Diseases of Vegetable Crops and Their Integrated Management

NEW INDIA PUBLISHING AGENCY

New Delhi – 110 034

About the Editors

Dr. Amar Bahadur is working as Asstt. Professor in the Department of Plant Pathology, College of Agriculture, Tripura, Lembucherra, Agartala-799210. He did his Ph.D. in Mycology and Plant Pathology, Banaras Hindu University, Varanasi, India. He is Life Member of different society. Dr. Bahadur has published more than 25 research papers in International and National Journals and more than 10 proceedings and book chapters in edited book. He has been Principal Investigator and Co-Investigator in DBT funded research project. He has participated in several seminar/symposia/conference in national and international and also visited Thailand. He was Zonal President of NEZ (North-Eastern Zone) in 2018 of Indian Phytopathological Society, IARI, New Delhi Organizing secretary and Organised a National Symposium in 2019. He has engaged in research, teaching and extension activities more than 15 years.

Dr. Pranab Dutta is a Associate Professor in Plant Pathology in the School of Crop Protection, College of Post Graduate Studies in Agricultural Sciences, Central Agricultural University, Umiam, Meghalaya. He served Assam Agricultural University, Jorhat, Assam, KVK-West Tripura (Now Khowai), CMER&TI (CSB) etc. He completed MSc and Ph.D. in Biological Control and Post Doctorate training at FERA, York, United Kingdom. He has been actively involved in teaching, research and extension for the last two decades. He has published 100 research papers, 45 research abstracts, 13 proceedings paper, 81 popular articles (English, Bangla and Assamese), 22 training manuals/hand books, 36 book chapters, 9 practical manuals /teaching aids, 12 books (Edited, authored or co-edited). Dr. Dutta has planned, participated in and given invitational lectures at numerous national and international conferences and workshops. He is the recipient of meritorious contribution award in sericulture (CMER&TI, Central Silk Board,) ICPP-2018 Bursary award (The American Phytopathological Society, USA), Young Scientist Award- 2018, Distinguished Scientist Award-2019, Prof. K S. Bilgrami Award-2016, Stanford Research Award-2015, Bharat Siksha Ratan Award, Fellow of SBER etc. Developed 7 bioformulations with native isolates, filled 5 patents and commercialized five bioformulation products. Pursued 12 externally funded project funded by DBT, DST, Osceola County, Florida, USA, (through CABI via University of Florida), PCIL, AYUSH, RKVY, BRNS etc. He is the life member of 8 professional societies and reviewer of 7 international and nationally reputed journals published by Elsevier, Springer etc. He is the Editor-In -Chief of Journal of Plant Health Issues. He organized two ICAR sponsored 10 days short course on Biological control during 2017 & 2019 and a national workshop on Biocontrol as Director. He has supervised 3 Ph.D. and 11 M.Sc. students as Major adviser and more than 100 students as member of advisory committee. His research area includes nanotechnology, biological control, ecotoxicology and plant defence.

Diseases of Vegetable Crops and Their Integrated Management

Amar Bahadur
Assistant Professor
Department of Plant Pathology
College of Agriculture
Lembucherra, West Tripura-799210, India

Pranab Dutta
Associate Professor (Plant Pathology)
School of Crop Protection
College of Post-Graduate Studies in Agricultural Sciences
Central Agricultural University (Imphal)
Meghalaya-793103, India

NEW INDIA PUBLISHING AGENCY
New Delhi – 110 034

NEW INDIA PUBLISHING AGENCY
101, Vikas Surya Plaza, CU Block, LSC Market
Pitam Pura, New Delhi - 110 034, India
Email: info@nipabooks.com
Web: www.nipabooks.com

For customer assistance, please contact
Phone: + 91-11-27 34 17 17 Fax: + 91-11- 27 34 16 16
E-mail: feedbacks@nipabooks.com

ISBN : 978-93-90591-09-1

Composed and Designed by NIPA.

Preface

The edited book on **"Diseases of Vegetable Crops and Their Integrated Management"** covered major diseases of nationally important vegetable crops with their integrated management practices. An effort has been made to compile the information on different aspects of diseases and their management. There is a record of huge losses of vegetables in terms of quality and yield, which is caused by different ailmnets. It has been felt necessary to compile the information related to the problem of important vegetable crops of India and their integrated management approaches.

This edited book contains chapters contributed by different authors throughout the country. The book will definitely help the students pursuing their degree in Agricultural Sciences, growers, teachers, extension personnel's and fellow researchers in their respective fields as well as will act as a ready reckoner for the readers.

Contribution and co-operation received from the contributing authors of the respective chapter is duly acknowledged. The editors express their sincere gratitude to all those who helped directly or indirectly in preparing of the book.

The editors feel great pleasure in expressing the deep sense of gratitude to New India Publishing Agency, Pitam Pura, New Delhi for publication of this edited book so efficiently and promptly.

Editors

Contents

1

Diseases of Potato (*Solanum tuberosum* L.) and Their Integrated Management

***N. Surmina Devi*[1], *R.C. Shakywar*[1], *P. Raja*[1] and *Salma Begum*[2]**

[1] *Department of Plant Protection, College of Horticulture and Forestry Central Agricultural University, Pasighat-791102, Arunachal Pradesh*

[2] *Agriculture Technology Management Agency, Churachanpur, Manipur*

Introduction

Potato is mostly considered as a vegetable crop in India, whereas it is a staple food crop in Europe, Canada, Australia, U.S.A, South America, Russia, Asia and other countries. In India, it is cultivated extensively in the hilly tracts during December-March or March-June and in the plains during September-January, as it requires cool and moist climate condition. As per FAOSTAT (2015), the potato production in India during 2013 was 45.34 mt from 1.99 mha area with a productivity of 22.76 t/ha. There are several varieties of potato grown in India and many diseases, some of which cause considerable damage to the crop.Potato crop can be affected by diseases caused by fungi, bacteria, virus, viroids and others by non-parasitic.

Late Blight

Economic Importance: Late blight is one of the most devastating diseases of potato throughout the world. It is recorded in area with frequent cool and moist weather. The disease is believed to have been introduced into Europe from South and Central America and spread to other countries which include the East Coast of Canada, Western Europe, Central and Southern China, South Eastern Brazil and the tropical highland. It created history by causing a famine in Ireland known as Irish Famine during 1845-46 resulting in death of many people due to starvation, dysentery and typhus spread among the farmers and subsequent migration of people from Europe especially from Ireland. Late blight has the tremendous potential to cause upto 70% reduction in the yield in a susceptible cultivar. Region wise economic importance of late blight shows that the disease takes highest toll of potato in Sub-Saharan Africa (44% crop losses) followed by Latin America (36%), Caribbean (36%), South-East Asia (35%), South-West Asia (19%) and Middle East and North Africa (9%) (CIP,1997). In India the disease was first introduced into the hilly regions of the Nilgiris during 1870-1880. The losses ranging from 20-25% in Punjab, 40-50% in Haryana, 15-20% in Uttar Pradesh and 5-10% in Bihar and West Bengal have been reported (CPRI,1987). In India, annual losses due to late blight have been reported to range between 10-75% (Dutt, 1979). Late blight disease also affects tomato crop equally.

Symptoms

The late blight may appear during any growth stage of the crops. It affects leaves, stems, petioles and tubers. Initially the disease is appear as small brown, water soaked lesions at the tip or margin of the leaves, which soon turn to brown to purplish black, necrotic lesions and enlarge rapidly under favourable weather condition. The infection spread inward and a whitish cottony or mildew growth can be seen underside of the infected leaves. Dark brown lesions may develop on stem which elongate later girdle the affected part and toppled down. In the advanced stage the entire potato may be killed in just few days giving a blighted appearance. Affected tubers also show some small irregular brown to purplish black tissues which extend into the flesh of the tuber. The infected tuber become firm, dry, sunken, is often invaded by secondary pathogen causing soft rot of tuber.

Causal Organism

Phytophthora infestans (Mont.) de Bary (Sub-Division: Mastigomycotina; Class: Oomycetes; Order: Peronosparales; Family: Pythiacea). The mycelium is of coenocytic, much branched, endophytic, hyaline which produced simple to club shaped haustoria. The sporangiophores arise from internal mycelium and emerge through the stomata on the leaves and tubers. The sporangiophores are slender and divide at the tip into 2-4 branches. Sporangia are formed at the tip of the branched which are of thin wall, hyaline and lemon shaped and multinucleate. Sporangia germinate almost entirely by releasing three to eight zoospores. The zoospores are reniform and biflagellate. After actively swimming for a few minutes they lose their flagella and enter the host by germinating the germ tube either by direct penetration of the epidermis or through the stomata. *Phytophthora infestans* is of heterothallic and requires two mating types A_1 and A_2 for sexual reproduction which is very rare in nature. In India although A_1 mating type is of common occurrence, prevalence of A_2 types has been recorded in Shimla hills after 1984 (Singh *et al.,* 1994). Antheridia & oogonia are produced after A_1 and A_2 mating types come in contact. Antheridia are amphygynous while oogonia are spherical. Oospores are 24-46μm in diameter, thick walled and developed after fertilization. Oospores formation varies depending upon particular A_1& A_2 combination and also on infected potato genotypes (Cohen *et al.,* 1997) but self-production of oospores in A_1 population is also reported.

Disease Cycle

The mycelium from infected tubers or from germinating oospores and zoospores spreads into shoots produced from infected or healthy tubers, causing discoloration and collapse of the cells. When the mycelium reaches the aerial parts of plants, it produces sporangiophores, which emerge through the stomata of the stems and leaves and produce sporangia. The sporangia, when ripe, become detached and are carried off by the wind or are dispersed by rain. If they land on wet potato leaves or stems, they germinate and cause new infections. A few days after infection, new sporangiophores emerge from the stomata of the leaves and produce numerous sporangia, which are spread by the wind and infect new plants. In cool, moist weather, new sporangia may form within four days from infection. Thus, a large number of asexual generations and new infections may be produced in one growing season (Agrios, 2005).

The second phase of the disease, the infection of tubers, varies between potato varieties and pathogen isolates. It begins in the field when, during wet weather, sporangia are washed down from the leaves and are carried into the soil. Emerging zoospores germinate and penetrate the tubers through lenticels or through wounds. In the tuber the mycelium grows mostly between the cells and sends haustoria into the cells. Tubers contaminated at harvest with living sporangia present on the soil or on diseased foliage may also become infected. Most of the blighted tubers rot in the ground or during storage.

Epidemiology

The development of late blight epidemics depends greatly on the prevailing humidity and temperature during the different stages of the life cycle of the oomycete. The oomycete grows and sporulates most abundantly at a relative humidity near 100% and at temperatures between 15 and 25°C. Temperatures above 30°C slow or stop the growth of the oomycete in the field but do not kill it, and the oomycete can start to sporulate again when the temperature becomes favorable, provided, of course, that the relative humidity is sufficiently high.

Integrated Management

Host Resistance: Growing of resistant varieties is the safest and regarded as the best means of disease management. Louwes *et al.,* (1992) mention that *Solanum circaefolium* sub sp.*circafolium* to be a good resistancesource against *Phytophthora infestans.*

Cultural Practice: Reduction of the primary sources of inoculums is the first step in management of late blight. Diseased haulms should be removed and destroyed by burning. Disease free tubers should be used as a planting material. Crop rotation with non solanaceous crops must be practice to avoid the disease. Choice of suitable cultivars, well aerated fields, pre-sprouting of tubers, early planting and use of resistant varieties are some of the measures against foliar blight while planting potatoes on large steep ridges, right time of mechanical weeding and harvesting, avoiding rapid shift of harvested tubers and long transports could minimize tuber blight (Meinck and Kolbe, 1999).

The Central Potato Research Institute, Shimla (2015) has developed INDO-BLIGHTCAST- a web based Pan-India model for forecasting potato late blight as an improvement over JHULSACAST. It predicts late blight appearance using daily mean temperature and RH data available with meteorological stations across the country without any calibration.

Chemical: Under congenial climatic conditions, application of fungicide is absolutely essential for the successful management of blight causing pathogen of potato. Spraying with an effective fungicide is a standard procedure for control of late blight. Ridomil MZ (metalaxyl + mancozeb) is one such mixture being commonly used against late blight (Alam *et al.,* 1991, Thind and Mohan, 1997). Newly developed fungicides like azoxystrobin, fenamidone, famoxadone, iprovalicarb and their combinations with contact fungicides like mancozeb have shown promising efficacy against potato late blight (Thind *et al.,* 2004)

Biological Control: Frequent and heavy dose uses of fungicide could pose threat to environment and human hazards. A few commercial fungicides have also been reported to induce oospore formation and phenotypic changes in *P. infestans* (Groves and Ristaino, 2000).

Early Blight

Economic Importance

Potato early blight occurs worldwide and is prevalent wherever potatoes are grown. It appears also on tomato, chilli, eggplant and many cultivated and wild host plants of India and other countries. The losses due to this disease have been reported from different parts of India to extent of 48-80 per cent (Datar and Mayee, 1981). Total aggregate losses caused by the various Alternarias on all of their hosts rank among the highest caused by any pathogen. (Gupta, 2006) estimated losses to the extent of 49.28 per cent due to this disease.

Symptom

The characteristic symptoms of the disease are dark brown to black concentric rings, which produce a target board appearance.This disease appears usually as leaf spots and blights, but they may also cause damping-off of seedlings, stem rots, and tuber and fruit rots.The leaf spots are generally dark brown to black, often numerous and enlarging, and usually developing in concentric rings with a yellow halo, which give the spots a target-like appearance. Lower, senescent leaves are usually attacked first, but the disease progresses upward and make affected leaves turn yellowish, become senescent, and either dry up and droop or fall off. Stem lesions developing on seedlings may form cankers, which may enlarge, girdle the stem, and kill the plant.

Causal organism

Alternaria solani (Ellis & Mart.) Jones & Grout. (Sub-division: Deuteromycotina; Class: Hypomycetes; Order: Moniliales; Family: Dematiaceae). It has dark-colored, septate mycelium. Short, simple, erect conidiophores that bear single or branched chains of conidia may arise from the mycelium present in the dead centers of spots. Conidia arelarge, dark, long, or pear shaped and muriform, with both transverse and longitudinal cross walls. The perfect stage of this fungus is not known.

Disease Cycle

The pathogen is overwinter as mycelium or spores in infected plant debris and in or on seeds. The pathogen also infects other hosts like tomato and capsicum as well as weeds like *Solanum nigrum* which may play an important role in its survival (Singh and Nagaich, 1977). The spores are produced abundantly, especially during heavy dews and frequent rains, and are blown in from infected debris or infected cultivated plants and weeds. The germinating spores penetrate susceptible tissue directly or through wounds and soon produce new conidia that are further spread by wind or splashing rain and cause secondary spread of the disease.

Epidemiology: Warm and humid climate with frequent rains followed by dry weather is most favourable for disease development. The optimum temperature for the germination of conidia is around 25^{0}C and the fungus can grow well at temperature between 23-25^{0}C (Tong *et al.,* 1994).

Integrated Management

Host Resistance: Resistance to early blight has been identified in *Solanum* species such as *S.phureja* and *S. chacoense* that has been exploited for breeding of resistant cultivars (Stewart *et al.,* 1994). A few varieties such as Kufri Sindhuri show good resistance to early blight.

Cultural Practice: Disease free planting material should be used. Crop rotation, removal and burning of plant debris and eradication of weed hosts help reduce the inoculum for subsequent plantings of susceptible crops.

Chemical: Spraying of three to four times of mancozeb @ 2gm/lt of water at an interval of 7-14 days interval found to be effective against early blight of potato.

Biological Control: Several mycoparasitic fungi are known to parasitize various species of *Alternaria,* but so far none has been developed into an effective biological control of the pathogen.

Wart Disease

Economic Importance

Potato wart disease is also one of the important and serious disease of cultivated potato. The disease was first reported from Hungary in 1896. Now, it is widely distributed in several countries of the world, including the United States of America, Britain, Canada, South Africa, Peru, South America, India etc. In India, the disease was reported for the first time by Ganguly and Paul (1953) from Darjeeling and is presently restricted to that area only due to strict domestic quarantine on the disease. The disease once established is difficult to eradicate since the resting spores of the pathogen may remain viable for 20 to 30 years or longer.

Symptom

The disease affects mainly the tubers during the growing season. There is no visible sign of wart disease on the above ground part of the plant. The characteristic symptom of brown or brownish-black dirty protuberances of cauliflower like outgrowths. They are produced as result of hypertrophy of epidermal cells surrounding the infected cell. The infected tubers may rot due to attack by secondary saprophytic soil microbes.

Causal Organism

Synchytrium endobioticum (Sub-division: Mastigomycotina; Class: Chytridiomycetes; Order: Chytridiales; Family: Olpidiaceae). The fungus is an obligate parasite, holocarpic and endobiotic in nature. The fungus lacks mycelium and has a thin walled summer sporangium stage & a thick walled 'winter' or resting sporangium stage. Both summer and winter sporangia produce with an extended vesicle called sorus from where zoospores are produced. The zoospores are pear shaped and possess a posterior flagellum. Potato is the main host although experimentally a number of *Solanaceous species* gets infected upon artificial inoculation (Phadtare and Sharma, 1971).

Disease Cycle

The resting sporangia can remain viable in the soil for many years in the absence of potato crop. It may be carried from one place to another through irrigation or rain water, through infected seed tubers, farm implements, feet of cattle and birds and even by man. The resting sporangia under wet soil conditions and temperaturebetween 10 to 27 ^{0}C germinate to release haploid uni-nucleate zoospores. The zoospores swim in soil, encyst and infect epidermal cells of meristematic tissues of growing buds, stolons tips or leaf primodia by means of an infection peg within 1 to 2 h of their formation. After successful infection a uni-nucleate thallus develops within the infected cell which enlarges to form a pro-sorus. A vesicle develops from prosorus and contents of the prosorus pass on to the vesicle to form a sorus within an infected cell. The sorus divides repeatedly to form several sporangia in which zoospores develop. Finally wall of the sorus breaks releasing sporangia and zoospores in soil. New infection results from the zoospores. This process continues throughout the growing season. Growth of the pathogen fungus within the host stimulates hypertrophy and hyperplasia of neighboring host cells without actively infecting them which result in an increase in meristematic activity and development of warts of variable size depending upon the degree of stimulation (Lapwood and Hide, 1971).

Epidemiology

The resting sporangia are very resistant to extremes of temperature, either high or low. A very high degree of soil moisture is necessary to bring about infection by the resting sporangia. Zoospore emergence is highest at very high humidity of 90-100% and low temperature between 14^{0}C -24^{0}C. A period of flooding, followed by draining and aeration of the soil is favourable for disease occurrence. The disease incidence is low at high degrees of alkalinity.

Integrated Management

The disease has been successfully managed by sanitation, long crop rotation, growing resistant and immune varieties and by enforcing strict quarantine legislation in countries of EPPO region (McNamera and Smith, 1998), Canada (Hampson, 1993), Maryland USA (Putnam and Sindermann, 1994) and India (Singh and Shekhawat, 2000). Periodic surveys are required to monitor viability of the pathogen in soil and efficiency of the quarantine measures.

Host Resistance: Some of the potato varieties, such as Adia, Ackersegen, Blanki, Capella, Mira, Pimpernel, Ronda, Ultimus and Voran are reported to be immune to wart disease. A number of varieties are developed by CPRI, Shimla like Kufri Kanchan, Kufri Sherpa, Kufri Jyoti etc. have been found to be resistant.

Cultural Practices: The pathogen is soil borne disease and it is difficult to eradicate, once introduced to a field. Strict domestic quarantine measures enforced in India have helped to restrict the disease in Darjeeling district of West Bengal only where it was first reported. Clean cultivation should be adopted and plant debris and infected tubers should be collected and destroyed by burning. Intercropping potato with maize, or growing rotational crops such as bean and radish has been found to reduce pathogen population (viable resting spores) in soil (Singh & Shekhawat, 2000).

Chemical Treatments: Application of fungicides and chemical to soil is costly and not a practical approach (Hodgson *et al.* 1974; O'Brien and Rich, 1976).

Black Scurf

Economic Importance

Black scurf disease also known as *Rhizoctonia* stem canker. The disease occurs throughout the world and in India, it is found in all potato growing areas, both in the hilly tracts and in the plains. Black scurf disease heavily affects the quality and reduces the market value of the potatoes and the affected tubers become unfit for human consumption. Lakra (1992) reported that black scurf reduced seed germination by 11 per cent and affected yields by 68.2 per cent in the cultivar Kufri Chandramukhi.

Symptom

The disease has two distinct phases *viz*, the stem canker phase and the black scurf phase. In the stem canker phase, the growing tip of the tuber sprouts is killed before emergence. When mature sprouts are severely affected, sunken, circular or elongated brown necrotic spots may be observed. The sprouts which escape infection, continue to grow and sunken or shallow, brown cankers are formed.The cankers may girdle the stem region and cause damage to the stem, as a result normal downward translocation of carbohydrates is impared and carbohydrates tend to accumulate in the top portions. This leads to excessive formation of anthocyanin and the leaves turn purple. The uppermost leaves tend to become bunchy and often the leaf edges may role abnormally(Henry and Devasahayam, 2011).

The presence of black crust on the tubers due to the formation of sclerotia of the fungus is the most prominent symptom of black scurf disease of potato. Black scurf phase is more common than the stem canker phase in India. In this phase, newly formed tubers show brown, discoloured, crusty areas extending from the affected stolons.Sclerotiaare typically formed on the tubers and less frequently on the affected stems or roots. They are formed at various spots on the tubers, especially on the underside to a large extent. Dry rot symptoms on tubers are also seen if the pathogen enters through wounds (Gupta and Thind, 2012).

Causal Organism

Rhizoctonia solani Kuhn; telemorph: *Thanatephoru-scucumeris* (Frank) Donk. The mycelium is light brown in colour, septate with multinucleate cells and produces branches that grow at proximately right angles to the main hypha and are slightly constricted at each junction and have a cross wall near junction. The branching characteristics are important for identification of the fungus as *Rhizoctonia.* The sclerotia of the fungus are quite distinct from sclerotia produced by other fungus. The sclerotia are small, dark brown to black and there is no differentiation of the sclerotial tissue into rind and internal medulla. Basidial stage of the fungus has also been reported in some parts of the country on dead potato haulms, which appears as flaky pellicle on the substrate under cool and humid environment. The basideospores appear on sterigmata arising from barrel shaped to clavatebasidia and are usually four in number.

Disease Cycle

The pathogen overwinter as sclerotia in the soil and on infected seed tubers. The sclerotia can remain viable for many years in the soil. The fungus spreads through rain, irrigation, through contaminated tools and implements and infected planting materials. The mycelium from germinating sclerotia, attacks sprouts, stem and young tubers. After entry into the host, the epidermal cells are rapidly destroyed.

Epidemiology

Soil moisture and temperature are the two important factors affecting scurf development. The fungus can grow at temperature between 8-35^{0}C with optimum at 25-30^{0}C. The sclerotia germinate between 8-30^{0}C with optimum at 23^{0}C. The disease is more severe in sandy to sandy loam and moderately wet soils.

Integrated Management

Cultural practices: Proper drainage facilities should be provided to keep the soil moisture level low. Wet and ill-drained areas must be avoided. Disease free seed material should be used. Soil amendment with neem cake @ 40kg/ac. or wood sawdust @ 40 kg/ac has been reported to give good control of the disease. The amendments are to be given, three weeks before planting. Bhattacharrya *et al.,* (1977) reported that green manuring with Diancha (*Sesbania cannabina*) reduced the black scurf on tubers by 40 per cent. Crop rotation for two to four years with cereals, brassicas and legumes is helpful for the management of this disease.

Chemical Treatments

Complete control of black scurf by dipping infected seed tubers in 0.25% solution of Monceren (pencycuron), a phenyl urea based fungicide, for 10mminutes before sowing were reported effective by Thind *et al* (2002). Brassicol (PCNB) @ 2gm/ lit. of water has been found effective in controlling the disease. Dipping of seed tubers with 3% boric acid has also reported effective against the black scurf (CPRI,2015). The treatment with boric acid is not only effective but also less toxic and it replaced the hazardous organomercurials chemical.

Biological Control: The fungus can be suppressed by soil application of its antagonists, such as *Trichoderma* species.

Powdery Scab

Economic Importance

Powdery scab is an important disease of potato in temperate climate and high altitude in the tropics (Hines, 1976). The disease was first reported from Germany in 1814. In India, the disease occurs in the hilly regions of Himachal Pradesh, Sikkim, Uttar Pradesh, West Bengal and Tamil Nadu. The disease causes scab like lesions on tubers and markedly reduces its market value.

Symptom

The disease attacks all the underground parts of the plant. The characteristic symptoms on potato tubers appear as purplish brown sunken lesions which later turn to scab like lesions but unlike common scab the lesions of powdery scab are round, raised, filled with powdery mass of spores and surrounded by ruptured remains of epidermis.The pathogen may also serve as soil borne vector of potato mop top virus (O'Brien and Rich, 1976).

Causal Organism

Powdery scab is caused by *Spongospora subterranea* (Wallr.) Lagerheim. (Class: Plasmodiophoromycetes; Order: Plasmodiophorales; Family: Plasmodiophoraceae). The fungus does not produce any mycelium during its entire life cycle.

Disease Cycle

The pathogen can remain in a viable state in soil for long periods of about five years or more. It can also remain survive on infected seed tubers. The spores may be carried from one place to another through irrigation or rain water, farm implements or through infected tubers.

Epidemiology

High soil moisture level of 90-100% and low temperature below 18^{0}C favour the disease occurrence. Generally prolonged rainfall, followed by cool and damp weather are conducive for the development of the disease. The disease incidence is higher in heavy clay soils than in sandy loam soils.

Integrated Management

Host Resistance: To some extent the varieties such as, Arran Chief, Arran Banner and Bintje are resistance to powdery scab.

Cultural Practices: Planting of disease free seed obtained from disease free area helps in management of the disease. The disease can be minimized in field by avoiding flooding through proper drainage. Crop rotation with non solanaceous hosts is effective.

Chemical Treatment: Seed tubers treatment with mercuric chloride @ 0.1% solution for 1½ hours or with formaldehyde @ 5% solution for 3 hours prior to planting helps to eliminate the spores in the tubers.

Common Scab

Economic Importance

The disease is occurred in most of the potato growing areas. It is widely prevalent and important in neutral or slightly alkaline and light sandy soils during dry periods. The affected tubers fetch low price in market due to their bad look and also because deeper peeling is required before consumption.

Symptom

Scab symptoms on tubers are of categories as a shallow scab and deep pitted. The common or shallow scab is characterized by the formation of concentric wrinkled layers of the skin around a central depression, while the pitted scab is characterized by the formation of deep depressions or pits bordered by torn skin. The tissues around the interior of the pits become corky and dark.

Causal Organism

The disease is caused by bacteria, *Streptomyces scabies* (Thaxter) Waksman and Henrici. (Class: Thallobacteria; Order: Actinomycetales; Family: Streptomycetaceae). The pathogen is gram negative and aerobic in nature. The thallus consists of a mass of simple or branched filaments and septate at irregular intervals. The spores, which are abstracted in succession at the apex are roughly cylindrical and measure 1.0-2.0x 0.6-0.7μ in size.

Disease Cycle

The pathogen is a soil inhabitant and can remain viable in the soil indefinitely as a saprophyte. The pathogen penetrates the host by way of stomata or young lenticels and through wounds.

Epidemiology

Soil conditions greatly influence the pathogen. Dry soils with alkaline reaction are favourable for the development of common scab. The disease is inhibited in clay soils and in soils with high moisture content. At temperature in the range of 20 to 30 ^{0}C with low soil moisture, the disease incidence is more.

Integrated Management

Host Resistance: Varieties Menominee, Russet, Rural and Sebago were reported resistance to scab.

Cultural practice: Disease free tubers should be used as planting material. Proper irrigation of the field should be followed from the time of tuberisation till the time of tuber maturity. Mulching during earlier part of season and frequent irrigation to maintain soil moisture approaching field capacity, from tuber initiating stage to maturity of the crop, has been practiced for control of common scab (Borowezak and Gladysiak, 1999). Four year crop rotation with wheat-oat followed by potato-onion-maize have also been observed to reduce incidence of scab in the infested field (Singh and Jeswani, 1987).

Chemical Treatment: The disease can be reduced by use of acidic fertilizers such as ammonium sulfate (Huber, 1980), single super phosphate (Grewal *et al.,* 1988); and potassium chloride (Heald, 1993). Application of magnesium and manganese sulfate, and sulfur to potato crop has also been found to reduce the scab (Lambert and Manzer, 1991). Application of gypsum is also reported effective to reduce common scab (Singh and Soni, 1987).

Bacterial wilt

Economic Importance

Bacterial wilt or brown rot is a destructive disease of potato especially in tropical and subtropical and warm temperature regions of Asia, Africa, Australia, Europe, America, West Indies, Indonesia, Peru and India. Losses up to 75% have been recorded under extreme conditions (Gadewar *et al.,* 1991).

Symptom

It causes the severe losses on potato, tomato; tobacco and eggplant in some warm areas outsite the tropics. The characteristics symptom of bacterial wilt is sudden wilt, vascular browning and infected young plants die rapidly. Older plants first show wilting of the youngest leaves, or one-sided wilting and stunting, and finally the plants wilt permanently and die. The infected tubers turn brown, and in cross sections they ooze a whitish bacterial exudate.

Causal Organism

Bacterial wilt is caused by *Pseudomonas solanacearum* Smith (Syn. *Ralstonia solanacearum*). (Order: Pseudomonadales; Family: Pseudomonadaceae). It is a rod shaped Gram negative bacteria measuring 0.5-0.6 x 0.8-1.2 µm and having 1- 4 polar flagella.

Disease Cycle

Ralstonia solanacearum is both a soil borne and seed borne disease. It overwinters in diseased plants or plant debris, in vegetative propagative organs, such as potato tubers, on the seeds of some crops, in wild host plants, and probably in the soil. Bacteria are spread through the soil water, through infected or contaminated tubers and by contaminated knives used for cutting tubers. Bacteria enter plants through wounds made in roots by cultivating equipments, nematodes, insects, and at cracks where secondary roots emerge. Bacteria reach the large xylem vessels and through them spread into the plant. Along the vessels they escape into the intercellular spaces of the parenchyma cells in the cortex and pith, dissolve the cell walls, and create cavities filled with slimy masses of bacteria and cellular debris.

Epidemiology

Optimum temperature of 28-30^{0}C and relative high soil moisture are favourable for wilt development. Singh (1980) observed that disease incidence was maximum during July-August when both soil moisture and temperature was comparatively higher.

Integrated Management

An integrated approach involving use of pathogen free seed potato, reduction of field inoculum, growing crop under right environmental conditions, chemical and biological control has helped in control of the disease (Lemaga, 2001).

Cultural Practice

Bacteria-free propagative material should be used, and tools, such as knives, should be disinfected when moving from one plant to another. Infested soils should be kept fallow for about a year and frequently disked during the dry season to accelerate the desiccation of plant material and the death of wilt bacteria. Crop rotation, the most generally used control measure, is employed world-wide to check bacterial wilt. By shifting the date of the sowing to avoid periods of high soil moisture due to rainfall and high temperature can also help to decrease the disease incidence.

Chemical Treatment: Application of stable bleaching powder @ 2 kg/ha has been found to reduce bacterial wilt by 80% when applied in furrows at the time of planting (Shekhawat *et al.,* 1988). Ojha *et al.,* 1986 reported that Plantomycine alone or in combination with Blitox-50 proved effective in controlling bacterial wilt up to 95%.

Biological Control: Biocontrol of bacterial wilt by use of antagonists such as *Pseudomonas flourescens*, *Bacillus* spp, avirulent *P. solanacearum* and actinomycetes have been found to be effective (Shekhawat *et al*.,1993).

Bacterial Soft Rot and Black Leg

Economic Importance

Bacterial soft rot and Black Leg of potato is found wherever potatoes are grown and cause considerable loss to the standing crop. The disease affect the crop at all stages of growth. It causes soft rot of tubers at harvest, transit, storage and blackleg of foliage during the crop growing season. Losses under poorly ventilated storage or transit may go up to 80% (Somani and Shekhawat, 1990).

Symptom

The disease symptom can occur at any stage of plant development. A soft black lesion appears at base of stem which extend to decaying soft rot affected seed tuber in soil and up to a little above ground level. Tissues in the lesion shrivel and rot. The affected plants become stunted and plants show tendency to grow erect. The affected plants exhibit yellow chlorotic foliage, wilt and die without producing fresh tubers. On tubers, the disease may appear as small soft water soaked spots around lenticels which enlarge under high humidity or shrivel and get sunken under dry conditions. The pith of infected tuber decay beyond the boundary of external lesion, turn cream to tan brown in colour and the tissues becomes soft and granular. A brown to black pigment may develop around the lesion.

Causal Organism

The disease is caused by *Erwinia carotovora.* It is gram negative bacteria, rod shaped with peritrichus flagella.They can grow both under aerobic and anaerobic conditions, produce pectolytic enzymes and degrade pectin in middle lamella of host cells, breakdown tissues causing soft rot and the decay.

Disease Cycle

The bacterial pathogen causing the soft rot and black leg are survive in lenticels, tuber wounds and even on their surface and spread to healthy tubers in stores. They also live on decaying plant debris left over in the field after harvest. Tripathi (1979) reported that water film on tuber surface leads to proliferation of lenticels and also creates anaerobic conditions as well as other injuries on tuber surface predispose potatoes to soft rot. The bacteria from tubers carrying latent infections or from soil infect the haulms at various times during the season. These pathogen can reach daughter tubers through stolonsand initiate tuber decay at the site of tuber attachment (Shekhawat *et al,* 1984). Decaying tubers in soil could serve as source of contamination for healthy tubers. The pathogen may also spread while washing of the produce with contaminated water (Dartz, 1999).

Epidemiology

The bacteria can survive for longer duration in cool and moist conditions than in warm climate. Infection of seed tubers is favoured by moist soil and relatively cool temperature lower than 18-19^{0}C.

Integrated Management

Cultural Practices: The control of bacterial soft rots is difficult and depends on proper sanitation, avoiding injuries, keeping storage tissues dry and cool, assuring good insect control, and practicing crop rotation. Infected plant debris and tubers should be removed and destroyed by burning. Healthy seed tubers should be used as planting material. Crop rotation with non solanaceous may reduce the soft rot and black leg incidence. Wales *et al.,* (1986) reported that hot water treatment also give some success in controlling the rot in seed tubers.

Chemical Treatment: Application of bleaching powder with the last irrigation to the crop also reduces soft rot of storage in storage (Parashar *et al.,* 1986). Tuber treatment with 3 % boric acid (Somani and Shekhawat, 1985) or 0.05 % copper sulphate (Zhang *et al.,* 1993) or 160ppm Kasugemycin also reduce the incidence.

Ring Rot of Potato

Economic Importance: Ring rot is one of the most contagious and most feared diseases of potatoes, especially among seed potato growers. The disease occurs and used to cause severe losses in North America and Europe.Through strict inspection and certification of potato seed tubers, the disease has almost been eliminated from seed lots, but occasional outbreaks still occur by contamination from handling or transportation equipment at the seed producers or on the farm.

Symptom

The infected plant become stunted growth and the interveinal areas of its leaflets turn yellowish, and the leaf margins roll upward and become necrotic. The leaves then begin to wilt, continuing until all the leaves of the stem wilt and the stem then dies. The disease also affects one or more tubers of each plant. Symptoms begin to develop at the stem end of the tuber and progress through the vascular tissue. The infected tubers show a ring of light yellow vascular discoloration and some bacterial ooze exudates from the tuber. As the disease advances, a creamy yellow or light brown crumbly or cheesy rot develops in the region of the vascular ring, and if the tuber is squeezed, soft, pulpy exudate oozes from the diseased areas while a more or less continuous ring of cavities is formed by the rotting of tissues in the vascular area. Secondary, soft-rotting bacteria often invade infected tubers, and these bacteria may cause complete rot of the tuber while producing a foul odor.

Causal Organism

Ring rot of potato is caused by *Clavibacter michiganense* subsp. *sepedonicum.* It belongs to gram positive bacteria.

Disease Cycle

Ring rot bacteria overwinter mostly in infected tubers and as dried slime on machinery, crates, and sacks. Bacteria enter plants only through wounds and invade the xylem vessels in which they multiply profusely and may cause plugging. Bacteria also invade the roots and cause them to deteriorate.

Integrated Management

Use of healthy seed tubers is important for avoiding introduction of ring rot disease in the field. Field sanitation should be done before planting.

Potato Leaf Roll Virus

Economic Importance

Potato leaf roll virus is worldwide in distribution and account for greater losses than many of the other diseases.

Symptom

The characteristic symptom of PLRV is the rolling up of leaves along the margins like a boat, with the midrib at the bottom and assumes an erect habit.The infected leaves turns leathery, brittle and pink-brown pigmented. Infected plants remain stunted, chlorotic and make a rattle noise, when the leaves strike against each other. Tuber formation is considerably reduced and tuber sizes also become small, resulting in poor yield and quality.

Causal organism: Leaf roll of potato is caused by Potato leaf roll virus (PLRV) or Potato Virus 1. (Genus: Luteovirus).

Disease Transmission: PLRV is transmitted through infected seed tubers and by some aphid vectors. *Myzus persicae* is the most efficient and important vector which transmit the virus in a persistant manner.

Mild mosaic/ latent mosaic/ faintmottle/ interveinal mosaic of Potato

Economic Importance

The Mild mosaic of potato disease is stable and occurs in high concentration in the infected plants and it is not as severe as leaf roll or some other virus disease in India. The disease occurs where the potato crops are grown.

Symptom

Negligible visible symptoms occur on the infected plants and hence it is called latent mosaic. The affected plants may also show interveinal mottling or mosaic. Slight stunting of infected plants or deformation of leaves may be seen.

Causal Organism

The virus disease is caused by Potato Virus X (PVX). (Genus: Potexvirus).

Disease Transmission

The disease is sap transmissible. The virus can also be transmitted by dodder plant (*Cuscuta campestris)*. The virus is not transmitted by any insect vector.

Severe Mosaic of Potato

Symptom

The symptoms are more severe than in the case of mild mosaic virus. The severe mosaic spots and blisters are the predominant symptoms of the severe mosaic virus. On the susceptible varieties, local lesion, consisting of black, necrotic spots appear on the underside of the leaves. Later, the uppermost leaves develop a wrinkled appearance, turn yellow or mottled, while the lower leaves collapse and remain hanging on to the stem or fall off. The diseased plants produce undersized tubers.

Causal Organism

The disease is caused by Potato Virus Y (PVY). (Genus: Potyvirus)

Disease Transmission

The disease transmitted through infected seed tubers and many aphid vectors. *Myzus persicae, Myzus circumflexus* & *Aphis rhamni* are the main vectors and they transmit the disease in a non-persistent manner.

Rugose Mosaic of Potato

Symptom

Severe mottling of leaves is the characteristic symptom of rugose virus. The disease is more severe than both mild mosaic and severe mosaic. The infected leaves become wrinkled, puckered and rough and are markedly reduced in size. The margins of the leaves roll downward and the plants are very much stunted. Tuber formation is considerably reduced and tuber sizes also become small, resulting in poor yield and quality.

Causal Organism

Rugose mosaic is caused by a combination of Potato Virus X (PVX) and Potato Virus Y (PVY) and both the virus components are necessary to cause the disease.

Disease Transmission

The disease is transmitted through infected seed tubers. Potato Virus X is transmitted by sap and Potato virus Y is transmitted by aphid vectors.

Integrated Management of Viral Diseases of Potato

The best way to control a virus disease is by keeping it out of an area through system of quarantine, inspection, and certification.

Cultural Practices

Eradication of diseased plants to eliminate inoculum from the field may, in some cases, help control the disease. The use of virus-free seed tubers is the most important measure for avoiding virus diseases especially those lacking insect vectors. Early planting may help to ward off the disease incidence to some extent by the way of disease escape. Alternate hosts of the viruses should not be allowed to grow in nearby cultivated field. Rogueing of diseased plants should be done from the field as soon as possible.

Chemical

Chemical control is not always completely effective when viruses are transmitted in a non-persistent manner, as the aphids can infect many plants before the insecticide is able to kill them.Application of systemic granular insecticide such as aldecarb 10G @ 4kg or phorate 10G @4kg or disulfaton 5G@ 8kg/acre, and be mixed thoroughly with 10 kg of sand and applied uniformly in the planting furrows at the time of planting to prevent the infestation of aphids, which transmit the viruses. Following this, spraying with dimethoate @2ml/lt of water on the 50th day protects the crop from aphid infestation for another 15 days.

Potato Spindle Tuber

Economic Importance

The potato spindle tuber disease occurs in North America, Russia, and South Africa. It causes severe losses, reduced the yield by 25% or more and in some regions it is one of the most destructive diseases of potatoes. It attacks all varieties and spreads rapidly. It also attacks tomato but seems to be of little economic importance in that crop.

Symptom

The characteristic disease symptom is the occurrence of erect, spindly, and dwarfed. The infected leaves are small and erect, and the leaflets are darker green and sometimes show rolling and twisting. The tubers are elongated, with tapering ends. Tubers are smoother, but tuber eyes are more numerous and more conspicuous.

Causal Organism

The disease is caused by Potato spindle tuber viroid (PSTVd), the first recognized viroid. PSTVd consists of 359 nucleotides. Under an electron microscope, purified PSTVd appears as short strands about 40 nanometers long and has the thickness of a double-stranded DNA.

Disease Transmission

The viroid is mechanically transmissible and is spread primarily by knives used to cut healthy and infected potato seed tubers and during handling and planting of the crop. PSTVd seems to also be transmitted by pollen and seed and by contaminated mouthparts of several insects not normally considered as virus vectors, e.g., grasshoppers, flea beetles, and bugs. After inoculation of a tuber with PSTVd by means of a contaminated knife or of a growing plant with sap from an infected plant, the viroid replicates itself and spreads systemically throughout the plant.

Integrated Management

Potato spindle tuber can be controlled effectively only by growing PSTVd free seed tubers.

Minor diseases of Potato

i) **Silver scurf** is caused by *Helminthosporium solani* Dur. & Mont. The most obvious symptom is development of a smooth, gray, leathery skin, especially near the heel end. Symptoms can develop prior to harvest or in storage. They are most conspicuous when tubers are wet at which time they exhibit a silvery sheen hence the name 'silver scurf'.

ii) **Dry Rot** is a storage rot disease and caused by *Fusarium* spp. small brown areas may be visible on the tuber surface. The surface of infected tubers become wrinkled and sunken, and the rolled tissue may turn brown, gray or black. Cavities frequently develop in affected tubers, which may become more or less filled with yellow, pink or red *Fusarium* molds.

iii) **Wilt of Potato** caused by *Verticillium albo-atrum.* Early symptom involved the wilting of the leaves and gradually the infected leaves turn dull green, then yellow and finally brown.

iv) **Charcoal Rot** caused by *Macrophomina phaseolina.* A soft, dark-colored shallow rot develops on the lower stem area, resembling black leg.Around the lenticels black areas appear and slowly spread all over the tuber surface. Inside the tuber also the flesh shows blackening.

v) **Crinkle Virus** disease caused by PotatoVirus X and Potato Virus A. affected plants show marked downward curling and puckering of leaves. The leaves become hard and brittle. Slight mosaic symptoms also appear on the leaves. The plants become stunted, bushy and chlorotic.

vi) **Black heart** is a non-parasitic disease due to lack of oxygen in storage godowns. Due to high temperature, the tissues break down, resulting in high respiration and failure of gas-exchange. When the affected are cut for examine, the cut surface turns pink, then dark brown to black. There is no involvement of any microbial agent with the disease. The black heart of potato can be avoided by providing sufficient aeration and storing the tubers in thin layers on racks (Rangaswami,G. and Mahadevan, A., 2004)

Future Approaches and Conclusion

Investigation of the disease incidence and severity of potato in different agroclimatic zones has to be carried out. Under changing climatic condition, the behavior of the pathogen should be thoroughly studied. Constant research should be continued to explore newer and newer strategies to manage these diseases. New information technology should be transfer to the farmer and aware them about diagnosis of disease so that they can approach the control measures in time. Forecasting model for late blight must be developed in hilly and plains areas, it may predicts the late blight wellin advance thereby helping the potato growers in protecting their crop by applying fungicides pro-phylactically. Growing of resistance variety is the only safest measure for disease management. Through biotechnology a high yielding variety and disease free plant with minimization of usage pesticides can be introduced to feed an expanding world population.

References

Arora, R.K. and Paul Khurana, S.M. (2004). Major fungal and bacterial diseases of potato and their management. In; *Disease management of fruits and vegetables (Fruits and vegetable diseases),* (Eds, K.G. Mukerji), Kluwer Academic Publishers, New York, pp. 189-232

Bhattacharrrya, S.K., Phadtare, S.G. and Sharma, V.C. (1977). Fungal diseases of potato. In: *Recent Technology in Potato Improvement and Production.* (Ed. B.B. Nagaich), CPRI, Shimla, pp. 414-424.

Borowezek, F. and Gladysiak, S. (1999). Disease infestation of potato tubers depending upon irrigation and cultivation system.*Progress in Plant Protection.*, **39**: 786-788.

Central Potato Research Institute (2015). Vision 2050.CPRI (ICAR), Shimla,pp.12

CIP (1997) Annual Report. The International Potato Centre, Lima Peru. pp.179

Cohen, Y., Farakash, S., Reshit, Z and Baides, A. (1997). Oospore production of *Phytophthora infestans* in potato and tomato leaves. *Phytopathology.*,**87**:191-196.

CPRI (1987). Annual Scientific Report. pp. 91-99.

Dartz, J.A. (1999). Washing fresh fruits and vegetables: lessons from treatment of tomatoes and potatoes with water. *Dairy, food and Environmental Sanitation.*, **19**: 853-864.

Datar, V. V. and Mayee, C. D. (1981). Assessment of loss in tomato yield due to early blight. *Indian Phytopathology.*, **34** : 191-195

Dutt, B.L. (1979). Bacterial and fungal diseases of potato. ICAR, New Delhi. pp.196

FAOSTAT. (2015) . Statistical database. http://faostat.fao.org/

Gadewar, A.V., Trivedi, T.P. and Shekhawat, G.S. (1991). Potato in Karnataka. Technical Bulletin No. 17, CPRI, Shimla, pp. 33

Ganguly, A. and Paul, D.K. (1953). Wart disease of potatoes in India.*Sci. & Cult.*, **18**: 605-606.

George N. Agrios. (2005). Plant Pathology. Academic Press, California, USA, pp. 425

George N. Agrios. (2005). Plant Pathology. Academic Press, California, USA, pp. 649-651.

Grewal, J.S, Sud, K.C., Singh, J.P., Sharma, V.C. and Singh, S.N. (1988). Phosphorus and potassium nutrition of potato. Annual Scientific Report. CPRI, Shimla, pp.60,67.

Groves, C.T. and Ristaino, J.B. (2000). Commercial fungicide formulations induce *in vitro* oospores in *Phytophthora infestans. Phytopathology.,* **90**: 1201-1208.

Gupta, A.B. (2006). Epidemiology and management of early blight of potato, *Ph.D, Thesis N.D.U & T.*, Kumarganj, Faizabad, India, pp.110.

Gupta, S.K. and Thind, T.S. (2012). Disease problems in vegetable production. Scientific publishers, Jodhpur, India, pp.28-43

Hampson, M.C. (1993). History, biology and control of potato wart disease in Canada.*Canadian Journal of Plant Pathology.,* **17**: 69-74

Heald, F.D. (1993). Manual of Plant Diseases. McGraw-Hill Book Co. Inc., New York and London, pp.953

Henry, L.D.C. and Devasahayam, T.H.L. (2011). Crop Diseases Identification, treatment and Management. New India Publishing Agency, New Delhi, pp-379-401.

Hines, M. (1976). The weather relationships of powdery scab disease of potatoes. *Annual Applied Biology.*, **84**: 274-275.

Hodgson, W.A., Pond, D. D., and Munro, J. (1974). Diseases and pests of potatoes. Canada Department of Agriculture, pp. 1492.

Huber, D.M. (1980). The role of mineral nutrition in defense. In: *Plant Disease - An Advanced Treatise*. Vol. 5 (Eds. Horsfall, J.G. and Cowling, Ellis, B.). Academic Press, New York, London, Sydney, San Francisco, pp.381-406

Kaur, S. and Mukerji, K.G. (2004). Potato diseases and their management.In; *Disease management of fruits and vegetables (Fruits and vegetable diseases),* (Eds, K.G. Mukerji), Kluwer Academic Publishers, New York, pp.233-280.

Lakra, B.S. (1992). Correlation of infection intensities of black scurf with yield components of potato. *Indian Journal Mycology and Plant Pathology*., **22**: 203-204.

Lambert, D.H. and Manzer, F.E. (1991). Relationship of calcium to potato scab. *Phytopathology.,* **81**: 632-636.

Lapwood, D.H.and Hide, G.A. (1971). Potato diseases, In; *Diseases of Crop Plants*, (ed. Western, J.H.) Wiley, New York, pp. 89–122

Lemaga, B, Siriri, D and Ebanyat, P.(2001). Effect of soil amendments on bacterial wilt incidence and yield of potatoes in south western Uganda. *African Crop Science Journal*., **9**: 267-278.

Louwes, K.M, Hoekstra, R. and Mattrheij, W.M. (1992). Interspecific hybridization between the cultivated potato *Solanum tuberosum* sub sp. *tuberosum* L. and the wild species *S. circaefolium* sub sp. *circaefolium* Bitter exhibiting resistance to *Phytopthora infestans* and *Globodera pallida*. *Theor. Appl. Genet*., **84**: 362-370.

McNamera, D.G. and Smith, I.M. (1998). National control measures for *Synchytrium endobioticum*. Bulletin-OEPP., **20**: 507-511.

Meinck, S. and Kolbe, H. (1999). Control of leaf and tuber blight in ecological potato cultivation. *Kartoffelbau.,* **50**: 172-175

O' Brien, M.J. and Rich, A.E. (1976). Potato Diseases. Department of Agriculture Handbook. U.S, pp. 474

Ojha, K.L., Yadav, B.P. and Bhagat, A.P. (1986). Chemical control of bacterial wilt of ginger. *Indian Phytopathology*., **39**: 600-601.

Parashar, R.D. and Sindhan, G.S. and Singh, J.P. (1986). Field application of Kloricin for the control of soft rot of potato in field and storage.*Journal Indian Potato Association.*, **13**: 81-82.

Phadtare, S.G. & Sharma, K.P. (1971). Additional hosts of *Synchitrium endobioticum*. *Indian Phytopathalogy*., **24**: 389-392.

Putnam, M.L. and Sindermann, B. (1994). Eradication of potato wart from Maryland.*American Potato Journal*., **71**: 734-747.

Rangaswami, G. and Mahadevan, A. (2004). Diseases of crop plants in India. Prentice-Hall of India, Private limited, New Delhi, pp.333.

Shekhawat, G.S., Gadewar, A.V., and Chakrabarti, S.K. (1993). Potato bacterial wilt in India.Technical Bulletin No. 38. C.P.R.I., Shimla, pp.55

Shekhawat, G.S., Kishore,V., Sunaina, V., Bahal, V.K., Gadewar, A. V., Verma, R.K. and Chakrabarti, S.K. (1988). Latency management of *Pseudomonas solanacearum,* Annual Scientific Report, CPRI, Shimla, pp. 117-121.

Shekhawat, G.S., Piplani, S. and Ansari, M.M. (1984). Endophytic bacterial flora of potato plant in relation to soft rot disease.*Indian Phytopathology*., **37**: 501-505

Singh, B.P. and Jeswani, M.D. (1987). Managing common scab of potato through chemical control and cultural practices. *J. Indian Potato Assoc.*, **14**: 26-32.

Singh, B.P., Roy, S. and Bhattacharrya, S.K. (1994). Occurrence of A2 mating type of *Phytophthopra infestans* in India. *Potato Res.,* **37**: 227-231.

Singh, D.S. and Nagaich, B.B. (1997). *Nicandra physaloides* Gaertn, an additional host of *Alternaria solani* (Ellis & Mart.) Jones & grout causing early blight of potato. *Sci. and Cult.*, **43**: 454.

Singh, H. and Soni, P.S. (1987). Chemical control of common scab of potato.*Pl. Dis. Res.*, **2**: 77-79.

Singh, P.H. & Shekhawat, G.S. (2000). Wart diseases of potato in Darjeeling hills. Technical Bulletin No. 19, Central Potato Research Instittue, Shimla, pp. 73

Singh, Rah Pal (1980). Studies on *Pseudomonas solanacearum* the incitant of bacterial wilt of solanaceous crops. *Ph.D. Thesis, Univ. of Gorakhpur*, Gorakhpur, India.

Somani, A.K. and Shekhawat, G.S. (1985). Evaluation of chemicals to prevent soft rot of potatoes. *Journal Indian Potato Association.*, **12**: 214-216.

Somani, A.K. and Shekhawat, G.S. (1990). Bacterial soft rot of potato. Tech. Bull. 21. CPRI, Shimla, pp. 323

Stewart, H.E., Bradshaw, J.E., Wastie, R.L., Mackay, G.R., Erlich, O., Livescu, L. and Nachmias, A. (1994). Assessing progenies of potato for resistance to early blight. *Potato Res.*, **37**:237-269.

Thind, T.S. and Mohan, C. (1998). Severity of late blight and assessment of yield losses in potato during 1997-98 epiphytotic in Punkab.*Pl. Dis.Res.*, **13**: 204-207.

Thind, T.S., Mohan, C. and Kaur, S. (2002). Promising activity of pencycuron, a phenyl urea-based fungicide, for effective management of black scurf of potato.*Indian Phytopathology.*, **55**(1): 39-44

Tong, Yen Hui, Liang JiNong, Ku-Jug You, Tung, Y.H. and Xu, Jy. (1994). Studies on the biology and pathogenicity of *Alternaria solani* on tomato.*J. Jiangsu Agric. College.*, **15**:29-31.

Tripathi, R.K. (1979). Mechanism of resistance in potato tubers to bacterial soft rot (*Erwinia caratovara* var *caratovara*) In: *Proc. of "International Symposium on Postharvest Technology and Utilization of Potato"*. Central Potato Research Institute, Shimla C.I.P., New Delhi, pp. 1-2

Wale, S.J., Robertson, K., Robinson, L. and Foster, G. (1986). Studies on the relationship of contamination of seed potato tubers with *Erwinia* sp. to black leg incidence and large hot water dipping to reduce contamination. In: *Crop Protection and Quality of Potatoes.* Wellsbourne, Warwick, AAB, pp. 285-291.

Zhang, X.J., Wang, J.S., Fang, Z.D. (1993). Control of potato soft rot (*Erwinia caratovora* (Jones) Dye) by copper sulphate and its effect on some enzymes in tubers. *Acta Phytopathologica Sinica.*, **23**: 75-79.

2

Diseases of Tomato (*Lycopersicon esculentum*) and Their Integrated Management

Manashi Debbarma, Pranab Dutta **and** ***Lipa Deb***

School of Crop Protection, College of Post Graduate Studies in Agricultural Sciences Central Agricultural University (Imphal), Umiam, Meghalaya-793103

Tomato (*Lycopersicon esculentum* Mill.) is one of the most important vegetable crop and is known as protective food due to its wide spread production and special nutritive value. Tomato is referred to be native of tropical America. In the 16th century, it got spread from America to other parts of the world and became popular in India within the last six decades. It is known as world's largest vegetable crop after potato and sweet potato. The overall world production of tomato is 1,180,000 tons with a productivity of 65 t/ha. Contribution of USA in tomato production is 13.2 million metric tons to the total world production (FAOSTAT). USA ranks third in position in the total world production of tomato after China and India (FAOSTAT). India produces 22,337MT of tomato from 801ha area.

Fungal Diseases

Fusarium Wilt

Economic Importance

Fusarium wilt caused by *Fusarium oxysporum* being so widespread, turns out to be a significant problem in many crops. It is one of the most widespread and damaging diseases of tomato wherever tomatoes are grown intensively. The pathogen also causes damage to many other crops likepotato, and pepper of the family Solanaceae. Yield losses of effected crops can goup to 45% in tomato in India.

Symptoms

Initially, the infected plants show clearing of veinlets and drooping of the petioles on young plants. Initial typical symptoms are yellowing of the foliage, starting from the lower leaves and moving upward. Yellowing mostlystarts on one side of the vine. Later, theinfected leaves show downward curling, followed by browning and drying. Although the top of the vine wilts during the day and recovers at night, but wilting becomes progressively worse until the whole vine is permanently wilted. Vascular browning is observed in infected stems and large leaf petioles. Affected plants become stunted. The extent of stunting depends upon time of root infection. Young infected are more severely stunted than plants infected at later stage. Ultimately, the whole plant show wilting .

Causal Organism

Fusarium wilt is caused by *Fusarium oxysporum* f. sp. *lycopersici*. This pathogen has septate, hyaline mycelium which later turns cream in colour. Initially the mycelium is whitish in colour, which becomes pinkish after formation of macro conidia. Macro conidia are hyaline, mostly 2-3 septate, fusiform and having size of about 25-33 x 3.5-5.5μm, whereas micro conidia are single celled, ovoid, hyaline, and measuring about 6-15 x 2.5-4 μm.

Disease Cycle

The pathogen is a soil dweller and survives in the infected plant debris in the soil as mycelium and spore and in the cooler temperate regions, as chlamydospore (Snyder and Hanson, 1940). The germ tube of spores or the mycelium enter root tips directly or through wounds or at the point of formation of lateral roots, when healthy plants come in contact with contaminated soil. The mycelium move up through the root cortex intercellularly, and when it reaches the xylem vessels it penetrates them through the pits. Branching of the mycelium along with production of micro conidia take place in the vessels, which are detached and carried upward in the sap stream. Micro conidia develop at the point where their upward movement is ceased, the mycelium get through the upper wall of the vessel.

Epidemiology

The fungus is both soil and seed borne in nature. The disease is severe in sandy soils at a temperature between 27°C to 32°C. Soil is infested by planting infected transplants, from movement of infested soil by wind and water erosion or on farm implements. *F. oxysporum* is a common soil pathogen and saprophyte that feeds on dead and decaying organic matter. The pathogendisseminates in two basic ways,in short distances through water splash, and planting equipment, and long distances through infected transplants and seeds. *F. oxysporum* infects a healthy plant by means of mycelia or by germinating spores penetrating the plant's root tips, root wounds, or lateral roots.

Integrated Management

- There is no cure for fusarium wilt disease.
- Use of healthy seeds and transplants is necessary.
- Since the pathogen can also spread through infected dead plant material, so cleaning up at the end of the season is essential.
- Seeds suspected of being infected should be treated with hot water before planting.
- Methods like planting resistant varieties, removing infected plant tissue to prevent over wintering of the disease, flood fallowing, and using clean seeds each year can be adopted as a control measure. The best method used for controlling *F. oxysporum* is planting resistant varieties.
- Other effective methods are fumigating the infected soil and raising the soil pH to 6.5-7 (Snyder and Hans, 2003).
- The fungus *Trichoderma viride* is a confirmed biocontrol agent to manage this disease in an environment friendly way. So, seeds should be treated with *T. viride* 1% WP @ 9g/kg seed.
- For root zone application, 2.5 kg of *T. viride* 1% WP should be thoroughly mixed with 150 kg of compost or farmyard manure and the mixture should be applied in the field after sowing/ transplanting of crops.

Early Blight

Economic Importance

Early blight of tomato is the most important disease of tomato in USA, Australia, Israel, UK and India economically, where significant reductions in yield have been observed from 35% to 78% (Datar and Mayee, 1982; Basu, 1974; Jones *et al.*, 1993). Likewise, 79% yield loss have been observed due to the diseasein India, Canada, United States, and Nigeria (Datar and Mayee, 1981; Sherf and MacNab, 1986).

Symptoms

Initially symptom develop as small necrotic spots that appear dry and papery on young tomato leaves. As lesion increases they mostly produces concentric rings giving the lesion a target-like appearance. Also, when multiple leaf spots come together entire leaflets break down. Lesions forming on the leaf rachis or petiolemay causes complete leaf to turn brown and shrivel. Fruit infections may start either in green or ripen stage which appear as a sunken decay and are mostly covered with a black mass of spores.

Causal Organism

The causal agent is *Alternaria solani* (Sorauer). Conidia are found singly or in chains of two on distinct conidiophores measuring about 12-20 X 120-296 um having 9-11 transverse septa and long beaks. The mycelium is haploid and septate, which turns darkly pigmented with age.

Disease Cycle

The fungus survives in the soil by forming resistant spores in association with diseased tomato debris capable of persisting for one year or more. The life cycle starts with the fungus over wintering in crop residues or wild members of the Solanaceae family in the form of dormant mycelium. Under favourable climatic conditions, fungus germinates and cause the primary infection. The secondary spread of the disease in the field is by rain splashed or by wind dispersal of conidia onto an uninfected plant. The conidia infect the plant by entering through small wounds, stomata or direct penetration. After penetration, lesions may form within 2–3 days or the infection can live in dormant condition waiting for favourable conditions.

Epidemiology

Infection occurs very quickly under warm and humid conditions. Spores are incapable to infect a perfectly dry leaf as itrequires free water to germinate. Over a wide range of temperatures, *Alternaria* spores germinate within 2 hours but at 26.0-29.0°C, it may take only half an hour. The best temperature for sporulation of *Alternaria* is about 26.6°C when abundant moisture (as provided by rain, mist, fog, dew, irrigation) is present.

Integrated Management

- Cultural practices like eradication of weeds, proper spacing and mulching of plants, and use of balanced dose of fertilizer should be practised.
- Infected lower branches and leaves should be trimmed and disposed off.
- Prolonged wetting of leaves from irrigation should be avoided or drip irrigationcan be used.
- Resistant or tolerant tomato cultivars should be used. Complete resistance to early blight does not exist in commercial tomato cultivars.
- Using wild *Lycopersicon* species which show a high degree of resistance in breeding programs has led to the release of a number of cultivars of tomato with a degree of resistance to early blight.

- Use of fungicides like Mancozeb, Chlorothalonil or Copper fungicides are recommended. A spray program using a recommended fungicide beginning at fruit set and continuing on a 7 to 14 day schedule should be continued where early blight problems are expected.

Late Blight

Economic Importance

Late blight, is one of the most disastrous disease of tomato causing significant economic losses worldwide (Foolad *et al.*, 2008 and Nowicki *et al.*, 2012). The pathogen caused loss of more than a million lives in the Irish potato famine, for which it is best known (Andrivon, 1996). An exposed tomato field can undergo yield losses reaching upto 100% because of the infection (Nowicki *et al.*, 2012).

Symptoms

Symptoms appear as greasy, gray spots on the upper surface of leaveswhich spread rapidly especially when leaves are wet or humidity is high. White moldgrows at the borders of affected areas. Infection on stems and petioles lead to infection in areas above them which ultimately causes wilt and die. In favourable environmental condition the whole plant defoliate rapidly. Infection on fruits start as brown, greasy spots that rapidly spreads to rot the entire fruit.

Causal Organism

Late blight of tomato caused by the fungus *Phytophthora infestans* (Mont.) de Bary is potentially a serious disease. Mycelium is hyaline and coenocytic, and the nuclei are diploid. The mycelium produces branched sporangiophores bearing lemon-shaped sporangia at their tips. Sporangia germinate completely at temperatures up to 12 to 15°C by releasing three to eight zoospores, although above 15°C sporangia may germinate directly by producing a germ tube. For sexual reproduction, the oomycete requires two mating types (Gallegly and Galindo, 1958). The female hypha develops through the young antheridium and establish into a globose oogonium above the antheridium, when the two mating types starts growing adjacently. The antheridium then fertilizes the oogonium, which grow into a thick-walled and hardy oospore. Oospores germinate through germ tube that produces a sporangium, even though at times the germ tube grows directly into the mycelium.

Disease Cycle

The fungus exists mainly in infected tomato transplants. Some of them may also remain in tomato vines. *P. infestans* produces thousands of sporangia per lesion on sporangiophores in asexual form. The disease cycle starts when sporangia settle on host plant tissue, which must be coated with a film of water, essential for the motile, germinated spore movement approaching towards a penetration site (Hardham and Blackman, 2010). Germination of sporangia occurs either through the direct extension of germ tubes or by zoosporogenesis. Direct germination of sporangium occurs at temperatures above 21°C, on host tissue whereas at temperatures below 21°C, upto eight biflagellate zoospores are discharged from the sporangia. Motile zoospores enter through the film of water, detach their flagella, and encyst until they produce germ tubes. Germ tubes differentiate into appressoria that penetrate the host through the leaf cuticle. Intercellular hyphae develop and migrate inside the host between cellsto form biotrophic feeding relationships in the mesophyll, using haustoria. Hyphae develop and sporangiophores finally come out from stomata. Instantly, thereafter, generally between five and ten days after inoculation Late blight symptoms become apparent. In sexual form, *P. infestans* requires the mating of individuals with opposite mating types, known as A1 and A2 (Gallegly and Galindo, 1958; Gisi *et al*., 2011). Mating hormones activate gametangial formation in the opposing mating types and develops in sexual reproduction, when mycelia of the two mating types interact. In this life cycle, an antheridium fuses with an oogonium to form a diploid oospore. Survival of oospores in plant debris or soil serve as a persistent source of inoculum in the field. Oospores can germinate under favourable conditions and release diploid progeny of A1 or A2 mating type (Judelson, 1997).

Epidemiology

Disease is more severe in cool and wet conditions. Temperatures above 30 °C are considered unfavourable for late blight development. Disease development is fast with extended periods of favourable conditions and stops when weather becomes hot and dry. Fungal spores are spread by rain and wind.

Integrated Management

- Cultural practices like rotation and fallow, elimination of volunteer tomato and potato plants, planting noninfected seedlings and tubers, and elimination of late blight sources such as potato cull piles are specifically employed to manage the disease. The latter is of particular importance being cull piles can serve as a living host on which *P. infestans* mycelia can live over the winter and produce enormous amount of airborne spores at the onset of the new field season.
- Planting of only the most resistant potato variety available is recommended.
- Two main groups of fungicides have been used usually, including protectants (e.g., chlorothalonil, dithiocarbamates and triphenyl tin hydroxide), which

are generally applied before or at the time of disease development, and systemic fungicides (therapeutic fungicides; e.g., phenylamides such as metalaxyl/mefenoxam, aliphatic nitrogen fungicides such as cymoxanil, and morpholine fungicides such as dimethomorph), which restrict or reduce disease progress once symptoms are probable.

Septoria Leaf Spot

Economic Importance

Septoria leaf spot, also known as Septoria blight, is a very common destructive fungal diseases of tomato. It does not affect only tomatoes but also other plants belonging to the Solanaceae family, especially potatoes and eggplant. Even though Septoria leaf spot is not necessarily lethal for tomato plants, it develops rapidly and quickly defoliate and weaken the plants, rendering them unable to bear fruit to maturity.

Symptoms

Initially the disease appears on the lower leaves after the plant has establish fruit. Next, the leaves turn yellow in some areas, that later become circular with graycenters and dark borders. Spots may reach 1/8 inch in diameter surrounded by a yellow halo. Tiny black specks may get established in the center of these spots. Badly infected leaves fall off. Defoliation advances from the base of the plant upwards resembling early blight from a distance. However, smaller septoria leaf spots are clearly different from the larger dark leaf spots with concentric rings of early blight.

Causal Organism

This devastating disease of tomato foliage, petioles and stems (fruit is not infected) is caused by the fungus *Septoria lycopersici*. *S. lycopersici*has two types of mycelium, *viz*., hyaline or brown. The hyphal walls are thin, hyaline and septate. Diameter of the hyaline mycelium ranges between 1.2 to 5.8 µm. The pycnidium consists of a hollow globose body, without definite ostiole. The pycnidia have diameter of 300 to 750 µm. Pycnidial wall breaks and pycnidio spores gets detached at maturity (Endrinal and Celino, 1940). Pycnidiospores are filiform, transparent and septate with pointed or blunt end measuring about 34.00 - 69.00 µm x 2.20 – 4.80µm (Sohi and Sokhi, 1973).

Disease Cycle

Most infection possibly arises from infested plant debris remaining in the soil from a previous tomato crop. Although the fungus is not soil-borne, it can overwinter on crop residue from previous crops, decaying vegetation and some wild hosts related to tomato. The fungus overwinters on infected tomato debris or on weeds in the night shade family. The fungus can also continue to live on equipment such ascages and plant stakes.

Epidemiology

Spores of the fungus are disseminated by rain splash. The disease is favoured by mild temperatures and extended periods of high relative humidity. Many spores are produced in the pycnidia and are exuded when the fruiting structures are mature, under wet conditions. The temperature range for sporulation varies between 15°C to 27°C with optimum being 25°C. Spores may be carried through windblown water, splashing rain, hands and clothing of pickers, insects such as beetles, and cultivation equipment. Spreading is followed by germination of spores within 48 hrs under moist conditions and favourable temperatures. Since free moisture is essential for spore infection through stomates, long-lasting dew and rainy days favour disease development.

Integrated Management

- Removal of initial sources of inoculum can greatly reduce the extent of disease. Since, seed has been implicated as a source, it should be acquired from disease-free seed-producing areas.
- Rogue the seedlings before transplanting them to the field, in case infected plants are found.
- Crop rotation of nearly 3 years, sanitation (removal of crop debris) and thorough shredding and incorporation of infested plant residue soon after harvest are recommended to reduce Septoria leaf spot.
- Drip irrigation along with water splashing is recommended to reduce periods of leaf wetness whereas Sprinkler irrigation and over-head irrigation should be avoided.
- Most fungicides registered for use on tomatoes would efficiently control Septoria leaf spot which include Maneb, Mancozeb, Chlorothalonil, and Benomyl.
- The disease will be in controlled if fungicides namely chlorothalonil or copper fungicide, or mancozeb are applied repeatedly. Spraying of mancozeb 75% WP @ 600-800 g in 300 litres of water/acre is also recommended.

Buckeye Rot

Economic Importance

Occurrence of Buckeye rot of tomato was first reported in Florida in 1917.The disease can result in huge losses in the yield of harvestable fruit. It destroys up to 40 percent of the grown tomato plants.

Symptoms

Buckeye rot of tomato is a devastating fruit rot of green or ripe fruit. The initial symptoms are grayish-green or brownish, watersoaked spots progressingoften at the point of contact between the fruit and the soil. The spot rapidly enlarges to cover up to half of the fruit diameter when temperatures are warm (more than 27°C). The rot then develop brown in colour with concentric rings resembling the markings of a buckeye chestnut. The disease is most serious in poorly drained areas.White, cottony fungal growth appearson the buckeye rot lesions under moist conditions. As the disease progress with time, the entire fruit will rot.

Causal Organism

Buckeye rot is caused by the fungus *Phytophthora parasitica* Dastur.

Disease cycle

Phytophthora parasitica is primarily soil-borne pathogen and infects fruit lying on or near moist soil. Heavy rainfall or frequent irrigation may result in the sudden appearance of the disease. Saturation of the soil helps the zoospores to get released from sporangia in the soil. Two types of spores are formed. The primary spore, or sporangium, which develops first are lemon-shaped spores, formed at the tips of simple sporangiophores that emerge through the stomata. Later, the sporangia give rise to motile zoospores. Unless the soil is wet and reaches above 18°C, sporangiophores and sporangia are not formed. The sporangia may be formed within 24 hours, at 21°C. During wet periods, zoospores are discharged from the sporangia and are easily splashed by rain from the soil to the fruit. The zoospores swim about in a film of water for a time, encyst, and affect the fruit. Penetration can occur over the unbroken skin. symptoms may appear within 24 hours and fruit rot develops rapidly.

Epidemiology

The disease is most prevailing during periods of prolonged warm, wet weather and in poorly drained soils. The fungus live in the soil and is spread by surface water and rain. Healthy plants can be affected by the spores dispersed by splashing rain and spores present in soil. Even though cold soil temperatures kill the spores,once the fungus is introduced into the field it may remain permanently. Soil temperatures of 18–30 °C are needed for development of disease, & 27 °C is optimal for fruit rot. If temperatures are 21 °C or above, the fungus may even spread among harvested fruit. The fungus is also seed-borne and it can be spread by contaminated seed.

Management

- Grow tomatoes on raised beds in well drained soil and avoid compacted, poorly drained soils.
- Rotation and proper sanitation is required for reducing the disease occurrence.
- Fruits should not come in contact with the soil which should be maintained by proper staking and mulching of plants.
- Fungicides containing mefenoxam (e.g. Ridomil Gold/Bravo, Ridomil Gold, Ultra Flourish) as a soil surface application can be applied under the vines 4-8 weeks before harvest.
- Alternatively, mefenoxam fungicides (e.g. Ridomil Gold/Bravo, Ridomil Gold/ Copper) can be applied as a foliar spray when crown fruit are 1/3rd of their final size. Spraying with fungicides like Mancozeb, or Copper fungicides will give good control of buckeye rot.

Anthracnose

Economic Importance

Tomato anthracnose is one of the most important fruit rot disease of India and worldwide, resulting in post-harvest decay, severely affecting marketability of the fruits, causing yield losses of 30-70%. It is a common and widespread rot of ripe or overripe tomato fruit.

Symptoms

As the fruit starts ripening, the symptoms first appear as small, circular indented areas, which later develop darkened centres. The infected spots continue to grow bigger with time as each infection site also spreads wide into the fruit. Symptoms are small, sunken, circular spots (may increase in size up to 1/2 inch in diameter) on ripe fruit whereas it is exceptional on green fruit. Later the centre of older spots turn blackish. Spots may increase in number in severe cases, and secondary rotting organisms may penetrate anthracnose lesions and show complete rotting of infected fruit. Green fruits do not show symptoms until ripening even though they are infected.

Causal Organism

Anthracnose of tomato is caused by *Colletotrichum coccodes* (Wallr.) S. Hughes.The fungus produces salmon-colored spores which emits from the black fungal material in the center of the spots.

Disease Cycle

In the center of fruit spots, the fungus forms small, dark survival structures called sclerotia. These sclerotia survive for up to three years in the soil, and cause infections either directly or by producing secondary spores. Senescent leaves with early blight infections and leaves with flea beetle injury are important sources of spore since the fungus can colonize and produce new spores in the wounded areas.

Epidemiology

This disease is favoured by warm, rainy, moist and humid weather (from rainfall or over-head irrigation) and heavy defoliation.Growth of *C. coccodes* is very quick at 27°C, even though the fungus can cause infections over a wide range of temperatures (13°-35°C). The fungus then spreads from infected to healthy fruit since spores are splashed by rain or overhead irrigation, or by pickers working with wet plants.

Management

- Removal and destruction of crop debris and as soon as the crop has finished bearing.
- Treatment of tomato seeds by soaking them in hot water (50°) for 25 minutes is recommended for destroying the fungus.
- Mulching helps creating a barrier between the soil surface and the fruit and ultimately reducing infections.
- Weekly spraying of products containing chlorothalonil can reduce the disease.

Seedling Disease (Damping-off)

Economic Importance

"Damping-off" is a general term for the death of seedlings, under damp conditions, either before or after emergence. Anexact statement of the loss from damping-off of tomato seedlings is crucial to make. Damping-off is common throughout Ohio, and most of the countries are affected by this disease. There is adequate time to replant and grow more seedlings if one seedling crop is lost, due to ravages of damping-off. In case, partial losses of the crop occur, diseased seedlings do not die immediately but, when transplanted into flats, may die or live and produce unwanted plants. As most of the farmers depend on producing early tomatoes for the market, a set-back of one or two weeks usually does not lower the yield, it greatly reduces the net returns, since the price of late tomatoes is low.

Symptoms

Damping off of tomato develops in two stages, i.e. the pre-emergence and the post-emergence phase.In case of pre-emergence phase, the seedlings are killed just before they reach the surface of soil.Also the young plumule and radicle are killed accompanied by complete rotting of the seedlings whereas the post-emergence phase is distinguished by the infection of the young, juvenile tissues of the collar at the ground level. Here, the infected tissues become soft and water soaked. Finally, the seedlings topple over or breakdown.

Causal Organism

The fungi *Pythium* and *Rhizoctonia* causes damping-off of tomato seedlings.

Disease Cycle

Soil, seed and water are the primary sources whereas conidia through rain splash or wind are the secondary sources of dissemination. The survival of the fungus in the soil suggests that it lives widely as a saprophyte on plant debris or humus in the soil, and turn into parasite when the suitable host and growing conditions are present. Dissemination of inoculum takes place through oospores in the plant debris and mycelium. The soil or compost used in making up the seed beds can be a very common source of contamination in seed beds. Occasionally, these materials are obtained from an infested source and the organism is introduced along with the soil or compost.

Epidemiology

High humidity (84-95%), high soil moisture (83-86.50%) and low temperatures below 24° C for few days are suitable factors for infection and development of disease. Dampness due to high rainfall, crowded seedlings, poor drainage and excess of soil solutes effects plant growth and increase the pathogenic damping-off.

Integrated Management

- Sowing of seedlings should be done in raised beds.
- Provide frequent irrigation for better drainage

- Seed treatment with *Trichoderma viride* @ 4g/kg of seed can reduce damping off.
- As a chemical treatment, seed should be treated with captan 75% WS @ 20-30 g/kg seed and soil drenching with captan 75% WP @ 1000g in 400 litres of water/acre is recommended.

Bacterial Diseases

Bacterial Wilt

Economic Importance

The bacteria *Ralstonia solanacearum* responsible for the diseasecalled **bacterial Wilt** is considered as one of the world's most important **pathogenic bacteria in plants** due to the vast losses (0-91%) that it provokes in a large range of crops. It is one of the major diseases of tomato and other solanaceous plants. The disease is known to develop in the sub-tropics, wet tropics and some temperate regions of the world

Symptoms

Rapid wilting and death of plants is observed without any yellowing or spotting of leaves. Brown discoloration occurs inside the stems of infected plants which ultimately collapse. The bacteria multiply and increases rapidly inside the water-conducting tissue of the plant, filling it with slime which results in a rapid wilt of the plant, although the leaves stay green. Brown and tiny drops of yellowish ooze may be visible, if an infected stem is cut crosswise. The disease is quickly diagnosed by suspending a clean, cut section of diseased stem in clear water. After a few minutes, a white milky stream of bacterial cells and slime flow from infected stems into the water.

Causal Organism

Bacterial wilt is a serious disease caused by *Ralstonia solanacearum* (formerly *Pseudomonas solanacearum*). This pathogen is a gram-negative, rod-shaped, aerobic bacterium measuring about 0.5-0.7 x 1.5-2.0 µm in size. It is very sensitive to dryness and is inhibited in culture by low concentrations (2%) of sodium chloride (NaCl). The optimal growth temperature is between 28 and 32°C, for most of the strains; however some strains have a lower optimal growth temperature 27°C.

Disease Cycle

R. solanacearum is both soilborne and waterborne pathogen. This bacterium lives in the soil or water for lengthened periods which can form a reservoir source of inoculum and penetrate the roots through wounds made by transplanting, cultivation or insects.The bacterium infects tomato plants through the roots (through wounds or at the points of emergence of lateral roots). Root-knot nematode can also cause injury to plant roots and favorentry of the bacterium. *R. solanacearum* can continue to live for days to years in infected plant material in soils, infested surface irrigation water, and infected weeds. From these point of supply of inoculum, bacteria can spread from infested to healthy fields by soil transfer on machinery, and surface runoff water after irrigation or rainfall. Infected semi-aquatic weeds may also play a mainpart in disseminating the pathogen by delivering bacteria from roots into irrigation waters. At low temperatures (<4°C), bacterial population densities fall very quickly but the bacteria still can survive, often in a physiological latent state.

Epidemiology

Disease development is favoured by high temperatures and high moisture content. High temperatures (29° - 35°C) play a big role in pathogen growth and disease development. Several other factors like soil type and structure, soil moisture content, organic matter in soil, water pH and salt content, and the presence of antagonist microorganismsthat may affect pathogen survival in soil and water may also favour disease development.

Integrated Management

- Bacterial wilt of tomato is hard to control, and no single strategy has shown 100% efficiency in control of the disease so far. Best strategy for control of bacterial wilt in the field is by practising cultural method like Phyto sanitation.
- Non-susceptible plants, such as corn, beans and cabbage, should be used for rotation for at least three years as a control measure.
- All infected plant material should be removed and destroyed. Only certified disease-free plants should be planted.
- Some level of controlling/managing bacterial wilt is possible using resistant or moderately resistant tomato cultivars, such as such as FL7514 and BHN 466.
- Bactericides (copper) and antibiotics (streptomycin, ampicillin, tetracycline and penicillin) are environmentally destructive and fairly expensive to apply in the field and have proved less efficient on suppression of *R. solanacearum*.
- As a result, a combination of different control methods, which includes cultural practices, host resistance and the use of chemical or biological control should be used in an integrated pest management approach to manage bacterial wilt of tomato in regions where the pathogen is well-established.

Bacterial Spot

Economic Importance

It is a severe disease of tomato and sweet pepperin New York occurring regularly. Bacterial spot lowers yield and quality of tomato by defoliation and spotting of fruit. In India, the disease was first recorded at Agricultural College Farm, Poona, in 1948 (Patel *et al.*, 1950).

Symptoms

The bacterial spot pathogen may produce lesions on all aboveground parts of the plant like leaves, stems, flowers and fruit. Initial leaf symptoms produced are small, circular to irregular, dark lesions, which may be encircled by a yellow halo. The lesions may increase in size upto a diameter of 3-5 mm and tend to concentrate on the leaf edges and tip. Infected leaves may develop a burnt appearance. When spot increases in number, foliage turns yellow and eventually dies, leading to defoliation of the lower portion of the plant. In case of fruit, the first symptoms observed are small, dark brown-to black, raised spots. The lesions also may have a white halo, which is similar to the bird's-eye spotting seen with bacterial canker. The white halos disappear, along with the aging of fruit.

Causal Organism

The bacteria that cause bacterial spot are called xanthomonads. New taxonomic studies have pointed out that these bacteria belong to one of four groups: A, B, C and D. The original taxonomic name given to these bacteria was *Xanthomonas campestris* pv. *vesicatoria*, but current work has indicated that each group may represent a separate species.

Disease Cycle

Bacteria enter the plant through natural openings like stomates and hydathodes or wounds caused by wind-driven soil, insects or mechanical damage (handling, wind whipping, high pressure sprayers). The bacteria have a very definite survival period of days to weeks in the soil, and therefore their survival is nearly always in association with debris from infected or diseased plants. The pathogen can continue to live in association with seed, either externally or internally. On externally infested seed, cotyledons may become infected when they come in contact with the seed coat and exhibit lesions soon after emerging from the soil. Bacteria are then quickly splashed to new foliage and to other plants. In a 24-hr period, the bacteria can increase rapidly and produce millions of cells.

Epidemiology

The major sources of infection for these bacteria are thought to be seed and infected crop debris. The bacteria are spread primarily by splashing water and wind-driven rain or mists produced during storms. Extended periods of high relative humidity favour infection and disease development. Even though bacterial spot is a disease of warm, humid regions, it can also develop in arid, irrigated regions.

Integrated Management

- Certified disease-free seed and plants should only be used.
- All diseased plant material should be removed.
- Drip or furrow irrigation should be avoided.
- Spraying of streptomycin sulfate 9% + tetracycline hydrochloride 1% SP solution (streptocycline) 40-100 ppm in fields after the appearance of first true leaves is recommended for twice, one before transplanting (seed beds) and another after transplanting (main field).

Bacterial Canker

Economic Importance

Bacterial canker is a vascular (systemic) and parenchymatal (superficial) disease with a wide collection of symptoms resulting in loss of photosynthetic area, wilting and premature death, and production of unmarketable fruit.

Symptoms

Symptoms can be split into two types: *superficial* symptoms which develops from bacterial colonization of the *surface* tissues and *systemic* symptoms developing from bacterial invasion of the vascular tissue. Necrotic leaf lesions appear on the upper leaf surface of mature leaves, whereas in other cases, circular, slightly raised white spots appear. Symptoms on fruit may be observed at any age superficially, but are mostly seen first on green fruit showing diameter of 1/2-2 inches. White spots develop on the most-exposed parts of the fruit. The spots have a dark brown centre, which becomes raised and surrounded by a distinct white halo; they have been termed "bird's-eye spots." Sometimes the spots lose the white halo and become necrotic merging with others. Wilting progresses until the whole leaflet dies.

Causal Organism

Bacterial canker is caused by *Clavibacter michiganense* pv. *michiganense*.

Disease Cycle

The organism is seedborne and can survive for short periods in soil, greenhouse structures, and equipment and for longer periods in plant debris. Therefore, seedborne inoculum may serve as one source of the bacterium. The bacteria can exist on the seed surface as well as within the innermost layer of the seed coat. This makes the canker organism difficult to eradicate with seed treatments. When the seed germinates, the bacteria get into the seedling through small wounds, possibly through broken trichomes. The bacteria advances systemically through the xylem from which it penetrates the phloem, pith, and cortex. Secondary spread take place through splashing water, on contaminated equipment. Contamination can take place during clipping, cultivation, vine-training operations, and other activities. In the field condition, such kind of spread generally results in local infections (i.e., leaf, stem, and fruit spots).

Epidemiology

Soil temperature of around 28 °C is favourable for the disease along with high humidity or persistent dew and moist weather with intermittent showers.

Integrated Management

- Only certified seed should be used along with disease-free seed from canker-free plants. The certified seed used should be disease-free transplants that have been produced under a vigorous inspection program.
- Diseased plants should be removed as soon as they are detected by cutting the plants off at the ground line and dispose them by placing them in plastic bag.
- In the field condition, if bacterial canker becomes severe, fields should be ploughed down to prevent spread to nearby healthy fields.
- There are no effective chemical controls for bacterial canker. So, preventive measures are the best defense.

Viral Diseases

Leaf Curl of Tomato

Economic Importance

Tomato leaf curl represents an important constraint to tomato production, as it causes the most prevailing and economically important disease affecting tomato.

It has been reported from India, Sri Lanka, Malawi and South Africa. Incidence of tomato leaf curl disease is recorded to be 83% in winter crop planted during October and 14% in summer crop planted during February (Tripathi and Varma, 2002).

Symptoms

Tomato leaf curl is identified by severe stunting of the plants due to downward rolling and puckering of the leaves. Freshly formed leaves show chlorosis whereas older leaves become leathery and fragile. The entire plant looks pale bearing more lateral branches giving a bushy appearance. Partial or complete sterility of the plant is observed.

Causal Organism

Tomato leaf curl is caused by Tobacco leaf curl virus (Genus Begomovirus, Family- Geminiviridae). The genome of gemini viruses is composed of circular, ssDNA, enclosed by various subunits of a single capsid protein.

Epidemiology

Tomato leaf curl is transmitted in nature by whitefly, *Bemisia tabaci*. Transmission of a virus by this insect was first demonstrated in tobacco leaf curl in 1930s. Factors that favour disease development are high temperatures, and low or no rainfall.

Disease Cycle

Tomato yellow leaf curl virus is not seed-borne,therefore it is not transmitted mechanically. The disease is spread by whiteflies *(Bemisia tabaci)*. Whiteflies have a wide host range. New plant growth attracts whiteflies, which feed on the lower leaf surface. The virus requires time period of 15-30 minutes to infect the whitefly. The incubation period is 21-24 hours, and the transmission period at least 15 min.

Integrated Management

- Yellow sticky traps should be kept @ 12/ha for monitoring the white fly population.Planting tomatoes away from other whitefly host crops, such as brassicas and peppers also help manage the disease.
- Removal of the infected plants and weed host is necessary.
- Imidachloprid 0.05% or Dimethoate 0.05% @ 15, 25, 45 days after transplanting should be sprayed to control vector.

- Biological controls such as lacewing larvae and lady beetle larvae are recommended to control whiteflies.
- Spraying of Neem Seed Kernel Extract (NSKE) 5% or Azadirachtin 5% W/W neem extract concentrate @ 80g in 160 litres of water/acre may also be effective against whiteflies.

Tobacco Mosaic Virus

Economic Importance

Tobacco mosaic virus (TMV) is distributed worldwide and may cause significant losses in the field and greenhouse. Tomato and numerous other crops and weeds are infected by many strains of TMV. Yield loss from TMV results in reduced number, size, and marketability of fruit. 24% yield losses have been reported for tomato (Johnson *et al.*, 1983).

Symptoms

Typical leaf symptoms observed are mottled areas of light and dark green colour. Infected leaves turns out to be small, curled and malformed. Susceptible varieties which are infected early gives stunted look and appear pale green in colour. symptoms are usually absent in fruits, but may include numerous forms of uneven ripening. Severe strains may cause different mottling symptoms, rendering fruit unmarketable.

Causal Organism

TMV is the member of a big group of viruses within the genus *Tobamo virus*. The rod-shaped virus particles of TMV measure about 300 nm x 15 nm. A single TMV particle is composed of 2,130 copies of the coat protein (CP) that coats the RNA molecule having about 6,400 nucleotides.

Disease Cycle

Tobacco mosaic virus continues to live on plant debris, infected seeds, and even in clothing for months or years. The virus penetrate plants through wounds sustained in transplanting or pruning. It spreads very quickly once it is in the host. The virus is extremely tolerant of very high temperatures (50°C) and can remain viable for 10 years on dried plant debris. The most common source of inoculum are the tobacco products; smokers can affect plants by handling them. The virus is not spread by aphids.

Epidemiology

The virus may be introduced on, infected crop plants, weeds, contaminated seed and tobacco products. TMV is very persistent and contagious. It is spread generally by humans handling infected plants and mechanically carrying the virus to healthy plants via the sap.

Integrated Management

- Since virus-infected plants cannot be cured, so effective management depends on prevention.
- Washing hands with soap and water before and during the handling of plants and after using tobacco products is necessary to inactivate the virus.
- Planting should be begun with healthy transplants which are produced from virus-free seed.
- Seed treatments for eradication of the virus from seed include soaking seed for 10 minutes in a 10 % solution of trisodium phosphate (Na_3PO_4) or dry heat treatment of seed for 2 days to 4 days at 65ºC.
- No chemicals can cure a plant infected with Tobacco mosaic virus.

Root-knot Nematode

Root-knot nematodes (*Meloidogyne* species) are microscopic worms that live in the soil and plant roots. Affected plants are generally stunted, discoloured and may die. Knots or galls gets established on the roots.To reduce the nematode population, resistant cultivars should be used, pulling up and dispose of roots immediately after harvest is necessary. *Paecilomyces lilacinous*, egg-pathogenic fungus is the most widely tested biological measure for plant parasitic nematodes, and has shown promising potential as an alternative to chemical control at both pre-planting and planting applications. *P. lilacinus* reduces *M. incognita* soil and root population and increases the production of tomato (Kalele *et al*., 2010). *Trichoderma harzianum* is found to be antagonistic against root-knot nematode, Meloidogyne javanicaandit is an appealing alternative to chemical use. *T. harzianum* is an effective egg parasite of *M.incognita* because *T. harzianum* were able to grow on the egg surface and penetrated the egg cell reducing the incidence of root- knot nematodes and increasing plant growth through enhanced root growth (Naserinasab *et al*., 2010).Chemical control of root - knot nematodes have primarily been accomplished through nematicides which can be non-fumigants or fumigants (Widmer and Abawi, 2000; Rahman, 2003). Non fumigants are applied at the time of planting and are systemic affecting the nematode's behaviour. Examples of non fumigants are are Nemacur ®, Aldicarp ®, Oxamyl ®, and Carbofuran® . Fumigants are volatile liquids which are fast in action and dissolve in the soil killing the nematodes and their eggs. Examples of fumigants are 1, 3 dichloropropene methyl bromide, ethylene, Metham sodium® and Dazomet ®.

Disorders

i) **Blossom End Rot:** Symptom develops as water-soaked spoton the blossom end of the fruit which enlarge and become black followed bysecondary infection by decay-causing organisms. This disorderin developing fruit is caused due to calcium deficiency. Other factors like cool temperatures, rainy or cloudy weather with high humidity extreme fluctuations in moisture, root pruning from nearby cultivation, and excessive ammoniacal (NH_4^+) nitrogen, potassium, or magnesium fertilization can also increase the chances of blossom end rot. For controlling blossom end rot, the soil pH should be 6.5 in order to provide adequate calcium levels in the soil. Limestone is should be applied 3 to 6 months prior to planting & tilled thoroughly. Excessive potassium or magnesium fertilization should be avoided as these nutrients will compete with calcium for uptake by the plants. Ammonical nitrogen fertilizers side dress applications should also be avoided, as ammoniacal nitrogen also will compete with calcium for uptake. A consistent supply of moisture should be maintained through irrigation and adequate soil mulches.

ii) **Sunscald:** Sunscald occurs during hot weather, when tomatoes are exposed to the direct rays of the sun. It is most commonly found on green fruit. Decay causing fungi mostlypenetrate the damaged tissue. The exposed fruit should be covered as a control measure for sunscald.

iii) **Growth Cracks:** Tomatoes crack when environmental conditions like drought, heavy rain or watering stimulate rapid growth during ripening. Some cracks may be deep, which allows decayed organisms to enter the fruit causing fruit rot. Uniform soil moisture should be maintained along with regular watering to prevent growth cracks.

iv) **Poor Fruit Set:** Poor fruit set occurs due to several reasons:

   a) **Extreme temperatures**: The blossoms decreases without setting fruit when temperatures are below 13°C or above 32°C for extended periods.

   b) **Dry soil**: When the plants do not get enough water, blossoms dry and fall off.

   c) **Shading**: When the plants receive less than six hours of sun in a day, few blossoms are produced.

   d) **Excessive nitrogen**: High nitrogen levels in the soil advances leaf growth at the expense of blossom and fruit formation.

v) **Catfacing:** This is a disorder caused due to cold temperatures during fruit set. The fruit is severely malformed and scarred, mostly at the blossom end. Fruits which develop later in the season will not be affected.

vi) **Leaf Roll:** Leaf roll may occurdue to high temperatures, extended periods of wet soil conditions, and drought. It may also develop when tomatoes are pruned severely. The symptom is mostly observed on older leaves, with an upward curling of the leaflets, but may continue to affect up to 75% of the foliage. The rolled leaves may look leathery and stiff. Leaf rolling symptom occurs once the plants are under the stress of a heavy fruit set. To prevent this disorder, tomatoes should be planted on well-drained soil and should be irrigated during periods of drought.

vii) **Herbicide Injury**: Drift from nearby sprays of broad leaf weed killers, such as 2,4-D and dicamba, and non-specific herbicides, such as glyphosate, used on turfgrass, may harshly damage tomato plants.

Future Approaches

Tomato breeders, in collaboration with plant physiologists, biochemists, and gene technologists, will continue to develop the commercial tomato. The tomato is an exceptional model system for research on the genetic regulation of the biochemical events associated with ripening. Large catalogued international collections of wild tomato species are being researched and maintained. There is also immense potential to enhance resistance to pests and diseases, yield of solids for processing, and sensory quality. Increased consumption of fresh and processed tomatoes selected for high levels of lycopene can be expected, if further medical research confirms a role for lycopene in reducing chronic diseases in people, A large amount of germplasm remains to be explained in details in species such as *L. peruvianum*, in which there are significant compatibility barriers with *L. esculentum*. Molecular biology provides the tools for overcoming these barriers.

Conclusion

This chapter has described several major diseases affecting tomato in almost all the stages. Certainly, not all diseases have been discussed in this chapter. There are many other minor diseases that may be of significant local importance in tomato crops that have not been discussed here. In addition, new diseases will likely be discovered or introduced that will affect tomato. Tomato disease management is a moving target and therefore agricultural researchers, breeders, extension agents and growers must continually adjust disease management strategies to adapt new developments. Write few line on the most dreaded dissaes and how o tackle it through IDM approaches.

References

Andrivon, D. (1996). The origin of *Phytophthora infestans* populations present in Europe in the 1840s: A critical review of historical and scientific evidence. *Plant Pathol.*, **45**: 1027–1035.

Basu, P.K. (1974). Measuring early blight, its progress and influence on fruit losses in nine tomato cultivars. Can. *Plant Dis. Survey*, **54:** 45-51.

Datar, V. V. and C. D. Mayee (1981). Assessment of losses in tomato yield due to early blight. *Indian Phytopath.*, **34** : 191–195.

Datar, V.V. and Mayee, C.D. (1982). Conidial dispersal of *Alternaria solani* in tomato. *Indian Phytopathol.*, **35:** 68-70.

Endrinal, D. M. and Celino, (1940). Septoria leaf spot of tomato. *Phillip. Agric.*, **29**: 593-610.

FAOSTAT Database. Available online: http://faostat.fao.org/site/339/default.aspx (accessed on 19 September 2017).

Foolad, M.R., Merk, H.L. and Ashrafi, H. (2008). Genetics, genomics and breeding of late blight and early blight resistance in tomato. *Crit. Rev. Plant Sci.*,**27**:75–107.

Gallegly, M.E. and Galindo, J. (1958). Mating types and oospores of *Phytophthora infestans* in nature in mexico. *Phytopathol.*, **48**: 274–277.

Gisi, U., Walder, F., Resheat-Eini, Z., Edel, D., and Sierotzki, H. (2011). Changes of genotype, sensitivity and aggressiveness in *Phytophthora infestans* isolates collected in European countries in 1997, 2006 and 2007. *J. Phytopathol.*, **159**: 233–232.

Hardham, A.R. and Blackman, L.M. (2010). Molecular cytology of *Phytophthora*-plant interactions. *Australasian Plant Pathol.*, **39**: 29–35.

Johnson, C. S., Main, C. E. and Gooding, G. V. (1983). Crop loss assessment for flue-cured tobacco cultivars infected with tobacco mosaic virus. *Plant Dis*., **67**: 881-885.

Jones, J.B., Jones, J.P., Stall, R.E. and Zitter, T.A. eds (1993). Compendium of Tomato Diseases. St Paul, Minnesota, USA: *Am. Phytopathol. Society*, 73.

Judelson, H.S. (1997). The genetics and biology of *Phytophthora infestans*: Modern approaches to a historical challenge. *Fungal Genet. Biol.,* **22**: 65–76.

Kalele, D. N., Affokpon, A., Coosemans, J. and Kimenju, J. (2010). Suppression of root knot nematodes in Tomato and Cucumber using biological control agents. *Afr. J. Hort.* Sci., **3**: 72-80.

Naserinasab, F., Sahebani, N. and Etebarian, H. (2010). Biological control of Meloidogyne javanica by *Trichoderma harzianum* BI and salicylic acid on Tomato. Tehran University, Iran.

Nowicki, M., Lichocka, M., Nowakowska, M., Klosinska, U. and Kozik, E.U. (2012). A simple dual stain for detailed investigations of plant-fungal pathogen interactions. *Vegetable Crop Res. Bulletin,* **77**: 61–74, DOI:10.2478/v10032-012-0016-z.

Patel, M. K. (1950).Bacterial spot of Tomato (*Xanthomonas vesicatoria*) and its management – A Review. *Indian Phytopath*., **3**: 95-97.

Rahman, I. (2003). Root knot disease and its control. Agfact1, 3rd edition, abstract National Wine and Grape Centre, New South Wales Department of Agriculture, State of New South Wales. U.S.A.

Sherf, A. F. and MacNab, A.A. (1986). Vegetable diseases and their control. John Wiley and Sons, New York. pp. 634- 640.

Snyder, W.C. and Hans, H.N (2003) Soil borne Plant Pathogen Class Project,*Fusarium oxysporum* f. sp. *lycopersici* (Sacc.) PP728 *Springer*, **72**: 254-259.

Sohi, H. S. and Sokhi, B. S. (1973). Morphological physiological and pathological studies in *Septoria lycopersici*. *Indian Phytopath.*, **17**: 666-673.

Synder, W. C. and Hansen, H. N. (1940). The species concept in *fusarium*. *Am. J. Bot.*, **27**: 64-67.

Tripathi, S. and Varma, A. (2002). Eco-friendly management of leaf curl disease of tomato. *Indian Phytopath.*, **55** (4): 473.

Widmer, T. and Abawi, G. (2000). Mechanism of suppression of *Meloidogyne hapla* and its damage by green manure of Sudan grass. *Plant Dis.*, **84**: 562-568.

3

Diseases of Brinjal (*Solanum melongena* L.) and Their Integrated Management

***Gunadhya Kumar Upamanya*[1] and *Pranab Dutta*[2]**

[1] *SCSCA, Assam Agricultural University, Dhubri, Assam*
[2] *School of Crop Protection, College of Post Graduate Studies in Agricultural Sciences Central Agricultural University, Umiam, Meghalaya*

Introduction

Brinjal, *Solanum melongena* L. a solanaceous and the native vegetable of India is very favourite among the Indians. It is an economically important vegetable among the small scale farmers and low income consumers of India. It ensures some assured source of income to the resource poor farmers throughout the year due to long fruiting and harvesting period. Brinjal fruits are of a low calorie value and have a mineral composition that is beneficial for human health; these fruits are rich source of minerals (potassium, calcium, phosphorus, sodium, chlorine, iron etc.) and vitamins and have some medicinal importance (Choudhary, 1967). In India, brinjal is grown on the wide range of soils and other variable environmental conditions. In production and productivity, India stands second in the world after China which is mainly grown in the states like West Bengal, Orissa, Bihar, Gujarat, Maharashtra, Karnataka, Uttar Pradesh and Andhra Pradesh. The world production of brinjal in the year 2016 was estimated to be 51.29 million MT in an area of 1.79 million hectare with productivity of 28.59 MT per ha China, the world leader of brinjal production produced 32.03 million tones of brinjal with a productivity of 40.96 Mt per ha in an area of 0.78 million ha. The productivity of brinjal as compared to China is very less (18.90 MT/ha) in India although we are the second largest producer of this popular vegetable (FAOSTAT Website (http://faostat3.fao.org/home/E).

Lower productivity is attributed to the attack of numbers of insect pests and incidence diseases during the different stages of crop growth. The important diseases which cause considerable loss in brinjal in India are damping off of seedling caused by *Pythium aphanideramatum* (Eds.) Fitz., *Phytophthora* spp. and *Rhizoctonia* spp., *Alternaria* leaf spot by *Alternaria melongenae* Rang. & Samb and *A. solani* (El. & Mart) Jones and Grout, *Cercospora* leaf spot caused by *Cercospora melongenae* Welles and *C. solani-melongenae* Chupp, *Phomopsis* blight and fruit rot, *Phomopsis vexans* (Sacc & Syd) Harter, *Sclerotinia* blight by *Sclerotinia sclerotiorum* (Lib.) de Bary, *Verticillium* wilt by *Verticillium dahlia* Kleb., Fungal wilt (*Fusarium oxysporum* f. sp. *melongenae* (Fomg.), bacterial wilt caused by *Ralstonia solanacearum* Yabuuchi *et.al.,* little leaf caused by mycoplasma and mosaic caused by Tymo virus group. All these may cause yield loss more than 70 per cent.

Important Diseases of Brinjal

Pathogenic diseases have been posing a great threat and adversely affect the yield and quality of Brinjal in India. Some of the important diseases associated with brinjal caused by fungi, bacteria, MLO and virus are discussed below.

Damping off

Damping off, the diseases of seedlings occur worldwide in valleys and forest soils under tropical and temperate climates. The disease affects seed, seedlings and roots of all plants. However, the greatest damage is done on the seed and seedling roots during germination either before or after emergence. Losses vary considerably with soil moisture, temperature and other factors.

Symptoms

The plants are attacked at all stages of growth, producing damping off symptom in nurseries and collar rot of young pants. In pre emergence damping off, the young seedlings are killed before they reach the surface of the soil. The radical and the plumule undergo complete rotting as they come out of seed. Since this happens under the soil surface, the disease is often not detected except for the resulting poor stand. The post emergence damping off is very conspicuous. In this phase, toppling over of infected seedlings is observed any time after emergence of the seedlings until stem has hardened sufficiently to resist the invasion. Infection usually occurs at or below the ground level, the infected tissue become soft, water soaked and ultimately collapse and die. As the disease advances the stems become constricted at the base and plants collapse. In most cases, cotyledons and leaves slightly wilt before seedlings are prostrated, although sometimes they remain green and turgid until collapse of the seedlings occur. The disease usually radiates from initial infection points resulting killing of seedlings in large areas.

Causal Organism

There are too many fungi that can cause seed and seedling rot but species of Pythium are generally known as damping off fungi. These include *P. debaryanum*, *P. aphanidermatum, P. irregulare* and *P. ultimum.* Among these *P. ultimum* is mostly associated with brinjal damping off (Gholve *et al,* 2014). *Pythium* produces white, rapidly growing mycelium. The mycelium gives rise to spoarangia which germinate directly by producing one to several germ tubes or producing a short hypha at the end which forms a balloon like secondary sporangium called a vesicle. In the vesicle, 100 or more zoospores are produced, which when released swarm about few minutes, round off to form a cyst and then germinate by producing germ tube. The germ tube penetrate the host tissue and starts a new infection. The mycelium also gives rise to spherical oogonia and club shaped antheridia. Oogonium is fertilized with the nuclei from antheridia and produced oospore. Oospores are resistant to adverse temperature and moisture and serve as the survival and resting stage of the fungus.

Development of Disease

Spore germ tubes or saprophytic mycelium of *Pythium* coming in contact with seeds or seedling tissues of host plant enter by direct penetration. There are certain enzymes responsible for the development of the disease. Pectinolytic enzyme produced by the fungus causes the maceration of the tissue. Proteolytic enzymes break down the protoplast of invaded cells and in some cases cellulolytic enzymes cause complete collapse and disintegration of the cell walls. As a result infected seeds and young seedlings are killed and converted into rotten mass composed of the oomycetes and other non degradable substance like suberin and lignin.

Favorable Condition

Damping off disease is most severe in ill aerated, ill drained soil. Compact heavy soil shows this type of properties. Loose soil having good proportion of sand exhibit less from the disease. the temperature 24^0-30^0C, high moisture, poor aeration prologation of juvenile stage of the seedlings and general weakening of the plants favors the disease severity.

Management

i) Cultural practices like good soil drainage, good air circulation among plants, planting when temperature are favorable for plant growth, avoiding application of excessive amount of nitrate forms of nitrogen fertilizer etc. are helpful in reducing the disease.

ii) Use of chemical and/or biological seed protectants to keep away the pre-emergence phase and to adopt sanitary precautions in the nursery to check the appearance of post-emergence damping off. The common chemical seed protectants used against damping off are Thiram, Captan or systemic like Metalaxyl. *Trichoderma harzianum*, *Penicillium oxalicum* and *Pseudomonas fluorescens* are the common biological seed protectant.

iii) Application of talc based bioformulation of *T. harzianum, or Pseudomonas fluorescens* in the nursery bed before sowing significantly reduce the damping off disease.

iv) Gholve *et al.,* (2014) suggested integrated schedule of managing the damping off of brinjal by seed treatment of captan (at 1.5 g/kg) + metalaxyl (at 3g/kg seed) + garlic extract (at 100ml/kg soil) +soil application of *T. viride* (at 25g/kg soil).

Alternaria Leaf Spot

The diseases caused by *Alternaria* are among the most common diseases of many kinds of plants throughout the world. They affect the leaves, stems, flowers and fruits of vegetable crops. Total aggregate losses caused by by the various Alternarias on all their host rank among the highest caused by any pathogen.

Symptoms

The disease first makes its appearance in young seedling. It attacks leaves and then spreads to fruits which subsequently rot and become unfit for consumption The disease causes characteristic leaf spots with concentric rings. Initially disease appears as small, dark brown and sunken spots, which subsequently gets converted into concentric rings and then become olivaceous dark brown lesion due to spore formation. The spots are mostly irregular and coalesce to cover large areas of the leaf blade. Severely affected leaves drop off. The symptoms on the affected fruits are in the form of large deep-seated spots. The infected fruits turn yellow and drop off prematurely.

Causal Organism

The disease is caused by *Alternaria melongenae* belongs to the sub-division Deuteromycotina, class Hyphomycetes, family Dematiaceae. The genus is widely distributed all over the world. The mycelium of the fungus is dull brown and septate, produced muriform, olive colored conidia with longitudinal and transverse septa (3-5 septa).

Development of Disease

Plant pathogenic species of *Alternaria* overwinter as mycelium or spores in infected plant debris and in or on the seeds. If the fungus is carried out with seeds it may attack seedlings usually after emergence and cause damping off or stem lesions or collar rot. However, in most cases the spores are blown from infected debris or infected cultivated plants and weeds under favorable environment condition. The germinating spores penetrate susceptible tissue directly or through wounds and soon produce new conidia and are further spread by wind, splashing rain etc.

Favorable Conditions

The fungus grows and sporulates on plant debris during periods of rain, heavy dew, or under good moisture conditions. Low temperature ($20^0 – 25^0$C), high humidity and cloudy weather favors the disease development.

Management

Field sanitation, crop rotation, burning of plant debris if infected and eradication of weed host help reduce the inoculums for subsequent planting of susceptible host. *Trichoderma koningii* and *Trichothecium* are found to be the natural enemy of Alternaria. Neem leaf extract was found to be the better alternative to control *Alternaria* leaf spot of brinjal (Nwogbaga and Wuagwu, 2015). The disease can be managed through using treated seeds and chemical sprays with appropriate fungicides among the fungicides, Dithane M-45 (0.2%), Zineb (0.2%) Carbendazim 50 WP (0.1%) are useful to control *Alternaria* disease.

Cercospora Leaf Spot

The *Cercospora* attacks mostly on the leaves causing the leaf spot disease. The spots either stay relatively small and separate or may enlarge and coalesce resulting in leaf blights. The diseases are generally wide spread among most cereals and grasses, many field crops, vegetables, ornamentals and trees. It is also important disease causing significant loss in brinjal.

Symptoms

The infection starts from the older i.e. lowest leaves and if unchecked can move upwards and infect young leaves and stems. Severely infected leaves curl and prematurely drop off the plant, often causing a reduction in yield (Chupp 2006). However, the pathogen does not infect the fruits but reduce the growth of plants as well as the yield. In severe condition the disease may be fatal to plants. Initially, small, circular to oval chlorotic spots appear that may develop angular or irregular shapes. The spots on leaves can easily be confused with spots caused by a bacterial disease. On closer inspection, however, the spots due to *Cercospora* show distinguishing features. They are generally circular with light to dark tan centers, and the stomata have black spots that can be readily observed using a hand-held magnifying lens. Elliptical to oval lesions may occur on the leaf blades, veins, and petioles. In case of bacterial leaf spots, irregularly shaped or circular spots with clear stomata has been observed. The leaf spots on brinjal caused by *C. melongenae* are 4-10 mm in diameter and visible on both leaf surfaces, though they may appear more abundantly on the lower surface. During the later stages of disease, lesions become grayish brown, with excessive sporulation at their centers. Concentric rings of diseased tissue may appear as the lesions gradually expand in size. As a lesion dries, the tissue in the center may crack and drop out.

Causal Organism

A plant-pathogenic fungus, *Cercospora melongenae*, causes *Cercospora* leaf spot of eggplant in Hawai'i. The genus *Cercospora* is a hyphomycete fungus comprised of many plant-pathogenic species. They produce leaf spot diseases on a wide range of agriculturally important plants. Conidiophores (fungal structures that bear the spores/conidia) appear in fascicles (clusters) of 3 to 12. They are pale to medium brown, paler towards the apex, occasionally septate, and unbranched, and they bear hyaline, mildly curved conidia (Mycobank Database 2012).

Development of Disease

Cercospora melongenae can survive for at least one year in plant debris or in soil as minute stromata. The disease process begins when the fungal spores are dispersed to susceptible plants by rain, irrigation water, or wind, or on agriculture equipment or by people (Nelson 2008).

Favorable Conditions

The fungus is favored by high temperature and therefore is most destructive in the summer month and in warmer climates. Leaf wetness and high relative humidity favor infection and disease development.

Management

Both organic and inorganic approaches can be used for the management of this disease.

- Maintain proper field sanitation: Collect and bury infected crop residues; remove and destroy severely diseased leaves.

- Use disease-free transplants.
- Control weeds to reduce relative humidity in the eggplant canopy.
- Irrigate in the morning to reduce humid and damp conditions overnight.
- Avoid over-irrigation to reduce relative humidity.
- Avoid overhead sprinkler irrigation in order to minimize leaf wetness and spread of the pathogen in splashing water droplets.
- Do not work with wet plants or move through a field of wet plants, as such movement can disperse fungal spores among plants.
- Increase the spacing between plants to improve aeration and drying of wet foliage.
- Intercrop eggplant with other vegetables to interrupt pathogen transmission between plants.
- Grow eggplant under cover (i.e., in greehouses or under shadecloth) where possible to minimize leaf wetness.
- A calendar-based spray program using a protectant fungicide, combined with cultural practices, can reduce losses from *Cercospora* leaf spot on eggplant.

Sclerotinia Blight

Sclerotinia sclerotiorum (Lib.) de Bary causing *Sclerotinia* stem and root rot of vegetables has been considered as a ubiquitous necrotrophic soil borne plant pathogen. The disease caused by *Sclerotinia* is also known as white rot, watery soft rot, cottony rot or white blight based on their symptoms developed in the host. The pathogen has a wide host range. Purdy (1979) has reported that *S. sclerotiorum* infects 64 plant families, 225 genera and 383 plant species.

Symptoms

The infection occurred on the stem, branches, leaves and fruits. Primary symptom generally appeared on main stem near the soil surface. Water soaked lesion appeared on the stem which gradually increase in length and reached to the base of one or more branches. Major portion of the infected area are covered with white cottony mycelial growth. Sclerotia are observed on the mycelial growth, which are white at initial stage, later became more or less black in colour. In the mean time, infection girdle the stem, plant wilted and died. In some plants, instead of stem infection, symptom appeared on the branches and leaves of those branches. On the branch and leaves white mycelial growth covered the lesion embedded with sclerotia. Branch infection resulted partial wilting of the plant. Occasionally, the host exhibited dry lesions on the stalk, stem or branches with a well-defined border between healthy and diseased tissues. On opening dry portion of the stem, pith was full of fungal sclerotia, medium to large in size, elongated or cylindrical and often attached to each other end to end. Fruits are infected directly from the soil surface or through the peduncle rot quickly. In advanced stages, white, cottony mycelium blankets covered the affected tissue, and sclerotia form on the surface of the fruit.

Causal Organism

The disease is caused by the fungus *Sclerotinia sclerotiorum.* The mycelium of the fungus is hyaline, much branched, consisting of closely septate large hyphae which are both intra and intercellular and invade all the tissues of the affected portions. The hyphae are 9-18 µm broad and filled with dense protoplasm. No true conidia are produced. When the vegetative growth has ceased the hypahe with thick granular protoplasm, and short cells collect in small dense masses which gradually become sclerotia. At first these are pink, but later turn dark brown to black and become smooth. The sclerotia germinate myceliogenically directly producing mycelium or caprogenically initiating the development of apothecia.

Development of Disease

The fungus survives from year to year as hard dark structures called sclerotia. White hyphal strands extend from sclerotia that have been stimulated to germinate by host plant exudates. The fungus is germinated by direct emergence of hyphae (termed ‘myceliogenic’ or ‘eruptive’ germination). These hyphae infect roots, and crowns and other plant parts that are touching the ground. Infection then spreads to aboveground plant parts.

Favorable Conditions

High soil moisture coupled with temperatures range of 16 to 21°C is favourable for disease development.

Management

The control of *Sclerotinia* disease depends on a number of cultural practices and on chemical sprays. Affected plants and debris should be collected and burnt immediately after harvest. Deep summer ploughing should be given in such a way that surface soil is buried deep. Soil should be kept free of weeds between crops. Close planting should be avoided. Rotation of cropping pattern with crops like beet root, onion, maize or paddy to eliminate the fungal inoculum in the field. The biological control agent like *Gliocladium roseum, Trichoderma viride, Coniothyrium minitans* may be used successfully against *Sclerotinia* diseases. Spraying Mancozeb 75% WP @ 3 g per litre or Carbandazim 50% WP @ 2 g per litre controls the disease.

Bavistin (0.1%), Vitavax (0.1%), and Topsin-M (0.15%) proved to be the most effective in inhibiting the growth of pathogen *in vitro* and controlling the diseases in the field. Biopesticide, Nimbidine was also proved effective, but slightly less effective than systemic fungicide except Chlorothalonil. Nimbidine being a safe eco-friendly and economical bioproduct which can be used in the management of the disease (Singh *et. al.,* 2008).

Phomopsis Blight

The disease is of worldwide occurrence but is more severe in tropical and sub tropical areas. In India it was reported in 1935 in Gujrat (Rangaswamy, 1972). The pathogen attacks both the foliage and fruits but the later phase is more destructive.

Symptoms

It is a serious disease, which may cause damping off symptoms if attacked at seedling stage. When the leaves are infected, small circular spots appear which become grey to brown with a light color centre. Several dark pycnidia can be seen on older spots. The infected leaves may turn yellow and die. Lesion may also develop on petiole and stem cause blighting of affected portions. In course of time, the spot enlarges and produces concentric circular area. On the fruits, small, pale sunken spots appear which on enlargement cover the entire fruit surface. These spots become watery leading to soft rot phase of the disease. A large number of dot like pycnidia also develop on such spots. The infection of fruits through calyx leads to development of dry rots and fruits appear black and mummified.

Etiology

The disease is caused by *Phomopsis vexans*

Conidiomata pycnidial, subepidermal, erumpent, dark, thick-walled, flattened to globose, varying in size, often 100-300 μm diameter, with or without a beak; beak to 76 μm. Phialides hyaline, simple or branched, sometimes septate, 10-16 μm long, arising from the innermost layer of cells lining the cavity. Alpha conidia hyaline, aseptate, sub-cylindrical, 5-8 x 2-3 μm. Beta conidia filiform, curved, hyaline, septate, 18-32 x 0.5-2.0 μm, non-germinating. Hyphae hyaline, septate, 2.5-4.0 μm diameter

Ascomata perithecial, in culture usually in clusters, 130-350 µm diameter, beaked; beaks sinuous, carbonaceous, irregular, 80-500 µm long. Asci clavate, sessile, 24-44 x 5-12 µm, eight-spored. Ascospores biseriate, hyaline, narrowly ellipsoid to bluntly fusoid, one-septate, constricted at the septum, 9-12 x 3.0-4.5 µm.

Development of Disease

The pathogen is seed borne and also survives in plant debris. On the seed the fungus is normally restricted as mycelium on the seed coat and cotyledons (Karuna *et. al*., 1994). Sometimes the fungus can also be found in the embryo. On the plant debris, the fungus survives both as mycelium and pycnidia. Spores of the fungus are released from the pycnidia. The major means of spread is by rain splashing. Wind dispersal is usually considered to be of minor importance. The seed infection lead directly to diseased seedlings.

Favorable Conditions

P. vexans requires hot and humid conditions for infection and disease development. Spore germination is optimal at 27°C, and pycnidial formation is greatest between 30 and 35°C. The optimum relative humidity for disease development is 55% RH and above. Fruit rot was maximal at 30°C and 50% RH.

Management

- The use of resistant varieties can be one of the most effective methods of control.
- Burning of crop debris and burying it by deep ploughing are some of the cultural practices that may help to reduce disease incidence.
- Three-year rotation can be useful in reducing initial inoculum.
- Use of an appropriate nitrogen source at a reduced level with higher rates of phosphorus and potassium fertilizer may increase yield without increasing disease.
- The pathogen also survives on and in seeds, therefore seeds should be collected from healthy plants and only disease-free seeds should be used.
- Antagonistic *Pseudomonas fluorescens* and *Trichoderma harzianum* seed treatment and spray treatment were found to be effective against *P. vexans*. Upamanya and Dutta (2018) found that the consortia of *T. harzianum,* Metarrhizium anisopliae and Beauveria can reduce the severity of the disease in field condition significantly.

- Chemical control, especially the use of fungicides, is largely practised for *Phomopsis* blight control in *Solanum melongena* The more common fungicides applied as foliar sprays are Bordeaux mixture, captan, carbendazim, carboxin, chlorothalonil, copper oxychloride, dithiocarbamates, maneb, mancozeb, thiophanate-methyl, tolclofos-methyl, ferbam and zineb.
- The newer systemic fungicide tebuconazole also provides a high level of control at a low concentration.
- Seed treatment with mancozeb, carbendazim and thiophanatemethyl has also produced a reduction in disease incidence.

Fusarium Wilt

Many *Fusarium* species are plant pathogen and cause vascular wilts, root and fruit rots diseases on various types of vegetable crops. Vascular wilt is an important disease of brinjal occurring in temperate, sub tropical and tropical areas. In India North and Central parts of country i.e. Punjab, Haryana, UP, MP etc. are mostly affected. This pathogen blocks the xylem transport system and causes severe wilting and death of brinjal plants which results in significant reduction in fruit yield.

Symptoms

Symptoms often appear later in the growing season and are first noticed on the lower (older) leaves. At the initial stage, yellowing of the leaves from lower to upper leaves followed by drying and defoliation are observed. Vein clearing symptom on the leaves can also be noticed. The plant as a whole is stunted due to attack of pathogen and the lateral roots showing the symptom of black rotting. As the infection spreads up into the stems and leaves it restricts water flow causing the foliage to wilt. Partial or complete wilting of plant takes place at the time of flowering and fruiting and thus causing almost complete loss in the fruit yield. Dense mass of mycelium and hyphae can be seen in xylem vessels which cause necrosis and death of vascular tissue. Plugging of xylem vessels takes place as a result of the presence of pathogen in form of conidia or mycelium. Pathogen is also reported to release fusaric acid, a toxin which is known to cause necrosis and death of xylem vessel. Even the root appears healthy externally, parallel brown stripe or rotting of vascular tissue may be seen upon splitting.

Etiology

Brinjal wilt is caused by the Deuteromycetes fungus *Fusarium oxysporum* f. sp. *melongenae*. The pathogen have dense mycelium with sporodochium.

Mycelium is branched, septate and hyaline which later on develops into a fruiting body called sporodochium. Two types of conidia are formed by the fungus, macro conidia and micro conidia. Macro conidia are sickle shaped, septate having 3-4 cells, measuring 25-33 X 3.5-5.5 µm. Micro conidia are single celled ovoid to elliptical, hyaline and measures 2.5 X 4.4-6 µm. Pathogen also produces thick walled structure chlamydospore which are formed from hyphal cell at the end of the crop season or during adverse condition.

Development of Disease

The pathogen is facultative parasite, grows on infected plant debris and colonizes dead host tissue saprophytically in moist soil during absence of host. Pathogen sporulates and produces micro and macro conidia and colonizes around the root surface before causing infection on host root tissues. Primary infection is established either through germination of conidia or chlamydospore by forming germ tube which later develops into appressorium and establishes infection on rootlets or root hairs. The pathogen is initially intercellular, colonizes tissues of cortical region and multiplied. In later stage the fungus establishes the infection in xylem tissue and become intracellular. Pectolytic enzyme is secreted by the pathogen that blocks the xylem vessel and ultimately produces the wilting symptom.

Favorable Conditions

The optimum temperature and soil moisture for disease development 25^0C-28^0C and 50-60% respectively. Diffuse light, short day length and deficiency of N favor disease development.

Management

Use of chemical, cultural and biological measures are some common practices followed to control this disease to some extent. Being a soil borne plant pathogen, it is difficult to control using a conventional chemical fungicides, because spores of this fungus survives for many years in the soil. Intensive use of chemical fungicides accumulates toxin in the environment and create residue problems. Rhizospheric microorganisms are the ideal control for soil borne plant pathogens. However, the following management strategies are employed against *Fusarium* wilt of brinjal.

- Crop rotation for 3-4 years raising from non host crop like okra, colocasia, mustard, cucurbits, maize, turmeric, ginger, cowpea etc. to reduce pathogen population.
- Selection of disease free field for brinjal cultivation. The field should be well drained and leveled. Low lying area should be avoided for brinjal cultivation.

- Deep summer ploughing is helpful to reduce the inoculums of the pathogen
- Irrigation followed by soil solarization using white transparent polythene sheet during summer days for a period of 15-20 days is found effective in eradicating soil borne inoculums. However, the method is feasible in nursery beds and smaller plots.
- Use of FYM or vermicompost based biocontrol agents like *Trichoderma harzianum*.
- Nursery treatment as soil drench with Carbendazim or Benlate @ 2 g/L of water.
- Seed dressing with Carbendazim or Benlate @ 2g/kg of seed. Seedling dip treatment with Carbendazim @ 0.2% or Hexaconazole @ 0.2% for 1-2 hours before transplanting in the field.

Bacterial Wilt

Bacterial wilt incited by *Ralstonia solanacearum* is a devastating disease distributed all over the world including tropical, subtropical and temperate regions (Hayward, 1991). This disease attacks over 450 plant species (Daughtrey *et al.,* 1995). In India the disease poses a constant threat to tomato and egg cultivation particularly in the states of Karnataka, Madhya Pradesh, Maharashtra and West Bengal. The loss in yield may vary between 10.8 and 90.6 % depending on the stage of the plant growth at which infection occurs and on environmental condition (Kisun, 1987).

Symptoms

Disease develops very rapidly in warm weather. Symptoms are very clear during morning or immediately after irrigation. Symptoms manifest initially as leaf drooping followed by wilting of entire plant within a few days. Wilted plants look green, a distinct symptom when compared to other vascular wilt diseases which develops yellowing of the leaves. Vascular discoloration (brown) can also be seen in the wilted plant. Development of adventitious roots from the stem is considerably enhanced. In many cases when nematode infection is also present the stem at the base becomes dark brown and constricted leading to collapse of the plant. A stream of milky white bacterial ooze can be noticed when the cut ends of the stem/root is kept undisturbed for few minutes in a clear container with water. This is a simple diagnostic method and could be used in the field to identify bacterial wilt.

Etiology

The causal organism of the disease, *Ralstonia solanacearum* (Smith) Yabuuchi *et al* 1996 was earlier known as *Pseudomonas solanacearum* (Smith) Smith, later renamed as *Burkholderia solancearum* (Smith) Yabuuchi *et. al.,* 1993. The bacterium is rod shaped measures 0.5-0.7 X 1.5-2.5 μm and are motile by 1-4 polar flagella. However, virulent strains are usually not flagellated and cells are surrounded by thick polysaccharide material. Stain reaction is Gram-negative.

Development of Disease

The pathogen is soil borne, persisting for long periods in some soils. Intact roots are not usually invaded. Infection always occurs through wounds caused by transplanting, cultural operations or nematode invasion. Rifts in root cortex and breaking of the stem for emergence of adventitious roots also provide openings for entry of the bacteria. The entry of the bacteria in the taproot is faster in susceptible than resistant cultivars. However, once inside the root the bacteria colonize, to some extent, almost all regions of the susceptible and resistant plants. The acidic extracellular polysaccharide and plant cell degrading polygalacturonases are the major pathogenicity factors in *R. solanacearum.*

Favorable Conditions

The optimum temperature and soil moisture for disease development 25^0C-28^0C and 50-60% respectively. Diffuse light, short day length and deficiency of N favor disease development.

Management

The bacterial wilt is very difficult disease to manage as the commercial cultivars are highly susceptible and no chemical control is feasible. Recently great emphasis has been laid on non chemical methods such as biological and cultural practices.

Since the pathogen is a soil-borne, the disease will occur only if the pathogen is present in the field and environmental conditions are favorable for its development. Fields that have no previous history of bacterial wilt and have not been planted with susceptible hosts during the previous season/year are less likely to have the pathogen. Regular rotation with other non-host plants will reduce disease incidence. However, rotation with paddy for only one season doesn't make the field disease free.

- Use resistant varieties like Surya, Swetha and Haritha and pathogen free seedlings are simple and most promising way to control the disease. Soil solarization or fumigation of nursery soil for 15 days prior to sowing. prevent the spread of the pathogen. Infected plants should be removed and destroyed and also the infected portion of the field should be isolated if possible by preventing the water flow into and from the field. Frequency of irrigation and quantity of water should be reduced. Flood irrigation in the infected field increases the disease incidence by two fold. Growing marigold (*Tagetes* spp.) in rotation or as intercrop will suppress the pathogen in addition to its anti-nematode effect. Growing *Brassica* spp. and incorporating the plants into soil at flowering stage will reduce the disease incidence.

- Application of organic manures like FYM/poultry manure in every year will increase the beneficial soil microflora and reduce the disease incidence. Soil application of biological control agent like *Pseudomonas fluorescens/ Bacillus subtilis* along with vermicompost or FYM in nursery as well as main field is recommended for management of bacterial wilt of brinjal.

Little Leaf

Little leaf caused by *Phytoplasma* is an important disease of brinjal which causes considerable economic losses. In case of early infection complete failure of the crop has been observed. The disease is found throughout India, Sri Lanka, and Bangladesh. In India it was first reported from Coimbatore (Tamil Nadu). In India 90 per cent of brinjal yield was recorded as aresult of infection by little leaf of brinjal (Joshi and Bose, 1983). There is also loss of germinability of seeds from fruits formed on infected plants.

Symptoms

The main symptom of the disease is production of very short leaves by affected plants. The disease starts with reduction of size of leaf lamina and shortening of petiole. Such leaves are narrow, soft, smooth and yellowish in color. Newly formed leaves are further reduced in size. The internodes of the stem are shortened and at the same time a large number of axilllary buds are stimulated to grow into short branches with small leaves. This results in stunted and bushy appearance of the plant. Normally the flowers are not produced but in case these are produced, these remain green. Formations of fruits are rarely seen.

Etiology

Prior to 1968 the causal agent of the disease was considered as virus. In the year 1988, Lee et al recognized *Phytoplasma* of the clover proliferation group as the causal of the disease. The *Phytoplasma* is ovoid to spherical measures 40-300 nm in diameter and lack a rigid cell wall. These structures are present in the phloem sieve tube cells of the stem, petiole, phyllodes, and roots and haemolymph of the fat bodies of the vector *Hishimonus* (*Cestius*) *phycitis*.

Development of Disease

In nature, the little leaf Phytoplasma is transmitted by its insect vector, leaf hopper (*Hishimonas phycitis*). *Empoasca devstans* also transmit the *Phytoplasma* but is less efficient. During the off season of the brinjal crop, the *Phytoplasma* survives in the weed host like *Datura fastuosa, Vinca roseus*, and *Catharanthus roseus* and is transmitted by its main host by insect vector.

Management

Effective control measures against little leaf of brinjal is not known. Treatment with tetracyclines were claimed to control the disease. Elimination of weed hosts, control of vector through insecticides is the effective to manage the disease. There are some brinjal variety which are moderately resistant to phytoplasma. Some of the moderately resistant varieties are Pusa Purple Cluster, Arka Sheel, Manjari Gota, Banaras Giant Black Beauty etc.

Conclusion

Lot of research efforts have gone into various aspects of diseases management of vegetable crops including the most popular vegetable, brinjal under the National Agriculture Research System comprising of Central Institutes and State Agricultural Universities. The popularity of management of disease with biological method has been tremendously increased with respect to some of the diseases of brinjal like damping off, fungal and bacterial wilt, *Phomopsis* blight. Inclusion of advanced technology like use of nano particle in integrated disease management system against brinjal disease might be useful for the growers to enhance the productivity with increased B:C ratio.

References

Chaudhury, B. (1967). Vegetables. National Book Trust, India, New Delhi. 48p.

Chupp, C. (2006). Manual of Vegetable Plant Diseases, Discovery Publishing House, India.

Gholve, V.M., Tatikundalwar, V.R., Suryawanshi, A.P. and Dey, U. (2014). Effect of fungicides, plant extracts / botanicals and bioagents against damping off in brinjal. *African J. of Microbiol. Res.* **8** (30): 2835-2848

Joshi, R.D. and Bose, K. (1983). Yield loss studies on little leaf of egg plant. *Indian Phytopathol.* **36**:604-607

Karuna Vishnuvat, Kumar, Sanjay and Kumar, S. (1994). Location and infection of Phomopsis vexans in brinjal seed. *Indian J. Mycol. Plan Pathol.* **24**: 226.

Kishun, R. (1987). Loss in yield of tomato due to bacterial wilt caused by *Pseudomonas solanacearum. Indian Phytopath.* **40:** 152.

Nwogbaga, A.C and Wuagwu, C.C. (2013). Effect of fungicide and neem leaf extract sprays on the management of Alternaria solani leaf spot disease of egg plant (*Solanum melongena* L.) *Journal of Global Biosciences* **4** (6): 2441-2445.

Purdy, L.H. (1979). Sclerotinia sclerotiorum, history diseases, symptomatology, host range, disease resistance, geographic distribution and impact. *Phytopathology*, **69**: 875-880.

Rangaswami, G. (1972). Diseases of crop plants in India Prentice Hall of India Pvt. Ltd. New Delhi 504 pp

Singh, R., Singh, P.C., Singh, N. and Alka (2008). Evaluation of different fungicides and biiopestricides against Sclerotinia blight of brinjal (*Solanum melongena* L.). *International Journal of Plant Protection*. **1** (2): 97-99.

Upamanya G.K. and Dutta Pranab, Development of IPDM module. PhD Thesis, Department of Plant Pathology, Assam Agricultural University, Jorhat, India (2018).

4

Diseases of Chilli (*Capsicum annum* L.) and Their Integrated Management

***Sakendar Ali*[1] and *Bandana Saikia*[2]**

[1] *Ex. Professor and Head, Department of Plant Pathology, Assam Agricultural University, Jorhat-785 013, Assam*

[2] *Department of Plant Pathology, Assam Agricultural University, Jorhat-785 013 Assam*

Chilli is reported to be an inhabitant of South America and widely scattered in all tropical and sub-tropical countries including India. Amongst all the important consumed crops, chilli occupies a dominant position and its economic losses worldwide due to few serious diseases like chilli anthracnose, Phytophthora leaf blight and fruit rot, bacterial leaf spot, bacterial wilt, Cercospora leaf spot, chilli leaf curl virus and so on. *Capsicum annuum* L., *C. frutescens* L. are the biological names of chilli. Bird's eye chilli is grown in Mizoram and some areas of Manipur and Assam. Chilli has got its own regional names. In Hindi, it is popularly known as *Lal Mirch*, in Gujrat as *March*, in Marathi as *Mirchi* and so on. Kanthari White is grown in Kerala and some parts of Tamil Nadu. Kashmir chilli is grown in temperate region such as Himachal Pradesh, Jammu and Kashmir and also in sub-tropical regions of North India. Red chilli held high amount of vitamin C and provitamin A, whereas green chilli too contain a considerable amount of both substances. In addition, peppers are a good quality source of most B vitamins, vitamin B6 in particular. They are very high in Potassium, Magnesium and Iron supplement.

Anthracnose of Chilli

Economic Importance

This is one of the serious diseases of chilli causing severe infection if fruiting coincides with rainy season. The pathogen requires high moisture for the germination of its conidia and could infect leaf, stem and also fruits. The disease was reported from various countries of the world such as China, Philippines,

Uganda, Ceylon, USA, Florida, India, Pakistan, Bangladesh etc. (Roy *et al.*, 1997). The loss is high owing to the post and preharvest involvement of the pathogen causing a loss of 10-80% of the marketable yield of chilli fruit (Than *et al.*, 2008). As reported from Punjab and Haryana (Bansal and Grover, 1969), the crop loss due in Anthracnose disease range from 10-35%.In India a calculated loss of 10-54% has been reported due to the disease (Lakshmesha *et al.*, 2005).Additionally chilli has got the ability to boost the immune system and lower the cholesterol levels (Grubben and Mohamed, 2004).

Symptoms

Die-back: symptoms appear as necrosis on tender twigs from the tip downwards. With the advancement of the disease other branches are also infected. In severe cases, entire plant is infected resulted withering. The infected twigs, stems become straw coloured where large number of black dots of acervuli formed. Subsequently, conidia will appear on those conidiomata to cause secondary infection.

The disease could be noticed both on green and dry fruits. The ripe fruits however more prone to the pathogen attack. In bell pepper, both green and ripe fruits exhibit symptoms. The spot are sunken, small, black and circular. The pathogen shows characteristics circular ring on the spots, which could be easily detected to be fruiting bodies of the fungus. Several spots coalesce to produce larger spots. Under congenial environment the pathogen produces pinkish mass of spores on these spots. The fungus can invade the fruits and in severe cases infect the seeds. Fruits become rusty in colour and subsequently shrivelled.

On chilli fruits, *Colletotrichum gloeosporioides* also can infect and produces sunken bleached lesions on the fruits where many light black brown coloured acervuli appear either in ring or in scattered fashion (Khodke and Gahukar, 1995).

Causal Organisms

The major pathogen of this anthracnose disease is *Colletotrichum capsici*. However, occurrence of *C. gloeosporioides* (Teleomorph: *Glomerella cingulata*) has also been reported (Khodke and Gahukar, 1995).

C. capsici produces dense whitish to dark green aerial mycelium,pale buff to salmon coloured conidia in mass, individual conidia hyaline, unicellular, falcate, fusiform in shape. Appressoria are produce by the germinating conidia and (Ali, 1989) in vitro could produce appressoria in chain of *C. capsici* (Plate 1. c), one of the important diagonostic characters of the organism. Setae are formed in abundance. The appressoria are medium brown, clavate and globose to sub-globose.

C.gloeosporioides Penz. also produces sunken lesions where acervuli are prominent with setae. The conidiophores are formed in acervuli, hyaline, bear unicellular conidia. The conidia are guttulate, cylindrical in shape with obtuse ends. The fungus also produces appressoria, but unlike *C.capsici* formed not in chain, which are clavate, broadly ovoid and brown in colour.

Disease Cycle

Major source of inoculums is infected debris and seed. The pathogen remains in the soil along with debris for nine months and act as primary source of inoculum. Conidia are the secondary source of infection that takes place through wind borne conidia. The pathogen *C. capsici* is seed borne. Seed borne nature of the anthracnose pathogens was already in record. (Jindal *et al.,* 1994).

Epidemiology

Prolong dew deposition causes severe infection resulting die-back. Average relative humidity around 78.5% is essential to cause serious damage and temperature 26°C along with free moisture and RH 100 percent was optimum for disease development. According to Prakasam and Jayarajan, (1989), the ripe fruits are more susceptible than 5 day old resistant green fruits.

Consideration of pathogen diversity and aggressiveness bear some logic to interrupt pathogen population structure that could be reduced and limit the potential of epidemic growth. Wide host range of *C. gloeosporioides* and *C. capsici* had been reported from North-East region of India (Ali, 1989); and total of twenty numbers of hosts were recorded against these two pathogens.

Integrated Management

Due to wide variation in the pathogen *C. capsici*, development of resistant variety is quite impractical. However, several reports indicated the resistant sources of the pathogen and these are California Wonder, Hungarian War, Pazardiska, Kapija, Soroksari, Merrimack Wonder, *C microcarpus* L etc are resistant (Ullasa *et al.,* 1981). Spraying of chemicals *viz*., Bavistin (0.1%), Dithane M-45 (0.25%), Dithane Z-78(0.25%), Difolatan (0.2%), Captan etc. were effective in the containment of the disease under field condition. As reported (Rashid *et al.,* 2015), the combine effect of plant extract of neem (*Azadirachta indica*), mahagoni *(Swietenia mahagoni*) and garlic (*Alium sativa*) had significant effect in reducing chilli anthracnose and found higher yield.

Future Approach

The use of resistant cultivars is the cheapest, easiest, safest and most effective means of controlling crop diseases. But no resistant cultivars of chilli have been developed and commercialized, it is very important to develop alternative means to manage this disease particularly the eco-friendly biological management strategies. Depending on weather factors the planting dates may be adjusted so that disease could be escaped particularly in endemic areas and areas where hot weather prevails with high rainfall. As the important pathogens *viz*., *C. gloeosporioides* and *C. capsici* have wide host range,cross-inoculation of the pathogens within and between the hosts maybe a challenging approach to explore their presence in the crop area of a particular region.It was apprehended that many areas are still unexplored in terms of host- pathogen interaction,its spread and effective management strategies (Saxena *et al.,* 2016).

Conclusion

Further studies leading to Integrated Disease Management (IDM) strategies need to be carried out so that eco-friendly, economically viable technologies could be evolved, using less and less chemicals for the management; and technologies could be fitted for organic agriculture. There remain many questions to be answered concerning characterization of *Colletotrichum* species associated with anthracnose in particular species present in different countries and regions; pathogenic or genetic diversity of *Colletotrichum* spp. worldwide; infection process and the disease cycle of *Colletotrichum* spp. leading to effective disease control and resistant plant breeding (Than *et al.,* 2008).

Cercospora Leaf Spot

Economic Importance

Based on favorable conditions the disease may assume serious proportions lead to heavy losses. Warm and moist weather is preferable for the disease. The leaf spot disease is found both in nursery and main field. The loss due to the disease may vary based on severity, under severe attack young leaves defoliate and cause significant loss. There was no report available about the exact loss that the disease can cause in the literature. However, Meon, (1990) reported that due to the disease, photosynthetic activity of the infected plant reduces and may cause substantial loss.

Symptoms

Symptoms appear as brown to black diffused spots, scattered later can form circular to ellipsoid, centre become grey. The leaves turn yellow and defoliates. Sometimes the central portion of spots drops off. The spots may appear on stems, twigs as dark brown irregular with whitish centre. Infection on fruit stalk and calyx is also very common. Dieback of the twigs may occur in severe cases.

Causal Organism

Cercospora is a genus falls under the teleomorph fungus Ascomycota division, class Dothidiomycetes. Most species have no known sexual stage and when the sexual stage is identified it is in the genus *Mycosphaerella*, anamorph *Cercospora.*

Division : Ascomycota

Class : Dothidiomycetes

Order : Capnodiales

Family : Mycosphaerellaceae or Dothidiaceae

Rank : Genus

Anamorph : *Cercospora capsici* Heald and Wolf

The anamorphic stage is characterized by dark brown stromata from where fasciculate yellowish brown conidiophore arises. The size of the stromata ranges from 34 - 39 x 28.5-45 µm and conidiophores 31.2 - 37.5 x 4.5 - 5.5 µm. The conidia are yellow to brown, 1-3 septate, aerogenous, tapering towards the apex measuring 34-77 x 3.5- 5.5 µm.

Disease Cycle

Dormant mycelium and stromata in the infected crop debris may serve as a primary inoculum. The pathogen mainly hibernates in seed and can initiate infection. Volunteer plant also may add inoculums if found available nearby field. Conidia are dispersed by wind and rain splashes to the host plant surface where they germinate to cause infection. They remain dormant as mycelium either on plant debris or in seed to carry over the disease for the next season.

Epidemiology

According to Mech and Khan, (1987), the disease is caused by *Cercospora capsici,* needs a temperature <than 28 °C, 92% relative humidity and pH 5-6 to manifest disease. Below 90% RH the disease cannot develop. The disease is more severe in wet weather than in dry weather and become destructive in high relative humidity (Ceekauskas, 2004). As reported by Suang *et al.,* (1984) the disease development is favoured by high temperature and humidity. Meon, (1990) further reported that the pathogen sporulates abundantly at 20 to 30 °C, poorly at 10 °C and ceases at 40 °C. Faster progress of the disease was also recorded (Kawagoe, 1998), when temperature was around 20 to 25 °C and prevailing relative humidity more than 95%. Sporulation found maximum when moisture is high and decline accordingly when moisture reduces. He further

reported that the optimum temperature for sporulation of the pathogen is 17 to 18 °C in night and 25 °C in the dark. He further observed that the spore dispersal within glass house environment condition and recorded that spore release and dispersal increased with an increase in air currents and temperature and a decrease in relative humidity under close environment. Mean temperature 22.5 to 23.5°C, relative humidity of 77 to 85%, sunshine hours more than 5 hours/ day and more number of rainy days favour the disease. Intermittent rains were found to be favourable for disease development (Anonymous, 2012).

Integrated Management

Resistant or tolerant variety if available may be grown to reduce the initial inoculums. Infected debris is to be burnt or destroyed. Healthy seeds to be used as far as practicable. Disease free seed to be collected from reliable source, 3 to 4 years crop rotation have been found effective as the practice reduces the initial inoculums of the pathogen.

The tolerant variety having low rate of spread of disease are to be selected in areas where severity of the disease is more. Ullasa *et al.,* (1981) screened several chilli variety collections and found few were free from cercospora leaf spot disease. According to Sharma and Sohi (1981), the disease could be managed by adopting the cultural practices like intercropping, adjustment of time of sowing, rouging etc. Physical methods like seed selection, hot water treatment, solar heat etc. and use of chemical treatment assured to be useful for controlling the disease (Islam, *et al.,* 2015). The disease could be managed by spray of chemicals 5 to 6 time *viz*., Bavistin (0.1%) and Difolatan (0.3%) at fortnightly interval. They further found (Sharma and Sohi, 1984), seed dressing of Bavistin (0.2%) is in combination with a seedling dip (3 hours in 0.1% Bavistin solution) together with spray at fortnightly interval give good control of disease.

The systemic fungicides propiconazole, hexaconazole and tebuconazole were found best for containment of the disease (Suresh, 2013). Further he found the effectiveness of Mancozeb and Propineb too against leaf spot disease. The treatment combination of Tilt 25% EC @ 0.05% + Eucalyptus @ 10% + *Pseudomonas fluorescens* @ 5g/litre was found to manage Cercospora leaf spot disease with a percent disease index (PDI) of 8 to 21.33 followed by Tilt 25% EC @ 0.05% + *Pseudomonas fluorescens*@ 5 g/lit (Devappa and Thejakumar, 2016).

Seed treatment with fungicide (53°C, 10 minutes) reduced seed borne infection and higher seedling vigour (Meah and Wick, 2006). Applications of Bavistin 50 DF@ 0.02% three times spray with 10 days intervals are most effective to control cercospora leaf spot of chilli (Mech, 2006).

Future Approach

Much emphasis has been given on protected cultivation with respect to disease management. As the disease appear frequently under protected cultivation, proper manipulation of temperature, humidity, air current etc. are very important and therefore the proper study on these aspects is warranted. Host plant resistance along with other methods of control might be integrated to achieve desired result for management of the disease. Organic way of management is also a great challenge in the present day in context of production and productivity of the crop compared to inorganic practices.

Conclusion

While trying to enhance their production and productivity, it is also equally important to see whether the technology is feasible, economically viable and free from health hazard. In this aspect it is quite acceptable to consider the technology that has management strategy without chemical application.

Phytophthora Leaf Blight and Fruit Rot

Economic Importance

As reported by Sohi *et al.,* (1971) the leaf blight disease of chilli *Phytophthora capsici* Leon. recorded first time from Solan, Himachal Pradesh. Chowdappa *et al.,* (2014),also reported severe foliar blight from Karnataka and Tamil Nadu in hot pepper. Warm humid climate is necessary for the development of leaf blight disease and it prevalence is world wild where congenial environment exist.

The losses may occur due to seedling mortality and also post-transplanting plants. Premature defoliation and premature fruit rot and drop are the major problem due to the disease. As flower get infection, fruit setting drastically reduced. Where congenial environment coincide with planting of capsicum plants, heavy reduction of yield is certain. Warm humid conditions followed by rain during June to September aggravate the leaf blight, necrotic lesions on stems along with rotting of fruits and subsequent reduction of potential yield. As reported (Liang *et al.,* 1992), the reduction of yield may occur to the tune of 43-100%, depending upon severity of the disease.

Symptoms

The pathogen can attack all the aerial parts of plant *viz*., leaves, stem, flowers, and fruits along with growing shoots. It may attack the collar region and may proceed downwards towards the root region. In that case sudden drop of leaves occur resulted sudden wilting symptoms.

Symptoms on leaves appear as water soaked to dull faded areas compared to healthy leaves. Premature leave fall occur due to lesion expansion during humid weather conditions. Some of the spots get bleached to light tan and margin become brown. Stems when infected turn dark brown to black. The infection of growing shoots resulted rapid killing of growing tips, as well as flower buds, progress downward covered more area towards the base. Complete defoliation may result due to severe infection.

Adult plants show whitish growth of mycelium of the pathogen on the crown region at the point of contact of soil, spread towards the root region.

On fruit the disease first appears in the form of dark green water soaked spots. The lesions coalesce and cover the entire fruit. When the infected fruit is cut open, profuse growth of mycelium inside the fruit cavity could be observed. The growth may cover the pericarp as well as the immature seeds, the fruit may rot completely.

Causal Organisms

There are three species of *Phytophthora* found associated with the disease *viz., P. capsici, P. nicotianae var. nicotianae* and *P. boehmeriae*. Among those, *P. capsici* Leon. has been reported from various parts of the world. *P. boehmeriae* was reported from South India (Chowdappa *et al.,* 2014) and association of *P. nicotianae* (Breda de Hann) var. *nicotianae* Waterhouse (Sharma and Bharadwaj, 1976) has also been found reported from Himachal Pradesh.

i) *P. capsici*: The fungus shows fluffy, cottony growth on oat meal agar. The hyphae hyaline, aseptate, typically variously swollen and tuberous growth. The sporangia are hyaline, papillate and non-papillate, ovoid to pyriform, lemon shaped, measuring 29.7-59.8x33.0-72.5 µm, could be germinated to release zoospores at temperature of 17-23°C in tap water. The shape of zoospores is reniform, biflagellate, motile for a while and encysted. Oospores are formed in medium, circular to spherical, average size of 21.28µm. The oogonia are hyaline, circular to spherical and antheridium amphygynous. The pathogen is homothallic.

ii) *P. nicotianae* var. *nicotianae*: Most of the cultural characteristics found similar to *P. capsici*. The sporangia are hyaline, globose to sub-globose, non-pedicellate, pappiliated and size within the range of 15-42 x 17-47µm which is a bit reduced in size of *P. capsici.* Oospores abundant, oogonia globose 23 x 21 µm. The oogonial stalk in engraved by a persistent amphigynous antheridium, oospores golden brown and measures 19-37 µm.

iii) *P. boehmeriae*: The pathogen is different from other two species as it has sporangia caduceus and pappilate and borne on simple sporangiophore. Sporangia are globose, ovoid to ellipsoid, nearly spherical, obpyriform measuring 32.76-38.8 x 18.21-28.57 µm (Chowdappa *et al.,* 2014). Pedicel length of sporangia was less than 5 µm, oospore size 21.9-25.3µm.

Disease Cycle

The pathogen overwinters in plant debris and seed of host plants. The oospore remains in the soil and infected plant debris are the major source of inoculums to carry over the disease from one season to the other. The survival of the pathogen is more in moist soil and could remain viable for 15 months without host. The oospore can remain viable for 7-8 months. As reported (Jindal *et al.,* 1994), the infected seeds are the major source of the pathogen to continue its disease cycle. The fungus is both internally and externally seed borne. Under congenial environment when temperature prevails around 24-26°C and with the onset of monsoon the oospore germinates directly or through formation of sporangia. Germination of sporangia releases biflagellate zoospores which encyst after swimming for sometime in presence of water. Such encysted zoospores act as primary inoculums and spread through raindrop splashes from soil falling on leaves and fruits; subsequently zoospores begin to germinate either directly or by germ tube and penetrate through the epidermal cell or less often through stomata. Sometimes the germ tube form appressoria where from infection peg originates and penetrate the epidermal cell. From initiation of infection to symptom development takes place within 24-72 hours based on congenial environment. Above 95% RH, sporangiophore and sporangia production begin from the lesions on fruits. Secondary infection takes place from the sporangia to nearby field and initiate new infection. During unfavourable conditions, the fungus goes for sexual stage and form oospores.

Epidemiology

Both the pathogens *viz*., *P. capsici* and *P. nicotianae* var. *nicotianae* are seed and soil borne. In soil they remain as oospores and survive for more than a year to cause re-infection. They can infect on leaves, fruits, stem and roots. Incubation period by *P.nicotianae var. nicotianae* is shortest on leaves (24 hours) followed by fruits 28 hours (Gupta and Jarial, 2014). Lower most leaves and fruits are generally infected first due to easy reach of inoculums from soil particularly during rainy season. Abundant rainfall, high RH and warm weather contribute for disease development and spread, but during dry weather disease is completely absent. Due to heavy rainfall, the inoculums in soil splashes and fall on host, the sporangial formation and zoospore germinate also enhance rapidly to cause infection. Subsequently the disease spread to nearby field. From infected field, the fruit act as supplier of inoculums to cause infection to both leaves and fruits.

It has been reported that incidence of the disease is more in clay loam soil than sandy loam soil (Yang *et al.,* 1991). However, optimum temperature of 25-30 °C with RH more than 70 % is conducive for maximum disease development (Shyam, 1969). As reported by Bharadwaj *et al.,* (1985) that high RH, rainfall and temperature has got positive correlation in disease development.

Integrated Disease Management

Due to persistence of pathogen in soil and seed borne nature, soil management through amendment is a practical way to manage the disease. Kim *et al.,* (1990), had reported about the management of soil with high C/N ratio material. Moreover, cultivation of non-host crop like sesame, groundnut, onion, garlic, pea and spinach along with capsicum crop had been found to reduce the disease (Lee *et al.,* 1993). Cultural practices like seed selection, rouging of diseased plants, removal of infected parts, mulching with straw, planting on top of furrow etc. could able to reduce the disease (Ferreyra *et al.,* 1984).

Host resistance also played vital role in disease severity to a great extent. Fruit rot tolerant lines *viz*., Feroz, Danish Wonder, HC-201 had been recorded, which could be exploited to be used as resistant source against the pathogen. The moderately resistant lines 'HC-201' could show slowest and 'Feroz' moderate rate of infection.

Future Prospect

Potassium phosphate (Axomine 3-4ml/lit), an antifungal compound was found to control root rot in black pepper and found compatible with the biocontrol agents like *Trichoderma* spp. and *G. virens* showed varying degrees of inhibition of *Phytophthora capsici* (Rajan and Sharma, 1996). Effective of axomine has already been suggested by the work of Verma *et al.,* (2006) against *P. nicotianae* var. *nicotianae*. The use of GRAS (Generally Regarded as Safe) chemicals may be taken into consideration while considering integrated approach for disease management. Generally, using a combination of different strategies like chemical control, physical control and intrinsic resistance have been recommended for managing the disease (Agrios,2005), but due to certain reasons the strategies failed to give desired result. It might be due to existence of significant amount of genetic diversity in *P.capsici* population and its rapidness of population build up on infected plants. Therefore, exploration of alternative approach is the need of the hour. External agents including bio agents, nano particles', organics like BABA and their analogs may be tried to enhanced induced resistance within the plants system to fight the pathogen. BABA applied to tomato plants as a foliar spray induced PR protein and provided 92 % protection against late blight pathogen of tomato (Cohen *et al.,* 1994).

Conclusion

As chilli is consumed in raw formed, therefore chemical management is very often discouraged. It is advisable to manage the disease through non-chemical methods and if at all required, one can resort for GRAS chemicals at recommended dose. The effectiveness of BABA (beta-amino butyric acid) as chemical elicitor has already been substantiated (Sanogo *et al.,* 2012) in enhancing induced resistance against the solanaceous crops. However, BABA can also impose growth stress (phytotoxicity) in some treated plants and therefore it is desirable to test with BABA analogs/ other elicitors to achieve desired goal.

Damping off

Economic Importance

Among the fungal diseases damping off caused by *Pythium aphanidermatum* (Edson) Fitz. in nurseries is a major constraint in chilli production causing 62% mortality of seedlings (Ramamoorthy *et al.,* 2002). However, based on susceptibility of host and prevailing environment, the mortality of the crop varies. Depending upon the mortality losses also varies from crop to crop and variety to variety.

Symptoms

The disease occurs in two phases as (i) pre-emergence and (ii) post emergence damping off.

Pre-emergence damping off: Seed rot and killing of young seedlings are the major problem on this phase of damping off disease. The seedlings are unable to emerge due to either seed rot or failure of seedling to emergence from the soil. In such cases the nursery bed becomes patchy in the early stage of their growth.

Post emergence damping off: This stage of the disease is characterized by the development of disease after seedlings have emerged out of soil but before the stems are hardened or lignified. The symptoms appear as water soaked lesion at collar region and subsequently the infected areas turn brown and rot; resulted softening of tissues, shrivelling and collapse of seedling. At the infection point on stem, constriction could be resulted toppling of seedling and severe cases all the seedlings may be killed.

Causal Organism

Pythium aphanidermatum (Edson) Fitz.

The pathogen is soil borne and its mycelium remains as intracellular in the host. It consists of coenocytic much branched slender hyphae. Asexual reproduction takes place by the zoospores formed within vesicle extended from zoosporangium. The zoospores are kidney shaped, biflagellate and swim in water film after release from sporangium. After swimming for a while, they encyst and germinate by germtube. In sexual reproduction both antheridium and oogonium are involved. Oogonium is plurinucleate, located terminally rarely intercalary on long and short hyphae where one nucleus migrates into the periplasm. Subsequently, one functional nucleus of antheridum will fertilize the egg. The branch bearing the terminal antheridium either arises at a distance from the oogonium on a separate branch or from the same branch. In due course, one male nucleus passes into the egg through a conjugation tube to form thick walled oospore.

Disease Cycle

Oospores are the primary inoculum resides in soil and carry over the new infection in newly sown crop. It can also remain as mycelium within plant debris to carry over the disease from one season to other. Oospore germination produces zoosporangia and under favourable condition the zoosporangia produce zoospores. The zoospores subsequently encyst and cause infection to germinating seedlings below the soil surface and at collar region near the soil. The infected seedlings topple down and from the infected tissues production of sporangia starts to initiate secondary infection. Under unfavourable climatic conditions the pathogen reverts to sexual cycle, producing oospores and proceeds to dormancy.

Epidemiology

The pathogen needs high soil moisture and relatively higher soil temperature for its development and more prevalent in poorly drained soil. Heavy rainfall, excessive and frequent irrigation provide favourable environment for its spread in temperature around 25-30° C.

Pre-emergence damping off is maximum at 20-25°C, while post emergence damping off is more severe at 30-40°C (Tripathi and Grover, 1976). As reported by Shyam, (1991), the rapid disease development also recorded at high soil moisture near 60-70% and temperature of 25°C (Kumar, 2017).

Integrated Management

Using the same nursery bed year after year is to be avoided to escape from damping off disease. Avoiding frequent irrigation was also suggested (Shyam, 1991). As reported (Raj *et al.,* 1997), the higher seed germination from solarised nursery bed could be obtained to the extent of 18.3 to 42%, lower incidence of post-emergence damping off with better seedling vigour during May-June. Mulching can raised the soil temperature to the tune of 13.5° C with polythene for 40 days and given effect to the depth of 8cm of soil.

Pre-sowing application of dazomet (Basamid G @ 30g/m^2) can manage damping off disease (Patel and Patel, 1998). Encouraging result could be obtained by seed treatment with chemical fungicides *viz*., Captan, Thiram, COC, carbendazim and combo fungicides like carbendazim+thiram which improve seedling stand. As reported (Champawat and Sharma, 2003), Captan when used as seed dresser (0.25%) and soil drenching (0.3%) 15 days after transplanting, total stand of seedling was found higher than control. Minimum pre-emergence and complete reduction of post-emergence damping off was recovered using fungicide metalaxyl as seed and soil drenching. Also other effective soil drenching fungicides were recorded, *viz*., Ridomil Gold, Apron XL, Curzate M-8 and Sectin.

Using bio-agents like *Trichoderma viride* and *T. harzianum* both as seed and soil treatment, could suppress damping off and root diseases (Basharat *et al.,* 2011). Using plant products like *Alium sativum*, garlic bulb extract maximum percentage of seed germination, growth and vigour of chilli seedlings was reported.

Future Approach

All effective non-chemical methods may be used to manage this disease, combining cultural, biological, using known ITK (indigenous technical knowledge) along with available resistant or tolerant cultivars known till date. Breeding for evolving resistant variety may be tried so that the use of harmful chemicals could be avoided.

Conclusion

Using toxic chemical may be avoided for soil sterilization. As far as predictable recommended polyethylene mulch to be used for management of the damping off disease of nursery.

Choneophora blight/ Wet Rot

Economic Importance

Choneophora cucurbitarum was first observed on chillies (*Capsium annum* L.) in India. It was found that up to 1990, host range of the pathogen very limited as reflected in Indian literature. Change in climatic situation, warmer and unpredicted pattern of rainfall distribution during monsoon season change the status of the minor pathogen as an emerging one attacking several numbers of hosts under Gangetic alluvial region of West Bengal (Das *et al.,* 2016). In a survey and surveillance study it was found that out of 30 isolates 10 isolates were very virulent and warranted its importance in near future to be considered as an important disease to cause loss of the crop. Twig blight is one of the major diseases of chilli during monsoon season. It is newly emerging pathogen, but a serious threat of the chilli cultivars and other vegetables under Gangetic alluvial region of West Bengal.

Symptoms

The vulnerable stages of the crop are the seedling stage, vegetative up to early flowering stages. Dead and dying tissues are generally more prone to the disease and could initiate to infect living tissues too. In case of fruit infection, calyx is first found predominantly attack by the pathogen. The pathogen attacks flowers through the senescing petals and overgrown on flowers resulting in brown or black mass of rotten tissue. Subsequently, the flower stalks, buds, leaves are also attacked by the pathogen. On infected tissues stiff silvery mass of whisker like or hairy strands of the fungus growth develops on which black mass of spores is produced. This type of sign gives us clue to identify the pathogen at ease. Individual branches may be infected and exhibits die back symptom. Stems of infected plants appear wet and green and the bark peels off into shreds.

Causal Organism

Choanephora cucurbitarum (Berk and Ravenel) Thaxt. Kingdom: Fungi, Order: Mucorales, Family: Choanephoraceae, Genus: Choanephora, Species: *Choanephora cucurbitarum*

The pathogen commonly occurs on fruits and flowers of higher plants and in soils of tropics. The members of Choanephoraceae are charactorized by their sporangia and sporangiola. Sporangiospores are brown or purple with a straite episore and posses stiff hairlike spines at both ends. The genus Choanephora is multispored, with presence of both columellate sporangia and monosporous sporangiola. The monosporous sporangiola resemble the sporangiospores in shape but are non-ciliate and have longitudinal striations with a short hyaline appendage at the base. Intercalary chains of chlamydospores are also formed by the pathogen. Masses of these sporangiola give the characteristic appearance like a pin cushion. The size of the sporangiola ranges from 8-30 x 5-10 μm. the pathogen is heterothallic and for sexual reproduction needs opposite strains that form zygospore.

Disease Cycle

The pathogen remains in soil as chlamydospores, subsequently produce mycelium and act as primary source of inoculum. As the pathogen is heterothallic, asexual reproduction is common and its secondary inoculums are the sporangiospores, monosporous sporangiola which are dispersed by rain splash and by wind.

Epidemiology

Warm, rainy and wet weather found favourable for the disease to develop and spread. Temperature 28°C, RH 97% along with extended period of rainfall followed by warm weather can cause quick spread of the disease. The crops infected by the pathogen are red pepper, taro, cucurbits, okra, common bean etc (Mehrotra and Aneja, 1990). Spore biology study revealed that temperature of 28±2°C coupled with 70-90% moisture enhanced rate of spore germination (Das *et al.,* 2016). Further it was found that 7.5% and 20.3% enhancement of average RH and rainfall respectively, during June-October might be associated with emergence of *C. cucurbitarum* as a major pathogen.

Integrated Management

Increase air movement in the crop canopy can be measure to reduce the humidity and may aid in the management of the disease. Increase plant spacing in endemic areas particularly help to check the disease. Adequate drainage also helps to reduce the soil moisture of cultivated field. Avoid excess nutrition particularly nitrogenous fertilizers which aggravate the disease quickly. Moreover, one should avoid overhead irrigation to take care for judicious use of water (Kucharek *et al.,* 2003). As reported by Chahal and Grover, (1974), Zineb could eradicate the disease to the tune of 80%. Good spray coverage where dense foliage occurs is important. Pepper cultivars where the petals of the flower fall soon after fruit set should be less susceptible to Choanephora blight.

Captan @ 0.15% was found to be the most effective fungicide in reducing the disease incidence of Choanephora twig blight of chilli (Chandrakala, 2016). She further found that combine effect of *Trichoderma viride*+captan+neem oil is controlling the disease. Meah and Mian, (1981) reported the efficacy of Carbon fungicide for the management of *Choanephora* blight.

Future Approach

A study on pathogen's isolates collected from different hosts is an important aspect whereby virulence of the pathogen on various crops could be explored. It has already been warranted that several isolates of the pathogens were virulent (Das *et al.,* 2016) and need attention of researchers.

Conclusion

Not much literature is available of the pathogen. However, the pathogen is one of the important enemies of Cucurbitaceae crop and now the chilli and need proper attention to combat the disease in future. As it had been reported that beta-amyno butyric acid was very effective in inhibiting (Cohen *et al.,* 1994) oomycetes group of fungi, one could try to explore wheather the chemical also effective against zygomycetes too.

Chilli Powdery Mildew

Economic Importance

Powdery mildew of chilli incited by *Leveillula taurica* is one of the most serious diseases of chilli. Due to the disease, the loss may occur to the extent of 14 to 30%, due to severe defoliation and reduction in photosynthesis, size and number of fruits per plant (Gohokar and Peshney, 1981). Powdery mildew disease of Capsicum was found more severe under protected cultivation, causing significant yield reduction by premature defoliation under warm and humid climatic condition (Gupta and Thind, 2018). In India the occurance of the disease was recorded in Himachal Pradesh under naturally ventilated polyhouse.

Symptoms

On the upper surface of the leaf the mycelial and reproductive fruiting bodies are seen forming irregular blotches. Leaves become yellow corresponding undersurface of the leaves also produces blotch and sporulate. The leaves rolled upward yellow and dropped prematurely. Stem generally not affected but leaves may fall off and fruits subjected to sunburn and become unmarketable. Generally older leaves are more prone to infection from where the infection may move upward and nearby leaves or adjacent capsicum plants.

Causal Organism

Oidiopsis taurical Leveillula taurica (Lev.) Arn.

Most powdery mildew pathogen does not produce sexual stage in tropical and subtropical areas and become a problem for taxonomy of his ascomycetous fungus (Monkhung *et al.,* 2011). Sontirat *et al.,* (1994) reported that *Oidiopsis* sp. is a causal agent of powdery mildew disease in Thailand, but identification at species level was not determined due to lack of teleomorphic stage on capsicum species. The phylogenetic analysis represented by maximum parsimony tree indicated that the five *Oidiopsis* specimens on chilli plants are located in a clade of *Leveillula* which confirms ananamorph-teleomorph connection of the fungus with *Leveillula* (Monkhung *et al.,* 2011).

Disease Cycle

For survival, the pathogen overwinters as cleistothecia or as mycelia and oidia/conidia. It has got wide host ranges which are coriander, tomato, brinjal, gram, cotton, onion etc. However host specificity had also been reported and needs critical investigation. The pathogen does not always produce cleitothecia and therefore has to perennates through mycelium and oidia produced on the cultivated and wild hosts.Oidia are wind borne or dispered with the splashes of rain. After landing of the oidia on host leaf surface, produce germ tube either enter directly through cuticle or stomata. The hyphae remify intercellularly, reaches spongy and palisade parenchyma. Pathogenesis starts and produces symptoms and continues to produce oidia/conidia for secondary infection.

Epidemiology

The disease is more in the humid or dry areas and more severe under shaded condition. Glawe, *et al.,* (2010) reported that the disease is restricted in warm semi arid regions. While Reuvani and Roten, (1973) found the conidia germination range within 8-35°C and relative humidity of 93 to 100%, but optimum temperature was 25 to 30 °C. But disease spread was found favouring at low relative humidity <50 (Diop Bruckler, 1989). Like other powdery mildew pathogens, they are too shy in spore germination on availability of free moisture. However, close canopy found to increase disease incidence where both other factors like warm and humid climate prevail.

Under protected cultivation maintenance of hygiene is an important measure to check the disease. Moens and Walvaert, (1981) reported that maintaining of wide spacing, sprinkler irrigation and misting air circulation are the effective means in reducing the disease severity.

Host resistance: Few cultivars/ lines are found resistant such as HV-12, S-32, Pondaicha, K-red and few identified as moderate resistant viz., B15, Pgedasali, Tinwari, Golden superior etc .

Integrated Management

In an integrated disease management, foliar spray with Sulphur alternating with dipotassium hydrogen orthophosphate (DHOP) or plant extract of *Azadirachta indica* was found effective as alternating with wettable sulphur fungicide. Among the chemical tested, DHOP significantly increased the protein, total phenol, phenylalanine ammonia lyase, polyphenol oxidase and peroxidase activity of the leaves (Sudha and Lakshman, 2009).

Propiconazole is the best fungicide to manage chilli powdery mildew incident by *Leveilulla taurica*. It was also reported that myclobutanil, triadimefon and hexaconazole were effective fungicides in reducing the disease. Another translaminar and systemic fungicide *viz*., Azoxystrobin was found effective due to its characteristic distribution from application side by diffusion (Vincelli, 2002). For organic cultivation, use of GRAS chemicals may be recommended. Fallik *et al.,* (1997), proposed the use of Sodium or Potassium bi-carbonate solutions for effective management of this disease. Some of the chemicals are found to have induced resistance on host and able to check the pathogen attack. Reuveni and Dor (1998), found both local and systemic resistance on green house grown pepper plants, subjected to application of mono-potassium phosphate (KH_2PO_4 @1%w/v).

Future Approach

Under protected cultivation use of GRAS chemicals to be recommended as these chemicals are user-friendly, ecological, sustainable and free from health problem. Many such chemicals could be tried integrating with non-chemical method for disease management. Sontirat *et al.,* (1994) reported that *Oidiopsis* sp. is a causal agent of Powdery mildew disease in Thialand but identification at species level was not determined due to lack of teleomorphic state in nature on *Capsicum* species. However, phylogenetic study revealed that the Powdery mildew anamorph was located in a clade of *Leveillula* which confirmed the anamorphic- teleomorphic connection of the fungus (Monkhung *et al.,* 2011) in Thailand. But it could be hypothesized that under certain environmental conditions, the teleomorphic state of the fungus might have formed and the future research could be carried out to establish the hypothesis in other parts of the globe. It is probable that a concrete management could be recommended after establishing the fact regarding the formation of teleomorphic state in nature.

Conclusion

Much stress has been given for organic production of crop and therefore management practices should be laid down to contain the disease with non chemical methods integrating with biological, use of GRAS chemicals, botanicals so on and so forth. High value pepper crop under protected cultivation needs more attention to obtain better control of the disease. To know about the perpetuation of the pathogen propagules along with its disease cycle are important consideration for management of a disease and therefore complete information about the pathogen yet lacking.

Chilli Leaf Curl Disease

Economic Importance

Leaf curl viruses have emerged as major treat to chilli crop and could cause loss to the tune of 80% (Singh *et al.,* 1979). The disease was first reported from India in 1954 and after 2005 the virus complex had been found emerging rapidly in India and sub-continent (Kenyon *et al.,* 2014). The vector of the disease is white fly and the transmission may occur through graft transmission too (Senanayake *et al.,* 2012) and severity of the disease is found where vector population is also more. Losses upto 80% have been reported from North India (Singh and Thakur, 1979).

Symptoms

Curling of the leaf is the major symptom of the disease followed by reduction of the size and yellowing of the leaves. Leaves curl towards midrib and become deformed. Stunted plant growth due to shortened internodes and leaves greatly reduced in size. Flower buds abcise before attaining full size and anthers do not contain pollen grains (Anonymous, 2016). The older leaves become leathery and brittle. Affected plants stunts and fruits that are formed are small and deformed.

Causal Organisms

The disease is caused by Tobacco leaf curl virus, which belongs to Begomovirus group (Circular dsDNA). The virus consists of geminate isometric particles measuring 18-20nm in diameter.

Disease Cycle

The virus has a wide host range. It can be transmitted by the vector white fly (*Bemisia tabaci*). Valend and Muniyappa, (1992) could experimentally able to transmit the virus to several other hosts, *viz., Lycoparsicon esculantum, Beta vulgaris, Sesamum indicum, Phaseolus vulgaris, Carica apaya* etc.

Epidemiology

To cause epidemic of the disease, source of inoculums, vector population, their activity etc. are important. Wet weather is not at all congenial for the vector activity whereby spread of the disease is reduced (Shivanathan, 1982). The virus survive from one season to another on various annual and perennial hosts and thereby transmitted by its vector i.e., white fly (*Bemisia tabaci*).

Integrated Management

As the host resistance is one of the most important component of IPM, the growing of resistant cultivars is paramount important to check the disease. The cultivars *viz.*, Perennial, Punjab Lal, Pusa Jwala , Jawahar Mircha, Surya Mukhi and Japani Loungi etc. were found resistant/tolerant to the virus. Resistance in different chilli cultivars has been reported to control by monogenic recessive gene (Bal *et al.,* 1995). Thakur *et al.,* (1987), further reported that Punjab Lal, a multiple resistant variety of chilli against several diseases including CLCV.

Besides growing resistant variety, the following measures may be adopted to contain the disease. Avoid monoculture, adjusting of planting dates, use of yellow sticky trap to reduce white fly population, use of neem oil (1lit/acre) etc for successful cultivation of the crop. Selective use of insecticide whereby natural enemies could be preserved was an another research carried out against Bt on cotton crop (Ellsworth and Martinez- Carrillo, 2001).

Future Approach

To contain the disease is difficult though several techniques are being followed starting from seed bed nursery to the main field, It is presumed that the techniques followed in cotton crop multi level, multi component approach that focuses *Bemesia tabaci* management may be tried on Capsicum/Pepper crop in future leaf curl management which had been tried in North America (Ellsworth and Martinez-Carrillo, 2001). Selective insecticides which could preserve natural enemies is another area of research where new generation chemicals particularly plant origin / bioagent or may be nano based formulation might be given due emphasis for sustainable practice for future use.

Conclusion

For viral disease management mere host crop management is not sufficient for containment of disease and it is important to manage the disease where both IPM and IRM are involved. It is also equally important that Insect Resistant Management (IRM) to be considered within the IPM framework so that cost effectiveness; surveillance with respect to change in pest behaviour, avoid monoculture etc might be focussed to accommodate strategy well sustainable.

Mosaic of Chilli

Economic Importance

As reported by several researchers the considerable loss may occur due to mosaic disease of chilli. Potyvirus and Cucumovirus groups mainly cause the disease. The incidence of CMV alone ranged from 20-50% in infected pepper field (Benner *et al.,* 1985), while mix infection could show yield loss of 66.7-74% (George *et al.,* 1995). Mix infection of Tomabovirus like Tobacco Mosaic Virus (TMV), Tomato Mosaic Virus (ToMV) and Pepper Mild Mottle Virus (PMMoV) are also formed associated with the disease in the world over. The Solanaceae family contains many species susceptible to infection by all the three viruses (TMV, ToMV, and PMMoV).

Symptoms: Caused by

a) ***Cucumovirus genus:*** The symptom appear at early stage of crop, leaves become light green, stunted growth, reduced fruit set and deteriorate quality, leaves become elongated, narrow and filliform. Associated symptom on bellpepper leaves are mosaic with leaf deformation, ring spots on leaves and fruits (Kapoor, 2012). It is a RNA virus and forms icosahedral particles of approximately 29nm in diameter.

b) ***Potyvirus genus:*** The viruses induce mosaic or mottled symptoms accompanied by dark green vein bending, necrosis of veins, petioles and stems of the crop. Pepper veinal mottle virus causes mosaic, mottling, filliform and distortation of leaves accompanied by bushy appearance of the plant. The virus causes systemic necrosis leading to death, while pepper severe mosaic virus produces vein clearing with deformed leaves along with mosaic mottle symptoms on infected plants. It is a RNA virus form flexous rod shaped particles ranging from 680-900nm in length.

c) ***Tobamovirus genus:*** TMV, ToMV and PMMoV are the type members of the Tobamovirus genus of RNA viruses. All are rigid, rod shaped and approximately 300nm in length. A general mosaic or mottle of dark and light green or yellow is observed on the leaves of plants infected with these viruses. PMMoV causes leaves to pucker and plants stunted. Some strains of TMV and ToMV can cause defoliation, premature flower drop, wilting, necrosis of the tissue inside the fruit wall.

Disease Cycle

All the members of the virus complex have wide host range. Among the Cucumovirus group *viz*., Cucumber Mosaic virus (CMV) is distributed over hundred families and 1200 species. Common hosts are tomato, cucumber, squash, melon etc.

Among the Potyvirus, PVY and PepMoV are primarily restricted to Solanaceous crops. Tomabovirus like TMV has wide host range, while ToMV and PMMoV have more limited host ranges. The Solanacae family contains many species susceptible to infection by all the three viruses.

Cucumo and Potyviruses are sap transmissible and also through aphids in a non-persistant manner. The most efficient vectors are *Aphis gossypii*, *Myzus persicae* and Green peach aphids and *M. persicae* of Cucumo and Potyvirus, respectively (Lopez-Moya and Garcia, 1998). But the Tobamoviruses are mechanically transmitted (Gibbs, 1977). The plant debris can harbour the infectious particules of tobamoviruses for many years. These are also carried on infected seeds.

Epidemiology

The severity of disease caused by Cucumo and Potyviruses depends upon the presence of their host, infected vector population. Most of the Solanaceous and weed host are reservoir of the viruses and act as primary inocolumn to initiate the disease. Secondary inoculum takes place through workers tools or during handling of plants by labourers or might be due to both the reasons. *Myzus persicae* and *Aphis caracivora* are the vectors of the pathogens and their population built-up takes place with the temperature range of 26 to 30°C, recorded corresponding enhancement of the disease (Sharma *et al.,* 1991).

Integrated Management

Barrier crops are one of the efficient means to check the aphid spread to the main crop. Aphids can transmit the virus in non persistent manner and therefore any cultural practice which can deter their activity and transmission capacity of the vector help to check the disease. Moreover removal of weed hosts, use of disease free seeds and avoid growing of susceptible solanaceous crops could reduced disease of CMV and PVY and able to check its spread too. Significant reduction of mosaic disease was found by growing the barrier crops like sunflower, snap bean, sorghum, pearl millet. Growing of marigold with pepper has also been found useful in reducing the incidence of viruses. Intercropping of onion with chilli resulted reduction of aphid population and also disease incidence (Basavarajappa and Rajasekhar, 2001). Silver (Polyester sheet) followed by white polythene mulching was found effective in controlling the disease (Kapoor, 2012).

Integration of cultural practices along with growing resistant cultivars/ varieties and using need based botanical pesticides might be good agricultural practices to contain the mosaic complex disease. As reported the chilli mosaic virus infection could be managed completely by application of leaf extract of *Datura stramonium* (Singh and Sharma, 1999). Neem seed kernel extract (5%) too can check the disease as well as contribute in enhancement of yield (Ragupati and Veeragavathatham, 2002). Among the insecticides, the following are found to check the disease viz., Methyl parathion, chloropyriphos, monocrotophos, dimethote etc. (Devi *et al.,* 1995).

Future Approach

Now-a-days, organic method of disease management has gained tremendous importance and in future research should concentrate in this aspect whereby use of synthetic chemical could be reduced. A modern method of prevention was found by a Japanese study on soil adsorption role in the PMMoV. The study found that increased humus content in soil will have an inhibitory effect on PMMoV adsorption (Yashimoto *et al.,* 2012).

Conclusion

Avoidance is the best means of controlling the disease because once a plant is infected; it cannot be treated. Only seeds that have been tested and treated for the pathogen should be planted (Anonymous, 2000). The Japanese strategy as proposed by Yashimoto *et al,* (2012) could be superimposed with the other non-chemical method of integration (IDM) to achieve desired goal.

Bacterial Diseases of Chilli

Bacterial Leaf Spot

Economic Importance

It is the most common widespread bacterial disease found to infect chilli crop. The causal organism is *Xanthomonas campestris* pv. *vesicatoria.* Due to infection, premature leaf falling occurs and could cause substantial yield loss. Besides, the infection fruits become dull, lack luster and unmarketable or receive less value than normal market value. For the disease, the loss may occur to the extent of 16% (Shekhawat and Chakravarti, 1977).

Symptoms

In this disease most of the plant parts, like leaves, fruits and stems are affected. The leaf exhibits small circular or irregular dark brown spots that become necrotic with brown centre, chlorotic borders. Spots may develop with straw colored centre. In lower surface, the lesion is raised. Severely spotted leaves turn yellow and drop. Raised brown lesions also found on the fruits, while narrow elongated lesion could be seen on stem. Disfiguration of the leaves and fruit may occur. In severe cases, the plant may die.

Causal Organism

Xanthomonas vesicatoria (ex Doidge) Dawson

The bacteria are strictly aerobic, motile, rod shaped, gram-negative. They possess polar flagellum, non-spore former measuring 1-1.8 x 0.5-0.7 µm. rarely do they form short chain. Colonies on NDA/ YDA are generally lemon yellow, convex, wet and shinning. The pathogen was isolated from different host *viz.,* tomato, chilli, bell pepper (Ramayashreedevi *et al.,* 2017) and undergone molecular study and characterized. They could able to group the isolates to pepper-tomato group.

Disease Cycle

The disease is both internally and externally seed borne and spread by rain splashes. It can overwinter in plant debris but cannot survive for long in soil or water alone. The bacterium is mainly wound parasite and enters the host through insect punctures, injuries hydathodes and stomata. Fruits are more prone to infection and develop lesions more frequently. The infected seeds overwinter for the next season as primary source of the pathogen.

Epidemiology

Moderate temperature, high relative humidity along with intermittent rain may cause epidemic of the disease, if other factors like susceptible host, inoculum pressure are high. Infection takes place at a wide range of temperature from 15 to 35°C but the optimum is 22 to 34°C. Forty to fifty days old plant is most susceptible to the disease. Prasad and Gupta, (2012) found that mean temperature 25 ±1°C, coupled with high humidity 60-80% and low sunshine hours with intermittent rainfall favored disease development.

Integrated Management

As the pathogen is seed borne, cultural practices integrating with other non chemical or chemical means could be used strategically to manage the disease. Well drained upland field may be an ideal field for growing crop to avoid congenial environment for the pathogen. Host resistance may be accommodated with disease free seed to contain the disease.Spraying Agromycin 100 alone or with 3g/lit copperoxycholride or copper sulphate or 2g/lit Mancozeb will control the bacterium. Disinfecting the seeds for one minute in 1.3 % sodium hypochlorite solution and drying in shade is the best way to reduce seed borne bacterium. Crop rotation can help to prevent bacterial leaf spot. It is better not to grow crops like peppers or tomatoes in location where either of these crops have been grown in the past four to five years. At the end of the season, it would be better to remove all the crop debris from garden and destroy it. Composting is even not to be tried with diseased plant debris. Once the area is cleaned of all visible debris, tilth the soil or turn it with shovel to burry any remaining bacteria. It is best not to save own seeds if ever had problem with bacterial leaf spot on pepper. Buy certified disease free seeds and seedlings.

Future Approach

Evolving of resistant or tolerant variety could be a solution for managing the disease, whereby farmers could be supplied with certified seed from the government or government undertaking organizations or agencies etc. However, other methods of disease management may be integrated for managing the disease. Characterization of the isolates across the country could be prime target so that effective management for the disease could be formulated.

Conclusion

Under protected cultivation generally disease problem is more. Utmost care should be taken to manage the disease from nursery, crop husbandry to fruit harvest. Breeding resistant variety is one of the means to combat the disease and partial success was achieved with two recessive genes (Stall and Lozano, 1982). Further works included that there genes were more potent to show resistance (Kim and Hartman, 1985), but none of the genes result in resistance to all strains of the pathogen.

Bacterial Wilt

Economic Importance

The disease is generally occurs in tropical, subtropical and temperate climates. Solanaceous crops are the good host of the pathogen. The disease is also important where the crop is grown under protected cultivation. Bacterial wilt is an economically important disease, distributed worldwide and has wide host range. Under favourable conditions the disease may occur where climate is congenial for its growth, multiplication and spread. The disease has been reported from different states of India *viz*., West Bengal, Kerala, Orissa, Uttarakhand and sporadically in Assam too. Khan *et al.,* (1980) reported that the disease is prevalent wherever tomato and capsicum is grown. In the sub humid areas, the pathogen appears endemically year after year.

Symptoms

Initial symptom showing dropping of youngest leaves but exhibited recovery in the next morning. Within 2 to 3 days another wilting occur causing death of the entire plant. Before wilting of the whole plants leaves drooping, followed by yellowing and discoloration of a part of the plant are seen. As the pathogen accumulates within vascular bundle, yellow to brown discoloration could be seen on splitting test may help to diagnose the disease very easily.

Causal Organism

Ralstonia solanacearum (Smith) Yabuuchi *et al.*

The bacterium is Gram negative, rod, 0.5 x 1.5 μm, monotrichous. In India the pathogen has three races namely 1, 3 and 4 and biovar II, III and IV (Mathew *et al.,* 2000 and Sunaina and Tomar, 2003). It has wide host range. Physiological properties and pathogenicity based upon which races and biovars have been identified. Among the host, the following crops are found infected by the pathogen *viz.*, potato, pepper, tobacco, tomato, eggplant and banana. The pathogen could survive in the rhizosphere, on the rhizoplane, inside the roots of weed plants (Samaddar *et al.,* 1998).

Disease Cycle

As the pathogen has wide host range including *Solanum nigrum, Solanum anguivi, Solanum torvum, Solanum xanthocarpum, Croton bonplandianum, Euphorbia hirta, Colocasia esculenta, Portulaca oleracea, Phyllantus niruri* and *Parthenium hysterophorus* (Samaddar *et al.,* 1998). It is probable that it can overwinter very easily from one crop season to the other. Moreover, the pathogen can remain in plant debris, considerable period in moist soil without debris. The bacteria colonize in the xylem vessel, spread throughout plant. It can enter through mechanical means, insect/ nematode injury or growth phases. It can invade intercellularly and multiply within the parenchyma cells in the cortex. The pathogen could disintegratethe cell walls resulting in pockets for bacterial cells and polysaccharides and cellular debris.From the infected plant debris or roots or both ,the bacteria released back to the soil for over seasoning to continue its life cycle (Bruchl, 1987).

Epidemiology

The organism can survive within the range of 15 to 37°C and optimum being 35 to 37°C. The pathogen is unable to survive and reproduce below 10°C. It can remain in soil and also has wide host range for perennation. High soil moisture combine with high temperature is congenial for its development and causes severe wilting. In general the disease may manifest in temperature above 20°C. The pathogen can grow well near neutral pH and range in between 6.2 to 7.4, but high pH within the alkaline range become unfavorable. Infected seed and soil may add as initial inoculum and spread through rain and irrigation water.

Integrated Management

Integration of growing resistant variety and adapting cultural practices are the most practical approach to check the disease. Moreover avoiding prone area and high soil moisture could be a strategy to avoid the disease.

Selection of variety is an important approach to contain this disease. Some of the commercial varieties having tolerance with low disease progress affinity, maybe chosen and integrated cultural practices will successfully suppress the disease. Even during selection of nursery bed, it is important to select disease free area with no history of occurrence of the disease. It is also required to disinfect the nursery soil through soil solarization or using recommended chemicals. Main field should also be disease free, planted healthy seedlings, weed free cultivation and to maintain proper hygiene of the field; so that no infested soil or planting material is introduced. Under protected cultivation, soil sterilization and other practices should be followed to avoid the disease.

Few of the resistant / tolerant Chilli cultivars may be planted to check the initial inoculum *viz.,*Pusa Jwala, Suryamukhi, Ujwala, Balijuri ,Krishna, Kulai, etc.

Under direct seeded condition Balijuri shows better result and could tolerate the disease to a considerable extent. Chola is a sweet pepper cultivar which is tolerant to the disease and shows maximum disease incidence up to 10% (Jimnez *et al.,* 1988). An organic input biosol was found effective in inhibiting the bacterium with 92.3%. In an integrated approach under protected cultivation in which mixture of biosol, fermented buttermilk, cow urine, vermiwash before and after transplanting along with seedling dip in biosol for 10 minutes before transplanting; provided 85% disease control (Ashlesha and Paul, 2012).

Future Approach

For management of the disease some organic products has been found effective *viz*., panchagavya, biodynamic compost tea, cow urine, vermiwash etc. where performance of biodynamic compost showed promising in inhibiting the bacterial growth (Ashlesha and Paul, 2012). It has also been reported that, drenching with mixture of biosol before and after transplanting with seedling dip for 10 minutes contain the disease upto minimum 85%. Growing of tolerant variety, followed by cultivation of crop and management of disease organically would be a practical proposition in the present day context of bacterial wilt management of chilli.

Conclusion

As the pathogen is soil borne, resistant host will be the best way to manage the disease. But in absence of highly resistant variety, moderately tolerant variety is a choice; where other methods of disease management could be integrated for management of the disease. Rigorous screening of cultivars for disease resistance, having low AUDPC values will be a practical proposition for containing the pathogen; while integrating all other effective practices for management of the disease.

References

Agrios, G.N. (2005). Plant Pathology.5th Ed. San Diego; Academic Press.

Ali, M. S. (1989). Taxonomic studies on Coelomycetes of Assam, Ph. D. Thesis, Assam Agric. Univ., Jorhat, Assam, India

Anonymous, (2016). Fungal disease protection. In: B Stress Disease,naipagropedia, Raichur, ICAR, NAIP.

Anonymous, (2000). Plant Pathology Fact Sheet: Pepper Mild Mottle Virus, Circular No. 400, http//www.fresh fromflorida.com

Anonymous, (2016). Crop Protection, TNAU Agri Tech portal, Horticultural Crops: Chilli. agritech.tnau.ac.in/crop_protection_disease_7.html.

Ashlesha and Paul, Y. S. (2012). Evaluation of organic inputs against *Ralstonia solanacearum*- A barcterial wilt pathogen of capsicum under protected cultivation. *Pl. Dis. Res.,* **27**(1):66-70

Bal, S. S., Singh, J. and Dhanju, K. C. (1995). Genetics of resistance to mosaic and leaf curl virus in chilli (*Capsicum annuum*). *Indian J. Virol.*, **11**(1):77-79.

Bansal, R. D. and Grover, R. K. (1969). Reaction of chilli (*Capsicum frutescens*) varieties to *Colletotrichum capsici. J. Res. Punjab Agric. Univ.*, **6**:345-348

Basavarajappa, M. P. and Rajasekhar, D. W. (2001a). Effect of intercropping in the management of Chilli mosaic transmitted by insect vector. *Insect Environ.* **7**:63-64

Basharat, N., Sobita, S., Das, S. and Soma, R. V. (2011). Comparative efficacy of *Trichoderma viride* and *T. harzianum* in management of *Pythium aphanidermatum* and *Rhizoctonia solani* causing root rot and damping off disease. *J. Plant Dis. Soc.,* **6**:60-62.

Benner, C. P., Kuhn, C. W., Demski, J. W., Dobson, J. W., Colditz, P. and Nutler, F. W. Jr. (1985). Identification and incidence of pepper viruses in Northeastern Georgia. *Plant Dis.,* **69**:999-1001

Bharadwaj, S. S., Sharma, S. L. and Dohroo, N. P. (1985). Role of meteorological factors in development of bell pepper fruit rot (*Phytophthora nicotianae* var. *nicotianae* Waterhouse). *Indian J. Mycol. Pl. Pathol.*, **15**:328-329

Bruchl, G. W. (1987). Soil borne plant pathogens. Mcmillan Publishing Co., 866 Third Avenue, New York, USA

Ceekauskas, R.C. (2004). Cercospora leaf spot published by AVRDC- the world vegetable centre. Shanhua, Taiwan,pp.741

Chahal, A.S. and Grover, R.K. (1974). Chemical control of soft fruit rot of chillies caused by *Choneophora cucurbitarum*. *Haryana J. Hortic. Sci.*, **3**: 190-194 Champawat, R. S. and Sharma, R. S. (2003). Integrated management of nursery diseases in brinjal, chilli, cabbage and onion. *J. Mycol. Pl. Pathol.*, **33**: 290-291

Chandrakala, J. (2016). Studies on choneophora twig blight *Choneophora cucurbitarum* (Berk and Rav.) Thaxt. of chilli (*Capsicum fruitescens* L.) and its management. M.Sc thesis, Dept. of Plant Pathology, College of Agriculture, Rajendranagar, Hyderabad-500030

Chowdappa, P., Madhura, S., Nirmal Kumar, B. J., Mohan Kumar, S. P. and Hema, K. R. (2014). *Phytophthora boehemeria* as dominant pathogen responsible for severe foliar blight of *Capsicum annuum* L., in South India. *Plant Dis.* **98**: 90-98.

Cohen,Y.,Niderman,T.,Mosinger, E. andFluhr,R.(1994).B(beta)-aminobutyric acid induces the accumulation of pathogenesis-related protins in tomato (Lycopersicon esculentum Mill.) plants and resistace to late blight infection caused by *Phytophthora infestans*. *Plant Physiol.* **104**:59-66.

Das, S., Mondal, B. and Dutta, S. (2016). Studies on variability, epidemiology and management of twig blight disease of chilli, caused by *Choanephora cucurbitum*. Conference: National Symposium "Imapct of climate change, biodiversity and good plant practices on crop productivity". AAPP, BCKV, Kalyani, WB, India, Sym. Abs:pp 57-58

Datar, V.V., Sontakke, M.B., Purandhari, N.D. and Shinde, N.N. (1990). Fungicidal control of anthracnose of chillies. *Indian J. Mycol. Pl. Pathol.,* **20**(2): 156-158.

Devappa, V. and Thejakumar, M.B. (2016). Integrated management of chilli leaf spot caused by Alternaria alternata and Cercospora capsici under field conditions. *Int. J. Adv. Res.*, **4** (4): 1468 -1474.

Devi, P. S., Reddy, H. R.and Devi, P. S. (1995). Effect of insecticides on aphid transmission on pepper vein banding virus and cucumber mosaic virus of chilli (*Capsicum annuum* L.). *Mysore J. Agric. Sci.* **29**(2):141-148.

Diop Bruckler, M. (1989). Effect of climate factors on development of *Leveillula taurica* and susceptibility of *Capsicum annuum* at different vegetative stages. *J. Phytopathol.* **126**: 104-114

Ellsworth, P. C. and Martinez- Carrillo, J. L. (2001). IPM for *Bemisia tabaci*: A case study from North America. *Crop Protection*, **20(9):**853-869. https:doi.org/10.1016/S0261-2194 (01)00116-8.

Fallik, E., Ziv, D, Grinberg, S., Alkalai, S. and Klein, J. D. (1997). Biocarbonate solutions control powdery mildew (*Leveillula taurica*) on pepper plants by foliar spray of mono-potassium phosphate. *Crop Prot.,* **17**(9):703-709

Ferreyra, E. R., Tosso, T. J. and Fernandez, M. C. (1984). Effect of irrigation water management on Phytophthora capsici Leon., causal agent of *Capsicum annuum* blight. *Agric. Tecnica,* **44**:319-324

George, T. E., Anand, N. and Singh, S. J. (1995). Influences of potato virus Y (PVY) and cucumber mosaic virus (CMV) on growth and yield of bell pepper (*Capsicum annuum* L.*). Indian Cocoa, Arecanut Spices J..* **19**:1-4

Gibbs, A. J. (1977). Tobamovirus group, CMI/AAB. Description of plant viruses. No.184

Glawe, D. A., Barlow, T., Eggers, J. F. and Hamm, P. B. (2010). First report of powdery mildew caused by *Leveillula taurica* of field grown sweet pepper in the Pacific Northwest. *Plant Health Prog*., **10**:1094

Gohokar, R.T and Peshney, N.I (1981). Chemical control of powdery mildew of chilli. *Indian J. Agric. Sci.,* **51**: 663-665

Grubben,G.J.H., Mohamed E.L.T.I.(2004).Capsicum annuum L.,in PROTAZ: Vegetables?Legumes, Eds.Grubben, G.J.H., Denton,O.A. (Wageningen, PROTA),154-163.(Google Scholar).

Gupta, S. K. and Jarial, K. (2014). Epidemiology and management of leaf blight and fruit rot of bell pepper. *Rev. Plant Pathol.,* **6**: 473-499

Gupta, S. K. and Thind, T. S. (2018). Disease Problems in Vegetable Production, Scientific Publishers, 2nd Edition, Ansari road, New Delhi-110002, India. pp.586

Gupta, S. K., Dohroo, N. P. and Shyam, R. M. (1998). Occurance of barcterial wilt of tomato in Himachal Pradesh. *Pl. Dis. Res.,* **13**:174

Hirata, T. and Takamatsu, S. (1996). Nucleotide sequence diversity of rDNA internal transcribed spacer exhibited from conidia and cleisthothecia of several powdery mildew of fungi. *Mycoscience,* **37**: 265-270.

Integrated disease management of powdery mildew *Leveillula taurica* (Lev.) Avn. of Chili (*Capsicum annuum* L.). *Archives Phytopathol. Pl. Protec.*, **42**(4): 299-317.

Islam, M. S., Fatema, K, Alam, K.M.B. and Meah, M.B. (2015). Daignosis and prescription for cercospora leaf spot of chilli. *J. Bangladesh Agril. Uni.*, 13(2):191- 196

Islam, M.R. (2006). An integrated approach for management of Phomopsis blight and fruit rot of eggplant. PhD dissertation, Department of Plant Pathology, Bangladesh Agricultural University, Mymensingh, Bangladesh.

Jimnez, J. M., Bustamante, E., Bermudez, W. and Cramboa, A. (1988). Response of four cultivars of sweet pepper to barcterial wilt in Costa Rica. *Manejo Tantegrado dePlagas*, **7**:19-28.

Jindal, K. K., Gupta, S. K. and Shyam, K. R. (1994). Study on the germination, vigour and microflora of bell pepper seeds from healthy and diseased fruits. *Indian J. Mycol. Pl. Pathol.,* **24**(3):327-328.

Kapoor, P. (2012). Detection of cucumber mosaic virus in bell pepper crop from Himachal Pradesh. *Pl. Dis. Res.,* **27**(1):71-80

Kawagoe, H. (1998). Influence of temperature and relative humidity on disease occurance of sweet pepper caused by *Cercospora capsici* in a plastic house. *Ann. Phytopthol. Soc. Jpn.,* **64**(2):137-138.

Kenyon,L.,Kumar,S.,Tsai,W.S. and Hughes J.A.(2014). Virus disease of peppers (*Capsicum* spp) and their control.*Adv. Virus Res*.90:297-354.

Khan, A. vvN. A., Shetty, K. S. and Patil, R. B.(1980). Occurance of barcterial wilt of Chilli in Karnataka and its relationship to the wilt of other solanaceous vegetables. *Indian Phytopath.* **32**:507-512.

Khodke, S.W. and Gahukar, K.B. (1995). Changes in biochemical constituents of chilli fruits infected with *Colletotrichum gloesporiodes* Penz. *J. Maharashtra Agri. Univ*., **20**: 142-143.

Kim, B. S. and Hartman, R. W. (1985). Inheritance of a gene (BS3) conferring hypersensitive resistance to *Xanthomonas campestris* pv *vesicatoria* in pepper (*Capsicum annuum*). *Plant Dis*., 69:233-235

Kim, C. H. (1989). Soil borne diseases: Phytophthora blight and other diseases of red pepper in Korea. Extension Bull. ASPAC, Food Ferti. Tech. Centre., **302**:10-17

Kucharek, T. A., Benny, G.L and Pernezny, K. (2003). Choanephora blight (wet rot), pp. 11-12, In K. Pernezny, P.D. Roberts, J.K. Murphy and N.P.Goldbay (eds.), Compendium of Pepper diseases. American Phytopathological Society Press, St. Paul. MN

Kumar, R. (2017). Studies on damping off of tomato caused by *Pythium* sp. M. Sc. thesis submitted to YSP University of Horticulture and Forestry, Solan.

Lakshmesha, K., Lakshmidevi, L., Aradhya, N. and Mallikarjuna, S. (2005).Changes in pectinase & cellulose activity of *Colletotrichum capsici* mutants and their effect on anthracnose disease on capsicum fruits. *Arch. Phytopath.Plant Prot.* 38:267-279.

Lee, H. U., Kim, C. D., Lee, Y. S. and Cho, C. T. (1993). Effects of soil chemical properties and soil microflora by anti plants on Phytophthora blight incidence of red pepper. *RDAJ Agril. Sci., Soil Ferti.,* **35**(2):305-310.

Liang, Y., Lu, J. D. and Ji, R. Z. (1992). Study on the occurance and yield loss of pepper due to Phytophthora blight in Shaanxi. *Plant Prot*., **1**:14-15.

Lopez-Moya, J. J. and Garcia, J. A. (1999). Potyviruses (Potyviridae). In: Encyclopaedia of Virology (Eds: A. Granoff and R. G. Webster). San Diego, Academic press, pp. 1369-1375

Majeed, M., Mirttassan, G., Hassan, M. and Mohulddin (2018). Biological management of damping off disease of Chilli (*Capsicum annuum* L.). *Eco. Env. Cons.*, **25**(1): 353-356

Monkhung, S., To-anun,C. and Takamatsu,S.(2011). Molecular approach to clarify taxonomy of Powdery mildew on Chilli plants caused by *Oidiopsis sicula* in Thialand. *J.Agril. Tech*.7(6).1801-1808.

Mathew, S. K., Girija, D., Devi, S. N., Sadhankumar, P. G. and Ranjan, S. (2000). Variability in isoltes of *Ralstonia solanacearum* affecting solanaceous vegetables in Kerala. *Veg. Sci.,* **27**(3):189-191.

Meah, M. B. and Wick, L. R. (2006). Diagnostic plant pathology, Dept of Plant, Soil and Insect Sciences, University of Massachusetts Amherst M. A., USA. pp.5-71.

Meah, M.B. and Mian, M. A. W. (1981). Pathogenicity of *Choneophora cucurbitarum* on chilli and possibility of its chemical control. *Acta. Phytopathol. Aead. Sec.Hung.*

Mech, M. B. and Khan, A. A. (1987). Checklist of fruit and vegetable diseases of Bangladesh. 1st edition. Department of Plant Pathology, Bangladesh Agricultural University, Mymenshing, Bangladesh, pp.13.

Mech, M.B. (2006). Plant disease diagnostic manual-plant disease diagnostic clinic. IPM laboratory, Department of Plant Pathology, Bangladesh Agricultural University, Mymensingh, Bangladesh, pp.6-11.

Mehrotra, R.S. and Aneja, K. R. (1990). An introduction to Mycology. Wiley Eastern Limited, Ansari Road, New Delhi, 110002. pp.764.

Meon, S. (1990). Infection of chilli by *Cercospora capsici*. *Pertanika*, **13**(3): 321-325.

Moens, M. and Walvaert, W. (1981).Capsicum varietal susceptibility to *Leveillula taurica* and the influence of different irrigation methods. *Meded. Fac. Landbouwwet. Rijksuniv. Gent.*, **46**: 805-811.

Monkhung, S., To-anum, C. and Takamatsu, S. (2011). Molecular approach to clarify taxonomy of powdery mildew on Chilli plants caused by *Oidiopsis sicular* in Thailand. *J. Agril. Tech.* **7(6):**1801-1806.

Patel, H. R. and Patel, B. N. (1998). Evaluation of Dazomet (Basamid G) against rove beetles, damping off, weeds and nematodes in bidi tobacco nursery. *J. Mycol. Pl. Pathol.,* **28**:134-139.

Prakasam, V. and Jeyarajan, R. (1989). Role of surface wax of chilli fruits in inducing resistance to fruit rot pathogens *Colletotrichum capsici* and *Helminthosporium rostratum. Curr. Sci.,* **58**(17): 976-977.

Prasad, D. and Gupta, S. K. (2012). Role of epidemiological factors on the development of barcterial spot (*Xanthomonas vesicatoria)* in bell pepper. *Indian Phytopath.,* **65**(4):412-414.

Raj, H., Bharadwaj, M. L. and Sharma, N. K. (1997). Soil solarization for the control of damping off of different vegetable crops in the nursery. *Indian Phytopath.,* **5**(4):524-528.

Rajan, P. P. and Sharma, Y. R. (1996). Biotechnology of spices, medicinal and aromatic plants. Proceedings of the national seminar on Biotechnology of Spices and aromatic plants, Calicut, India: pp.150-155.

Ramamoorthy, V., Raguchander, T. and Samiyappan, R. (2002). Enhancing resistance of tomato and hot pepper to Pythium diseases by seed treatment with fluorescent Pseudomonads. *European J. Pl. Path.,* **108**:424-441.

Ramayashreedevi, G. S., Singh, D., Srivastava, A., Biswas, K. S. and Gupta, A. K. (2017). Characterization of Xanthomonas species causing barcterial leaf spot disease of pepper (*Capsicum annuum*) in India. *Indian J. Agric. Sci*., **87**(12):1679-1686.

Rangaswami, A. (1979). Diseases of vegetables. In: Diseases of crop plants in India: 2nd edition. McGraw Hill Bose Company, New Delhi, pp.345-347.

Rasid, M.M., Kabir, M.H., Hossain, M.M., Bhuiyan, M.R. and Khan, M. (2015). Eco-friendly management of chilli anthracnose (*Colletotrichum capsici*). *Intn.J.Pl.Path*.6:1-11.

Reuveni, R. and Dor, G. (1998). Local and systemic control of powdery mildew (Leveillula taurica) on pepper plants by foliar spray of mono-potassium phosphate. *Crop Prot.* 17(9): 703-709.

Reuvani, R. and Roten, J. (1973). Epidemics of *Leveillula taurica* on tomatoes and peppers as affected by conditions of humidity. *Phytopath.Z.* **76**(2): 153-157.

Roy, A., Bordoloi, D.K. and Paul, S. R. (1997). Screening of resistant genotypes of red pepper (*Capsicum annuum* L.) against leaf curl infection. *PKV Res. J.,* **21:**175-176.

Samaddar, K. R., Chakraborty, M. and Kanjilal, S. (1998). Identification of the races of *Pseudomonas solanacearum* causing wilt of solanaceous vegetables in West Bengal and its survival. *J. Mycol. Res*., **36**:51-58.

Sanogo, S. and Ji, P. (2012). Integrated management of *Phytophthora capsici* on Solanaceous and Cucurbitaceous crops: Current status, gaps in knowledge and research needs. *Can. J.Plant Path*. 34(4) 479-492.(https:// doi.org/10,1080/07060661.2012.732117.)

Saxena, A., Raghuwanshi,R.,Gupta,V.K. and Singh,H.B.(2016).The epidemiology and management.*Front.Microbiol*.7:1527.(doi:10.3389/fmicb.2016.01527).

Senanayake, D.M.J.B., Varma, A., Mandal, B.J. (2012). Virus –vector relationships ,host range, detection and sequence comparion of chilli leaf curl virus associated with an epidemic of leaf curl disease of chilli in Jodhpur. *Indian Phytopath*. 160:146-155.

Sharma, P. N., Chowfla, S. C. and Andotra, P. S. (1991). Epidemiology of mosaic disease complex of blue pepper in Himachal Pradesh. *Indian Phytopath.,* **44:** 112-114.

Sharma, S. L. and Bharadwaj, S. S. (1976). A fruit rot and leaf blight of bell pepper caused by Phytophthora nicotianae var nicotianae. *Indian Phytopath.,* **29:** 93-94.

Sharma, S. R. and Sohi, H. S. (1981). Studies on non-chemical control of Cercospora leaf spot of chilli. *Indian J. Mycol. Pl. Pathol.,* **11**(1): 30-34.

Sharma, S. R. and Sohi, H. S. (1984). Control of Cercospora leaf spot of chilli with Bavistin. *Pesticides*, **18**(5): 21-23.

Shekhawat, P. S. and Chakravarti, B. P. (1977). Assessment of loss, symptomatology and occurance of barcterial leaf spot of chillies caused by Xanthomonas vesicatoria in Rajasthan. *Indian J. Mycol. Pl. Pathol.,* **7:**11-14.

Shivanathan, P. (1982). Epidemiology of chilli leaf curl disease in Sri Lanka. *Trop. Hortic.,* **138:**99-109.

Shyam, K. R. (1969). The fruit rot of bell pepper (*Capsicum annuum* L.) caused by *Phytophthora capsici* Leon. M. Sc. Thesis, Punjab Univ., Chandigarh, pp.80.

Shyam, K. R. (1991). Important diseases of summer and winter vegetable crops and their management. In: Horticultural techniques and post harvest management (eds. Jain et al.), Directorate of Extension Education, UHF, Nauni, pp.62-69.

Singh, S.J., Sastry, K.S. and Sastry, K.S.M. (1979). Effecacy of different insecticides and oil in the control of leaf curl disease of chilli. *J.Pl. Dis.Prot.* **86**:253-256.

Singh, S. A. and Thakur, D. P. (1979). Reaction of chilli (*Capsicum fruitesencs*) varieties to *Colletotrichum capsici* (Syd.) Butler and Bibsy. *Curr. Sci.*, **48(11):**512-513.

Singh, S. S. and Sharma, R. K. (1999).Control trial on chilli mosaic virus after inhibiting their infectivity by leaf juices of some angiospermic plants. *J. Living. World.*, **6:** 18-21.

Sohi, H.S., Shyam, K. R. and Sharma, S. L. (1971). Fruit rot of Capsicum annuum caused by *Phytophthora capsici* Leon. *Int. Symp. Pl. Pathol.,* Bangalore. pp. 31(Abstr.).

Sonitrat, P., Pitakpaivan, P., Khamhangridthirong, T., Choobamroong, W and Kueprekone, U. (1994). Host index of plant disease in Thailand. Plant pathology and Microbiology Division, Mycology Section, Department of Agriculture, Bangkok, Thailand (3rd edition).

Stall, R. E. and Lozano, J. C. (1982). Selection for component of horizontal resistance to barcterial spot of pepper. Proc. Fifth. Int. Conf. Pl. Pathogenic. Bacteria. pp.511-517.

Suang, J. M., Cho, D. J. and Kand, S. N. (1984). Undescribed fungal leaf spot disease of pepper caused by *Cercospora capsici* in Korea. *Korean J. Mycol.*, **12(2):** 75-77.

Sudha, A. and Lakshman, P. (2009). Integrated Disease management of powdery mildew (*Leveillula taurica*) Lev. Arn. of Chilli. (*Capsicum annuum* L.) *Archives of Phylopath. Pl.Prot.*, **42(4):** 299-317.

Sunaina, V. and Tomar, D. K. (2003). Distribution of barcterial wilt of potato and its races/ biovars in Uttaranchal hills. *Bionotes*, **5**:13-14 Sunderban region of West Bengal. *J. Mycopathol. Res.,* **30(1):** 31-35.

Suresh, K. R. (2013). Studies on leaf spot of chilli (*Capsicum annuum* L.) caused by *Cercospora capsici* Heald and Wolf. MSc thesis, UAS, Bangalore.

Thakur, M.R., Singh, J., Singh, H. and Sooch, B.S. (1987). Punjab Lal- a new multiple disease resistant chilli variety. *Prog. Fmg.* **23**(9): 11-16.

Than, P.P., Prihastuli, H.P., Sithisack, Taylor, P.W.J. and Kevin, D. Hyde (2008).Chilli anthracnose desease caused by *Colletotrichum* species. J. Zhejiang Uni. Sci.(B). Zhejiang University Press.

Tripathi, N. N. and Grover, R. K. (1976). Inoculum potential and soil factors affecting the pathogenesis of Pythium butleri in causing damping off of tomato. *Proc. Indian. Natl. Sci. Acad. B.,* **41(5):**466-474.

Ullasa, B. A., Rawap, R. D., Sohi, H. S., Singh, D. P. and Joshi, M. C. (1981). Reaction of sweet pepper genotypes to anthracnose, Cercospora leaf spot and powdery mildew. *Plant. Dis.,* **65(7):**600-601.

Valend, G. B. and Muniyappa, V. (1992). Epidemiology of tobacco leaf curl virus in India. *Ann. App. Biol.*, **120**: 257-267.

Verma, S., Shyam, K. R. and Gupta, S. K. (2006). Fungicidal management of Phytophthora leaf blight and fruit rot of bell pepper (*Capsicum annuum* L.). *Pl. Dis. Res.,* **21(2)**: 128-131.

Vincelli, P. (2002). QoI (Strobilurin) Fungicides: Benefits and Risks. *The Plant Health Indicator.* doi: 10.1094/PHI-I-2002-0809-02.

Yang, S. S., Kim, C. H., Cho, E. K. and Lee, E. J. (1991). Distribution and Characteristics of supressive soil to Phytophthora blight of red pepper in Korea. Res. Rep. Rural Dev. Adm., *Crop Prot.*, **33(3):**18-22.

Yashimoto, Ryota; Sasaki,Hirotaka;TakahashiTadashi; Kanno, Hitoshi; Nanzyo, Masami (2012). "Contribution of soil components to adsorption of Pepper Mild Mottle Virus by Japanese Soils". *Soil Biol. Biochem.*, **46**:96-102.

5

Diseases of Okra (*Abelmoschus esculentus* L.) and Their Integrated Management

Raghubir Kumar Patidar, Pranab Dutta, Mahesh Pathak **and** ***Lipa Deb***

School of Crop Protection, College of Post-Graduate Studies in Agricultural Sciences Central Agricultural University (Imphal), Umiam, Meghalaya-793 103

Introduction

Okra (*Abelmoschus esculentus* L., Family: Malvaceae) also known as lady's finger has been originated in Africa and introduced to India during early 20th century. As an economical vegetable crop, it can be grown in tropical as well as sub-tropical parts of the world. During 2018-19, the total area under cultivation of okra was 511 thousand ha while the production stood at 6219 thousand MT in India (Horticulture Statistics Division, 2018). Okra can be grown on wide range of soils, but higher yield can be obtained in well-drained fertile soils with adequate organic matter. It is utilized for various purposes in terms of fresh leaves, buds, flowers, pods and stems providing a rich source of protein, carbohydrates, vitamins, calcium, potassium, enzymes, and total minerals in human diet. The mucilage has medicinal applications when used as a plasma replacement (Nilesh, *et al*., 2012), and the seeds contain 20 to 40% oil content (Sorapong, 2012).

Okra pods (per 100 g edible portion) are composed of water (88.6 g), energy (36 kcal), protein (2.10 g), carbohydrate (8.20 g), fat (0.20 g), fibre (1.70 g), Ca (84.00 mg), P (90.00 mg), Fe (1.20 mg), β-carotene (185.00 μg), riboflavin (0.08 mg), thiamine (0.04 mg), niacin (0.60 mg) and ascorbic acid (47.00 mg) ((Priya, *et al*., 2014).). Therefore, okra has been called as "a perfect villager's vegetable" due to its robust nature, high dietary fibre as well as distinct seed protein with a balance of both lysine and tryptophan amino acids. Okra production is subjected to various diseases caused by fungi, virus and nematodes as a major biotic constraint. The mandates of integrated disease management of okra are to impart knowledge about symptoms, causal organism, disease cycle and epidemiology of major diseases. The present review will give a brief highlight about major diseases of okra *viz.* yellow vein mosaic, wilt, powdery mildew, cercospora leaf spot, damping off, collar rot and root knot nematode.

Yellow vein mosaic- Bhendi *Yellow Vein Mosaic Virus (BYVMV)*

Economic Importance: Yellow vein mosaic disease is one of the very serious constraint in bhendi production throughout the country. It was first reported in India from Bombay in 1924 (Kulkarni, 1924). Estimates show that plants affected with diseases within 35 days of germination exhibits retarded growth, few leaves and malformed fruits with yield loss as high as 94% (Pun and Doraiswamy, 1999). The extent of damage declines with the delay in infection of plants in field condition, whereas, plants infected at 50-65 days after germination suffer a yield loss of 84-49% respectively (Sastry and Singh, 1974).

Symptoms: The diseased plants can be recognized from a distance due to the yellowing of entire network of veins. The characteristic symptoms consist of alternate green and yellow patches exhibiting vein clearing and vein chlorosis of leaves. The yellow network of veins are very conspicuous due to thickening of vein and veinlets. In severe cases, the chlorosis may extend to the interveinal area that may results in to complete yellowing of leaves.

Distortion of leaf stalks and stems occur at an advance stage of infection. Infected plants are stunted bearing very deformed, small and yellow green fruits causing heavy loss in yield.

Causal organism

Taxonomic position of yellow vein mosaic virus of okra:

Domain : Virus

Order : Geplafuvirales

Family : *Geminiviridae*

Genus : *Bhendi Yellow Vein Mosaic Virus (BYVMV)*

Begomoviruses consists of spherical bipartite particles and ssDNA as genome with two similar sized DNA components (DNA A and DNA B). The DNA A component encodes for replication-associated protein (Rep), replication enhancer protein (REn), coat protein (CP) and transcription activator protein that are essential for viral DNA replication and gene expression. While, DNA B component encodes nuclear shuttle protein (NSP) and movement protein (MP) essential for systemic infection of plants (Gafni and Epel, 2002).

Disease Cycle

The continuous disease cycle of yellow vein mosaic virus in okra is maintained through its infected wild and cultivated hosts. The primary infection takes place by viral particles in infected plants and weeds as collateral hosts such as *Hibiscus tetraphyllus, Croton sparsifolia, Malvastrum tricuspidatum and Ageratum* spp.

Virus is transmitted from diseased to healthy plants via vectors *viz.* whitefly (*Bemisia tabaci: Hemiptera: Aleyrodidae)* and leafhopper (*Empoasca devastans: Hemiptera: Cicadellidae),* facilitating secondary spread of disease. The minimum acquisition and inoculation access period of 15-29 minutes are required for effective transmission of *BYVMV.*

Epidemiology

The disease incidence *OYVMV* is favoured by environmental factors such as temperatures, relative humidity and whitefly population. The congenial temperature and relative humidity for multiplication and spread of yellow vein mosaic virus lies between 25°-36°C and 40% respectively. Temperature above 38°C in May-June and low temperature in January-February causes low vector population. The viral disease incidence can be determined by estimation of whitefly population as per existing weather parameters mainly during the crop season. Warm humid weather favours insect population and they reproduces significantly during the summer season, catalysing the transmission of virus between okra plants. Crop sown in April-June or later shows higher disease incidence while 30-50 days old standing crop shows maximum rate of increase in disease development (Bhagat *et al.,* 2001).

Wilt of okra- *Fusarium oxysporum* f. sp. *vasinfectum*

Economic Importance

Fusarium wilt of okra is one of the most widespread disease of okra. It was first reported from Brazil in 1935 infecting cotton variety "Texas" in Paraiba state (Silva *et al.,* 2007). The disease causes significant yield losses in both cotton and okra.

Symptoms: The characteristic symptom of okra wilt includes typical wilt beginning with yellowing and stunting of the plant followed by wilting and rolling of the leaves as if the roots were unable to supply sufficient water. The leaves hang down during daytime and recover again in the night, eventually wilt and die. However, before wilting, infected plants show epinasty and vein clearing symptoms along with loss of turgor. The necrotic spots appear on leaves along their veins, while, the affected leaves fall and plants dries out.

If a diseased stem is split lengthwise, the vascular bundles appear as dark streaks. When severely infected, nearly the whole stem is blackened.

Causal Organism

Taxonomic position of wilt fungus of okra

Domain	:	Eukaryota
Kingdom	:	Fungi
Phylum	:	Ascomycota
Class	:	Sordariomycetes
Order	:	Hypocreales
Family	:	Nectriaceae
Genus	:	*Fusarium*
Species	:	*Fusarium oxysporum* f. sp. *vasinfectum*

Fusarium oxysporum f. sp. *vasinfectum* appear as white, pinkish to violet fungus with hyaline, sparse to floccose mycelia developing intracellularly in to the host. The fungus produces 3-5 septate macroconidia and 0-1 septate microcondia on sporodochia and pionnotes. Conidia appear as buff to salmon orange, fusiform-falcate in shape, pedicellate base and curve inwards at both ends. Microconidia measures about 5-12 x 2-3.5 microns, while macroconidia measures 40-50 x 3-4.5 microns in size. Chlamydospores of broadly ovate shape measuring about 7-13 microns in diameter are formed both intercalary and terminally.

Disease Cycle

The fungus *F. oxysporum* f. sp. *vasinfectum* is soil borne in nature that survives as a saprophyte on colonized roots and as chlamydospores at the depth of about 1 metre in soil. After coming in contact with root exudates from root surface of 7-16 days old plants, germination of conidia and chlamydospores takes place at rhizospheric region. They form germ tubes and hyphal strands on the root surface that proliferate inside the xylem vessels and multiplies rapidly. Wounds by nemtodes also facilitate the ingression of root surface by the fungus. Thereofre, the wilt fungus enter plants through roots, interfering water conducting vessels by restricting water flow. The host range of *Fusarium* wilt includes okra, cotton, tomato, potato, eggplant, pepper, lucerne, lupin, soybean, tobacco and weeds. Usually symptoms of wilt may appear in the plants of about 30-45 days old. Fusaric acid is known to produce by fungi in the infected plant tissues or rhizosphere or in the soil. Fusarium wilt can survive for years in the soil, seeds as well as infected okra vegetation residues and spread by water, insects and garden equipment.

Epidemiology

F. oxysporum f. sp. *vasinfectum* grows and develops at temperatures ranging from 10 to 35°C. Optimum growth conditions includes temperature between 18-27°C, soil humidity of 40-70% and acidic environment near to pH-5.3. Temperatures higher than 35^{0}C are inhibitory to the wilt development. The disease affects more intensively in places with high fraction of okra in field crop rotations. Fusarium wilt of okra is often interrelated with root knot nematode as they provide a venue for vascular wilt diseases. Therefore, the incidence of wilt significantly increases in the presence of root knot nematode. Excess use of nitrogenous fertilizers and deficiency of potassium in soil favours wilt disease incidence.

Powdery mildew- *Erysiphe cichoracearum*

Economic Importance: Powdery mildew dieases is an emerging potential threat to okra cultivation, which affects plants at all growth stages and may resuly huge losses in yield up 17-86.6 % (Ghanem, 2003).

Symptoms: Powdery mildew is characterized by grey-coloured powdery growth that appear on both upper and lower surface of the leaves, disrupting photosynthesis activity in plants. The leaves turn yellow, dries and finally drop, thus, reducing the yield of the crop due to severe infestation. The problem of re-sowing of okra crop may come during favourable weather condition for development of fungi.

Causal Organism

Taxonomic position of powdery mildew fungus of okra:

Domain	:	Eukarya
Kingdom	:	Fungi
Phylum	:	Ascomycota
Sub-division	:	Pezizomycotina
Class	:	Leotiomycetes
Order	:	Erysiphales
Family	:	Erysiphaceae
Genus	:	*Erysiphe*
Species	:	*E. cichoracearum*

E. cichoracearum DC ex. Merfat produce well-developed, evanescent mycelia bearing ellipsoidal or barrel-shaped conidia in chains measuring 34.8 x 15.2 µm.

Cleistothecia are gregarious or scattered, globose, becoming depressed or irregular measuring 34.8 x 15.2 µm having distinct walls of 10-20 µm wide. Appendages are numerous, basally inserted, myceloid, interwoven with mycelium, hyaline to dark brown and 1-4 times as long as the diameter of cleistothecia. Asci are ovate to broadly ovate, rarely sub-globose, more or less stalked, 60-90 x 25-50 µm in size having 10-25 asci per ascocarp. 2-3 ascospores measuring 20-30 x 12-18 µm in size are present per ascus.

The teleomorphic stage i.e. *Sphaerotheca fuligena* (Schlecht ex. Fr.) Poll produces hyaline, occasionally brown when old, usually evanescent but sometimes persistent mycelia bearing ellipsoid to barrel-shaped conidia in long chains with distinct fibrosin bodies measuring 27-31 x 15-18 µm in size. Cleistothecia are either rare or scattered to densely gregarious, 66-98 µm diameter and 25 mm wide peridial cells. Appendages are variable in number, usually as long as the diameter of ascocarp, myceloid, tortuous, brown, interwoven with the mycelium but sometimes long nearly straight and brown. Ascocarp are broadly elliptical to subglobose, 50-80 x 30-60 µm in size having one ascus per ascocarp. Each ascus contains eight ascospores, which are ellipsoid to nearly spherical measuring 17-22 x 12-20 µm in size.

Disease Cycle

Powdery mildew fungus tends to grow in fine tissue layers on infected plants surface. This white fungus is mainly composed of spores in chains that travel by wind, causing them to spread to new host plants in the area. Germination for all species of powdery mildew can occur without water. The disease cycle is initiated by wind blown conidia, which infect okra over a range of humidity (50-95%) during moderate to warm (20-27°C) temperatures. *E. cichoracearum* is more active at lower temperatures than *P. fuliginea*. Penetration of host cell by germ tubes from ascospore is entirely mechanical which result in dislocation and inward invagination of epidermal cells. Honeydew also favours the penetration by the germ tubes. The conidia of powdery mildew fungus are dispersed by wind and insects and usually overwinters on wild cucurbit hosts, weeds, and dead vines.

Epidemiology

Powdery mildew fungus has the to thrive under environmental condition as it prefers warm and humid weather. It is one of the most common leaf diseases in areas with dry hot summers. Temperatures between 16-27^{0}C are the most advantageous for this disease. Shady areas are also highly beneficial for spore development. Growth will become challenging for powdery mildew if temperatures exceed 33^{0}C, especially under direct sunlight.

Cercospora Leaf spot- *Cercospora malayensis, C. abelmoschi*

Economic importance: Cercospora leaf spot is a major disease of okra damaging leaves, resulting huge yield loss of the crop. It has been reported in tropical and sub-tropical countries of Asia and is commonly present in Western Africa wherever okra is cultivated, especially during the warm rainy season (Farrag, 2011).

Symptoms

The characteristic symptoms of cercospora leaf spot of okra differs on the basis of Cercospora species as well as age of the plant. In India, two species of cercospora causes leaf spot of okra. *C. malayensis* causes brown, irregular spots with grey centre and dark coloured margin, whereas, *C. abelmoschi* causes sooty black and angular spots as the first sign of infection in the lower surface of leaves.

Generally, symptoms first appear on the lower, older leaves and progress as the newer lesions appears on the younger leaves. As the disease advances, the leaf spots enlarge and eventually merge to cover the entire leaf surface, which then turns necrotic and often rolls as it dries, but remains attached to the stem. Both the leaf spots cause severe defoliation and are common during humid season.

Causal Organism

Taxonomic position of cercospora leaf spot fungus:

Domain	:	Eukarya
Kingdom	:	Fungi
Phylum	:	Ascomycota
Sub-division	:	Pezizomycotina
Class	:	Dothidiomycetes
Order	:	Capnodiales
Family	:	Mycosphaerellaceae
Genus	:	*Cercospora*
Species	:	*C. malayensis, C. abelmoschi C. malayensis* F. Stevens and Solheim

Disease Cycle

The disease is most damaging to okra, watermelon and other cucurbitaceous crops. The fungus survives through the conidia and stromata on diseased plant debris in the soil. The disease cycle begins when spores (conidia) are deposited onto leaves and petioles by wind or splashing water. Conidia germinate during moderate to warm temperatures in the presence of free moisture. New cycles of infection and sporulation occur every seven to ten days during warm and wet weather. The pathogen is readily disseminated within and among fields by wind, splashing rain, irrigation water and survives between okra crops as a pathogen on weeds and in infested crop debris in the soil.

Epidemiology

Rainfall and high humidity favour infection, disease development and sporulation of the pathogen on okra leaves. Rain splash or overhead irrigation creates a conductive microclimate for pathogen to multiply in the field and favourable conditions for conidia (spores) to move from plant to plant. Conidia rapidally germinate at 25-33^{0}C temperatures. Wind spreads conidia from one field to another. Infected okra seedlings,movement of people and animals also spread the disease within the field and over longer distances.

Damping-off of okra- *Pythium aphanidermatum, Phytophthora nicotinae* and *Rhizoctonia* sp.

Economic Importance

Damping off of okra have been considered as one of the most important and destructive disease of okra with the disease incidence reaching as high as 50-75% (Matny, 2013). Damping-off of okra was first reported to be caused by *Phytophthora nicotinae* bu Matny (2013) from Iraq (Matny, 2013). Three pathogens have been identified as causal agent of damping-off of okra *viz. P. aphanidermatum, P. nicotinae* and *Rhizoctonia* sp.

Symptoms

Plants are most vulnerable to infection by *P. aphanidermatum* during germination and juvenile stages. Damping-off of seedlings occurs in two stages *i.e.* the pre-emergence and post-emergence phase. The initial symptoms may be poor or uneven germination and complete rotting of seedlings killing young radicale and plumule during pre-emergence of seedlings. Seedlings that do germinate are susceptible to post-emergence damping-off. The post-emergence phase is characterized by the infection of young, juvenile tissues of the collar at ground level. The roots, crowns and hypocotyl region of infected seedlings become soft, water-soaked, and and topple over or collapse.

Causal organism: Taxonomic position of damping-off pathogen of okra:

Domain	:	Eukarya
Kingdom	:	Chromista
Phylum	:	Oomycota
Class	:	Oomycetes
Order	:	Pythiales
Family	:	Pythiaceae
Genus	:	*Pythium*
Species	:	*P. aphanidermatum*

P. aphanidermatum produces branched, hyaline, coenocytic hyphae up to 5 μm in diameter bearing globose to oval, terminal to intercalary sporangia measuring about 15-26 μm in diameter. Zoospores are kidney-shaped (reniform) with two flagella, encyst and germinate by germ tube. The size of zoospore is up to 8 μm in diameter. Oogonium are globose, terminal or intercalary with multinucleate ooplasm surrounded by a layer of periplasm measuring about 15-28 μm in diameter, while, anthredium are elongated or club-shaped. Oosphere are smooth, thick-walled, aplerotic measuring about 12-20 μm with 1.5 μm thick wall. The oospore germinate by producing a germ tube or by functioning as a sporangium.

Disease Cycle

Pythium is a weak sprophyte, can infect young seedlings or succulent plants as mature plant tissues become resistant to its attack. *P. aphanidermatum* overwinters in the soil as oospores and it germinates to produce germ tube. Under favourable environmental condition, oospores may produce sporangia, which in turn produce motile, biflagellate zoospores that swim to the host plant and germinate. Secondary infection takes place by asexual spores leading to rapid spread of the disease. Once infected, mycelium growth takes place throughout the plant tissue and kill plants over period of time.

Epidemiology

P. aphanidermatum can cause disease at temperature ranging between 13-18 ^{0}C but damping off is more severe at 24-30^{0}C. Disease is most brutal in poorly aerated and drained soil. Motile zoospores need to swim in order to infect the host, therefore moist conditions facilitate the rapid spread of the disease. Temperature also has an effect on the rate of pathogen propagation. Factors that may cause stress in plants such as high saline conditions, drought, nutrient deficiencies, and excessive moisture around the plant also increases the likelihood of infection. High saline content in the soil, low temperature and humidity also provides an ideal condition for promotion infection. Excessive nitrogen fertilization also increases the chance of infection as nitrogen decreases the function of the plant's innate defence response and damages the ends of the roots serving as a primary mode of infection.

Collar rot- *Macrophomina phaseolina*

Economic Importance

Collar rot disease of okra is a widespread, non-specific disease that infect more than 500 hosts in about 100 families including crops and weeds. Economic crop hosts include cotton, groundnut, jute, maize, millet, potato, sesame, soybean and other beans, sunflower, sweet potato, tomato and tobacco. Charcoal rot is an important disease during hot, dry weather or when unfavourable environmental condition.

Symptoms

Infected seedlings shows reddish brown water-soaked lesions on the hypocotyls near the soil-line are characteristic feature of charcoal rot or collar rot of okra. Fungus are commonly found in seeds and can be redistributed. Once the hypocotyls are girdled the seedling will die. Water-soaked lesion appear first on stem but then turn into dull light brown. Once the stem is infected, fungus rapidly colonizes the remaining tissues. The dead tissues turn black as microsclerotia are abundantly produced and infected seedlings may die in hot and dry weather. Sclerotia can also be produced on roots and can lead to yellowing and wilting of the plant. The sclerotia resemble a sprinkling of finely powder charcoal. When split open the root and base of stem shows black streak in the woody tissues and sclerotia formed in the pith of the stem.

Causal Organism

Taxonomic postion of collar rot fungus of okra

Domain : Eukarya

Kingdom : Fungi

Phylum : Ascomycota

Class : Dothidiomycetes

Order : Botryosphaeriales

Family : Botryosphaeriaceae

Genus : *Macrophomina*

Species : *M. phaseolina*

M. phaseolina (Tassi) Goid. produces white to brown coloured culture that darken with age having hyphal branches generally formed at right angles. Bears pycnidia of 100-200 μm in diameter, dark to grayish, globose or flattened, membranous to sub carbonaceous with inconspicuous or definite truncate ostiole. The pycnidia bear simple, rod-shaped conidiophores, 10-15 μm long and conidia are single celled, hyaline, elliptical or oval and measuring about 14-33 x 6-12 μm in diameter. Microsclerotia are jet black in colour, appearing smooth and round to oblong or irregular.

Disease Cycle

M. phaseolina is a soil borne fungus that causes charcoal rot. The microsclerotia serve as the primary source of inoculum and can survive as microsclerotia in the soil and infected plant debris for three years and. The microsclerotia are black, spherical to oblong structures that are produced in the host tissue and released in to the soil as the infected plant decays. These multi-celled structures allow the persistence of the fungus under adverse conditions such as low soil nutrient levels and temperature above 30^{0}C. Microsclerotial survival is greatly reduced in wet soils and can't survive more than 7 to 8 weeks and mycelium more than 7 days. Seeds may also carry the fungus in the seed coat. Infected seed do not germinate or produce seedlings that die soon after emergence.

Epidemiology

Germination of microsclerotia occurs throughout the growing season when temperature ranges between 28-35^{0}C. Microsclerotia germinate on the root surface intogerm tubes forming appresoria that penetrate the host epidermal cell walls through natural openings or mechanical or enzymatic digestion leading to disease development. The hyphae grow first intercellularly in the cortex and then through the xylem colonizing the vascular tissue. Through vascular tissue, *M. phaseolina* spreads throughout root and lower stem of the plant producing microsclerotia that plug these vessels. The rate of infection increases with higher soil temperatures and low soil moisture further enhancing disease severity. Hot, dry weather promotes infection and development of charcoal rot. The population of *M. phaseolina* in soil will increase when susceptible hosts are cropped in successive years and can be redistributed by tillage practices.

Root-knot Nematode- *Meloidogyne incognita, M. javanica*

Economic Importance

Root-knot nematodes (*Meloidogyne* spp.) are serious and economically most important pests of all the cultivated crops around the world. According to Sikora and Fernandez (2005), root-knot nematodes are particularly damaging vegetables in tropical and subtropical countries of the world and cause losses up to 80% in heavily infested fields. Okra ranked high amongst the economically important vegetables of the world. The avoidable yield loss in okra due to root-knot nematode, *M. incognita* was recorded to the extent of 27.02 per cent (Shendge, *et al.*, 2010).

Symptoms

Root-knot nematodes are sedentary endoparasites of roots. The physical symptoms are *viz.* yellowing of leaves, stunted growth, wilting during hot and dry weather conditions; undersized fruit and reduced yield are common above ground symptoms. Below ground symptoms are typical and diagnostic as they form the root galls/knots. The intensity and size of galls are variable depending upon the root knot nematode species, population intensity, crop vulnerability and age of the crops.

Okra is highly susceptible to root knot nematode and form heavy galling. The okra plants remain pale and stunted and pod set is very low. The leaves are yellowish green or yellow in colour. Drooping, sudden wilting together with a scorching of leaves from the margins inward can be observed. Formation of knots or galls in the root system is a characteristic symptom. The maximum number of galls per plant and number of females and egg masses per gram of root occur when plants are infected at two week stage. The main root and the laterals have spherical or elongated galls of various sizes. In advanced stages of infection, tissues decay and are attacked by other pathogenic and saprophytic organisms.

Disease Cycle

Root knot nematode is sedentary endoparasites. Reproduction is generally parthenogenetic. The rectal glands in female secrete gelatinous substance on the root surface in which eggs are laid. Oviposition continues for 10-12 days. Each female laid about 200-400 eggs held together in an egg mass. Embryogenesis takes about 10-15 days and first moult occurs within the eggs. Second stage juvenile (J_2) is the only infective stage and the hatched J_2 move freely in soil, for search of new roots. This stage is also known as pre-parasitic J_2. The J_2 penetrates the roots just behind the root cap through repeated thrust of stylet and head located near the vascular tissues while rest of the body remain completely inside the cortex.

As the feeding process begins J_2 start assuming swollen shape, now called parasitic J_2. Second moult occurs within a week and J_3 is formed than the third moult follows quickly. The pointed tail of J_2 still visible hence called "spike tail stage". The J_3 and J_4 retain their old cuticle. J_3 and J_4 are non-feeding stages due to absent of stylet. At the last moult adult female become sac like or rounded, stylet reappear and reproduction system gets fully developed. The whole life cycle is completed in about 25 days at 25-30°C.

Epidemiology

Meloidogyne incognita is abundant in warmer areas whereas *M. javanica* is common in cooler areas. Root-knot nematodes begin their lives as eggs and J_2 hatch out from eggs and it is the infective stage but eggs hatches under suitable environmental conditions. Damage and yield losses caused are generally more severe in coarse-textured sandy soils. The pore size of soil must be more than width of nematode body (20µm). *Meloidogyne* spp. are generally intolerant of flooded soil conditions. J_{2s} attack growing root tips and enter roots intercellularly, behind the root cap. Optimum temperature for nematode invasion is 15-16⁰C. The most of nematodes can stand against upon 10 atmospheric. They move to the area of cell elongation where they initiate a feeding site by injecting oesophageal gland secretions into root cells. These nematode secretions cause dramatic physiological changes in the parasitized cells, transforming them into giant-cells. For successful completion of development period of its life cycle it requires 18-24⁰C. They are introduced through infected seedlings and by shifting soil from neighbouring infested fields.

Integrated Disease Management Approach in Okra

Cultural Approach

- Before planting: Plant in sunny places, and, if possible, choose areas with good air circulation.
- Plant only high quality certified seed from government agencies.
- Do not plant successive crops of okra on the same land; practice crop rotation.
- Remove the affected plants from the field and burn them.
- Remove weeds from within and around the crop and collect crop debris after harvest and burn it.
- Two to three deep summers ploughings at 10-15 days interval in May/June month are found effective against the dormant stages of nematodes.
- Plastic mulch can be used to manage the nematodes by increasing the soil temperature. The moist soil should be cover with plastic mulch in small areas like nursery bed for 3-4 weeks before sowing.
- A 15-20 cm thick layer of rice husk (20 kg/m^2) is spread over soil surface and burnt to destroy the nematodes. The ash is mixed with top soil and a week later seed can be sown.

- Crop rotation with cereal crops like wheat, rice, maize, sorghum, pearl millet is found helpful in reducing the root knot nematode population.
- Follow okra-rice-fallow or okra-cucumber-mustard cropping sequence for the management of root-knot nematode (*M. incognita*) in vegetables.
- Trap crop are those plant species, which are highly susceptible to nematodes and can be used against nematode e.g. *Crotalaria* spp.

Chemical Approach

- Spray of Azadirachtin 0.03 % (300 ppm) Neem Oil Based WSP can be used against white fly and leaf hopper @ 2500-5000g dissolve in 500-1000 litre of water.
- Flupyradifurone 17.09% w/w SL 250g a.i./ha or 1250g of insecticides dissolve in 500 litre of water to control the jassids, whitefly.
- Spray of Sulphur 80 % WP @ 2.5kg a.i. /ha or 3.13kg fungicide dissolve in 750-1000 litre of water to manage powdery mildew disease. CIB & RC
- Spray the crop with Copper oxychloride (Blitox-50) @ 3g/litre of water or Dithane Z-78 or Dithane M-45 or Captan @ 2g/litre of water at regular interval for managing cercospora leaf spot disease.
- Post damping off okra can be manage by fungicidal spray Thiophanate Methyl 450g/l + Pyraclostrobin 50g/l w/v FS @ 15g a.i./ha or 30 g fungicide can be dissolve in 500 litre of water and spray it. CIB & RC
- In direct seeded crops like okra and cucurbits seed dressing with carbosulfan @ 3% a.i. (w/w) for root knot nematode management.
- Application of neem/eupatorium @ 15 t/ha two weeks before sowing for controlling root-knot and reniform nematodes.

Biological Approach

- Treat the seed with *Trichoderma viride* 1.5% WP @ 20 gm/kg of seeds and apply *Trichoderma viride* 1.5% WP @ 5 kg/ha enriched FYM* @ 5 tons/ha to the soil before sowing. CIB & RC
- Treat the seed with *Pseudomonas fluorescens* 1.0% WP @ 20 gm/kg of seeds and apply *Pseudomonas fluorescens* 1.0% WP @ 5 kg/ha enriched FYM* @ 5 tons/ha to the soil before sowing.
- Treat the seeds with *Verticillium chlamydosporium* 1.0% WP @ 20 gm/kg of seeds and apply *Verticillium chlamydosporium* 1.0% WP @ 5 kg/ha enriched FYM * @ 5 tons/ha to the soil before transplanting. CIB & RC

- Soil application of *Purpureocillium lilacinum* @ 2.5 kg/ha in 2.5 t FYM to manage *M. incognita* and enhanced fruit yield of okra. CIB & RC.

Resistant Varieties

- Plant resistant varieties like Parbhani Kranti, Arka Anamika, Arka Abhay, Punjab-7, Hybrid-6, VRO-5, VRO-6 and Pusa A-4 they remain free from viral disease under field conditions.
- Select *Fusarium* wilt resistance varieties like Pusa Makhamali, Okra I.S. 9273, 9857, C.S. 3232, 8899, I.S. 6653, 7194 and Pusa Sawni.
- Resistant varieties can be used to reduce the initial nematode population as well as development of root knot nematode e.g. Okra: Long Green Smooth, IC-9273 and IC-18960.

Future Approaches

Historically, more than 90% of the worldwide okra cultivation has been localized in the developing countries of Asia and Africa but very little attention has been paid to its genetic improvement and genomic information on *Abelmoschus.* Also, there is scarcity in the development and implementation of molecular techniques in okra breeding in comparison to other vegetables. Attempts should also be made for the integration of both coat protein genes and antisense RNA technology for the effective viral resistance in okra. With current and fast emerging technologies like chromosome engineering, RNAi, marker-assisted recurrent selection (MARS) and genome-wide selection (GWS), targeted gene replacement using zinc-finger nucleases, next generation sequencing (NGS), nano-biotechnology and genome editing; the future seems promising for the development of designer okra having improved features for viral disease resistance. To avoid the disease incidence during the cultivation okra the following information one should keep in mind such well drained-sandy loams are the most suitable for okra, it grows best in neutral to slightly alkaline soils with optimum pH 6.5, Overnight soaking of seeds are speed up germination, maintain the time of sowing, spacing should be maintain, primarily 3 to 4 seeds per point should be planted and after germination weak plants are removed by thinning. Avoid over doses of nitrogen fertilizers that favour the insects and diseases. Therefore, soil testing and follow recommendations are necessary before planting. The growers can also conduct small scale trials with minimum expenses for solving their problems by their own.

The project proposal on okra should be focused on a particular location, for a specific time period, often with outside resources, demonstrate techniques and methods that could be extended and sustained after the project period. In the last but not in least, it is also suggested that researchers must conduct various experiments on other aspects of non-chemical and bio-control methods etc, against diseases to reduce risk of health hazards and environmental pollution. Insecticide, fungicides, biopesticide, herbicides and plant growth regulators registered under the Insecticides Act, 1968, India should be used to control the plant diseases as per guideline of Central Insecticide Board & Registration Committee.

Conclusion

Abelmoschus esculentus is a medicinal plant of immense nutritional and medical applications. It has been used as an ingredient of many herbal formulations, which are used for the cure of various ailments, in particular the regulation of blood pressure, fat, diabetes, chronic dysentery genitourinary disorders, simple goiter and ulcer.

A well planned and long-term planning is required to manage the diseases of okra and it requires public sector investment jointly by Government of India. The word has started coming up on the biotechnological interventions in okra and it is expected that, in years to come, advance in okra improvement is going to happen. Any breakthrough in the transgenic or highly linked marker(s) with viral disease resistance QTLs will pave a new way for the improvement of viral disease resistance in okra. Hence, it is required to use interdisciplinary approaches to tackle the serious challenges of disease management in okra. Lastly, the outcomes of plant-breeding should reach the farmers as seeds of improved varieties.

References

AESA Based IPM Package No. 23 (2014). AESA based IPM-Okra. Department of Agriculture and Cooperation Ministry of Agriculture Government of India. Pp 81.

Akhtar, S., Khan, A.J., Singh, A.S. and Briddon, R.W. (2014). Identification of a disease complex involving a novel monopartite begomovirus with beta and alphasatellites with okra leaf curl disease in Oman. *Arch. Virol.,* **159**: 1199-1205.

Ali, M., Hossain, M.Z. and Sarker, N.C. (2000). Inheritance of yellow vein mosaic virus (YVMV) tolerance in a cultivar of okra (*Abelmoschus esculentus* (L.) Moench). *Euphytica.,* **111**: 205-209.

Amadi, J.E., Nnamani, C., Ozokonkwo, C.O. and Eze, C.S. (2014). Survey of the incidence and severity of okra (*Abelmoschus esculentus* L. Moench) Fruit rot in Awka South lga, Anambra state, Nigeria. *Int. J. Curr. Microbiol. App.Sci.,* **3**(4): 1114-1121.

Amit K., Verma, R.B., Ravi, K., Saksham, K.S. and Randhir, K. (2017). Yellow Vein Mosaic Disease of Okra: A Recent Management Technique. *International Journal of Plant & Soil Science.,* **19**(4): 1-8.

Arora, D., Jindal, S.K. and Singh, K. (2008). Genetics of resistance to yellow vein mosaic virus in inter-varietal crosses of okra (*Abelmoschus esculentus* L. Moench). *SABRAO J. Breed. Genet.,* **40**: 93-103.

Asare-Bediako, E., Van der Puije, G.C., Taah, K.J., Abole, E.A. and Baidoo, A. (2014). Prevalence of Okra Mosaic and Leaf Curl Diseases and *Podagrica* spp. Damage of Okra (*Albelmoschus esculentus*) Plants. *Int. J. Curr. Res. Aca. Rev.*, **2**(6): 260-271.

Czosnek, H., Ghanim, M., Morin, S., Rubinstein, G., Fridman, V. and Zeidan, M. (2001). Whiteflies; vectors and victims (?) of geminiviruses. *Adv. Virus Res.*, **57**: 291-322.

Farrag, E.S.H. (2011). First record of Cercospora leaf spot disease on okra plants and its control in Egypt. *Plant Pathology,* **10**: 175-180.

Ghanem, G.A.M. (2003). Okra leaf curl virus: A nanopartite begomo virus infecting okra crop in Saudi Arabia. *Arabian Journal of Biotechnology,* **6**: 139-152.

Gilmar, P. Henz., Carlos, A. Lopes & Ailton, Reis (2007). A Novel Postharvest Rot of Okra Pods Caused by *Rhizoctonia solani* in Brazil. *Fitopatol. Bras.,* **32**(3): 237-240.

Gray, S. M. and Banerjee, N. (1999). Mechanisms of arthropod transmission of plant and animal viruses. *Microbiol. Mol. Biol. Rev.,* **63**: 128-148.

Gyan P. M., Bijendra, S., Tania, S., Achuit, K.S., Jaydeep, H., Nagendran, K., Shailesh, K.T. and Prabhakar, M.S. (2017). Biotechnological Advancements and Begomovirus Management in Okra (*Abelmoschus esculentus* L.): Status and Perspectives. *Frontier In Plant Science*, Volume 8, Article 360.

Jambhale, N.D. and Nerkar, Y.S. (1986). "Parbhani Kranti", a yellow vein mosaic resistant okra. *Hortic. Sci.,* **21:** 1470-1471.

Kamalpreet, K. and Prabhjot, K. (2018). Agricultural Extension Approaches to Enhance the Knowledge of Farmers. *Int. J .Curr. Microbiol. App. Sci.,* **7**(2): 2367-2376.

Lamont, W.J.J. (1999). Okra- A versatile vegetable crop. *Hort. Technol.,* **9**: 179-188.

Matny, O.N. (2013). First report of Damping-off of Okra caused by *Phytopthora nicotinae* in Iraq. *Phytopathology,* **97**(4): 558.2-558.2.

Mustansar, M., Yasir, I., Muhammad, I.U., Qaiser, S., Muhammad, A., and Iram B. (2017). Incidence of okra yellow vein mosaic disease in relation to insect vector and environmental factors. *Environment & Ecology***., 35** (3C): 2215-2220.

Priya, S., Varun, C., Brahm, K.T., Shubhendra, S. C., Sobita, S., Bilal, S. & Abidi, A.B. (2014). An overview on okra (*Abelmoschus esculentus*) and its importance as a nutritive vegetable in the World. *International Journal of Pharmacy and Biological Sciences.,* **4**(2): 227-233.

Saha, L.R. (2002). Hand Book of Plant Diseases. Diseases of okra vegetables. Published by Kalyani Publication, Rajendra Nagar, Ludhiyana-141 008. Pp. 333-403.

Sanwal, S.K., Venkataravanappa, V, and Singh, B. (2016). Resistance to bhendi yellow vein mosaic disease: a review. *Indian J. Agr. Sci*., **86**: 835-843.

Sathish, K.D., Eswar, T.D., Praveen, K.A., Ashok, K., Bramha, K., Srinivasa, R.D. and Ramarao, N. (2013). A Review on: Abelmoschus Esculentus (Okra). *Int. Res. J. Pharm. App. Sci*., **3**(4): 129-132.

Schafleitner, R., Kumar, S., Lin, C.Y., Hegde, S.G., and Ebert, A. (2013). The okra (*Abelmoschus esculentus*) transcriptome as a source for gene sequence information and molecular markers for diversity analysis. *Gene.,* **517**: 27–36.

Series of Crop Specific Biology Documents (2011). Biology of *Abelmoschus Esculentus* L. (Okra). Department of Biotechnology Ministry of Science and Technology & Ministry of Environment and Forest. Pp. 26.

Sharma, D.K. (2019). Enumerations on seed-borne and post-harvest microflora associated with okra [*Abelmoschus esculentus* (L.) Moench] and their management. *GSC Biological and Pharmaceutical Sciences.*, **8**(2): 119-130.

Singh, B., Rai, M., Kalloo, G., Satpathy, S. and Pandey, K.K. (2007). Wild taxa of okra (Abelmoschus species): reservoir of genes for resistance to biotic stresses. *Acta Hortic.*, **752**: 323-328.

Singh, R.S. (1998). Plant Diseases. Published by Oxford & IBH Publishing Co. Pvt. Ltd. 66 Janpath, New Delhi-110 001. Pp. 653.

Singh, S.J. (1996). Assessment of losses in okra due to enation leaf curl virus. *Indian J. Virol.*, **12**: 51-53.

Synonyms (2019): National Horticulture Board Ministry of Agriculture and Farmers Welfare Government of India 85, Institutional Area, Sector - 18 Gurugram - 122015 (Haryana), www.nhb.gov.in.

Varshney, R.K., Bansal, K.C., Aggarwal, P.K., Datta, S.K. and Craufurd, P.Q. (2011). Agricultural biotechnology for crop improvement in a variable climate: hope or hype? *Trends Plant Sci.*, **16**: 363–371.

6

Diseases of Cabbage (*Brassica oleracea* var. *capitata* L.) and Their Integrated Management

Nongthombam Olivia Devi, Manashi Debbarma **and** ***R.K. Tombisana Devi***

School of Crop Protection, College of Post-Graduate Studies in Agricultural Sciences Central Agricultural University (Imphal), Umiam, Meghalaya-793103

Blackrot of Cabbage

Economic Importance

Blackrot is an important bacterial disease of cabbage and other crucifer crops worldwide. The disease was first described in New York on turnips in 1893, and has been a common problem for growers for over 100 years (Smart and Lang, 2013). Blackrot has been given many names such as bacterial blight,black stem, black vein, stem rot or stump rot (Pfeufer and Davis, 2019). These pathogen is particularly damaging to cabbage and cauliflower, but turnip, rutabaga, collard, kohlrabi and Chinese cabbage are also susceptible. Broccoli and radish aregenerally resistant to Blackrot (Flynn,1999).

Symptoms

Seedling stage is particularly susceptible to Blackrot. Initial infection of the bacterium starts from cotyledon and systemically spread through leaf veins. As the disease advances seedlings becomes stunted , chloratic, lower leaves drop and eventually die. Leaves may be affected on only one side of a seedling. Plants infected because of contaminated seed may not develop symptoms for many weeks (Pfeufer and Davis, 2019).

In mature stage symptoms of Blackrot generally begins with yellowing at the leaf margin, which expands into the characteristic "V" shaped lesion with the bottom of each "V"pointing inward which enlarges as the disease progresses resulting in yellowing, wilting and falling of entire leaves from the plantdr. Leaf veins and veinlet turn from green to brown and finally black which is the characteristic of the disease. The vascular discolouration may extend to the main stem and petiole causing systemic infection. Bacterial ooze may also be seen on diseased portions and cut open vascular tissues. The heads of the infected plants remains small and its qualityis reduced making it unfit for marketing.

Causal Organism

Blackrot is caused by a gram negative,rod shaped bacteria having a single polar flagellum, *i.e. Xanthomonas campestris* pv. *campestris* (Pammel) Dowson. When grown in nutrient agar medium it produces yellow, mucoid, glistening colonies(Vicente and Holub, 2013). The yellow colonies is due to a pigment called xanthomonadins and the mucoid culture due to an exopolysaccharide i.e xanthan (Vauterin *et al.,* 1995).

Epidemiology

Warm and humid condition favours the development of the disease and can spread rapidly from rain dispersal and irrigation water (Vicente and Holub, 2013; Meenu *et al.,* 2013). The optimum temperature for the growth of the bacteria is 26.5-30°C, minimum being 5°C and maximum 36°C with thermal death point of 50°C.Symptoms use to develop only 7-10 days after infection and develops more slowly at lower temperature.

Disease Cycle

The pathogen survives in diseased plant debris as saprophyte and in the seeds also. The bacteria also survive on weeds like *Brassica rapa*, *Brassica nigra, Cardaria pubescens*, *Sisymbrium officinale*, *Brassica juncea*, *Lepidium* spp., *Raphanus raphanistrum* etc and crop plants without showing symptoms. Seed is the primary source of inoculum of the bacteria where it can survive for three years. During the time of germination, the bacteria inside the infected seed gain entry into the cotyledons or young leaves through stomata, hydathods,wounds and spread inter-cellularly to reach the vascular region and then systematically to all the parts of the plant. Secondary spread in the field occurs through irrigation, rain water splash, wind, soil containing the Blackrot bacterium can be spread on machinery and footwear. It can also be spread on insects including cultural operations. After the harvest of the crop the debris left in the field harbours the pathogen for several years which become active again when a suitable host crop is grown in the same field.

Integrated Management

1. Selection of well-drained fields and planting on raised beds.
2. Use only certified disease free seeds.

3. Maintenance of proper field sanitation, cleaning of the greenhouse prior to transplanting even if the disease has been absent the previous year.
4. Removal of infected plant debris and alternate host from the field.
5. Crop rotation with non-crucifers crop for a minimum of three years since the bacterium can survive in plant debris in soil.
6. Sprinkling the foliage is one of the most common way of disseminating of the bacterium so it should be avoided.
7. Avoid overcrowding the plant by maintaining proper spacing and planting in well-drained soil.
8. Tools and equipment's should be decontaminated before using them.
9. Prevent injury to plants by insects such as cabbage worms, cutworms, root maggots which can serve as infection point by Blackrot pathogen by controlling them.
10. Use of plant cultivars that are resistant or tolerant to Blackrot. Cabbage varieties which are resistance to Blackrot are Guardian, Defender, Hancock, Gladiator, Bravo, Supermarket and Blueboy.
11. Use of biocontrol agents like various strains of yeast (Celetti *et al.*, 2002), *Bacillus subtilis* R14, *B. pumilus* C116, *B. megaterium pv. cerealis* RAB7 and *B. cereus* C210 due to their antibiotic effect against *X. campestrispv. campestris* bacterium (Luna *et al.*, 2002).
12. Hot water treatment of seed at 50°C for 25minutes or soaking the seeds in 0.5% sodium hypochloride for 30 minutes.
13. Drenching of nursery bed with formalin 5% 3-4 weeks prior to sowing.
14. Spraying of the crop with Agrimycin 100 or streptocycline @0.6g/litre of water 2-3times at fortnightly intervals.

Clubroot of Cabbage

Economic Importance

Club root also known as 'finger and toe disease' causes significant yield and quality loss of mustard family specially cabbage and it's closely related plants (Henry and Devasahayam, 2011). Clubroot causes substantial yield and quality loss in Brassica family crops all around the world. The disease was first reported from Britain in 1736. In India, club root has been present on cabbage and cauliflower crops for nearly 80 years in the Eastern Himalayan, Darjeeling Hills of West Bengal and South Indian Nilgiri Hills in Tamil Nadu (Bhattacharya *et al.,* 2014). In 1878, Mikhail S. Woronin was first to recognize plasmodiophorous organism as causal organism for clubroot disease and gave the name *Plasmodiophora brassicae* (Tommerup and Ingram, 1971).

Symptoms

It is a disease of the root as the name suggest i.e. club root and does not show any visible symptoms on the above ground parts until the disease advances to a considerable extent. When infection occurs at early stage, the plants become stunted, wilting of entire plant in hot sunny days, as if the plant is suffering from water deficiency and may die,whereas plants infected in later stage fail to make marketable heads or growth (Zitter, 1985). When such plants are pulled out from the soil, characteristic club root symptoms are seen where the roots are usually swollen and distorted due to hypertrophy of the root system which enlarges rapidly to form 'Clubs'. Clubbing of the roots are of different types: (a) clubbing of entire main and lateral root system, (b) presence of clubs only in main root, (c) clubs are present in only in lateral roots, (d) clubs appear as tumor, and (e) dark decomposed spots in root system (Singh, 2009). The gall formation affects the vascular system which eventually hinders the absorption of water and minerals from soil. In advanced stage the galls enlarges and results in disruption of the outer tissue which rots and turns black due to the invasion of secondary organism such as soft rot bacteria.

Causal Organism

Club root of cabbage is caused by soil borne obligate phytoparasitic pseudofungi, *Plasmodiophora brassicae* Wor. (Gahatraj *et al.,* 2019). Its vegetative phase consists of a multinucleate mass of protoplasm (plasmodium or slime mold). The resting spore are produced inside the host cell and are released into the soil which are spherical in shape and 2.0-4.0μ diameter. Such spore on germination give rise to single zoopores,which are naked,pear shaped, devoid of cell wall, uninucleated protoplasm with two unequal whiplash type flagella.

Epidemiology

Club root occurs at a temperature of 18-25°C with an acidic soil having a pH less than 6.5. The incidence of the disease is 100% at pH 5.7. The disease favours soil moisture of 60-70% for its development but low level of infection can occur at low soil moisture and vice versa (Hamilton and Crete, 1978). These are disseminated from field to field through irrigation water,transport of infested soil by means of tools, equipment, animals, and humans.

Disease Cycle

The fungus can persist in the soil and crop debris as resting spores for as long as for 3-6 years without any susceptible plant ever being grown there, once a susceptible crop is grown the pathogen becomes prevalent. The fungus can also attack several weeds commonly found in the field.Other important features of its life history includes means of spread, and its reaction to soil pH. When susceptible host are planted in the field.The chemicals from root of the susceptible plant triggers the germination of the resting spores in the soil, which releases zoospores that swim in the soil water and entre into the root hairs. Infection can also take place through wounds on roots and stems and also natural opening at the point of emergence of lateral roots. Inside they form a cell having many nuclei, called a "plasmodium". The roots produces characteristic swellings - the gall due to the formation of zoospores. As the roots decay, resting spores form and are released into the soil, where they can remain viable for 10 or more years (Jackson, 2017c).

Integrated Management

1. Use of disease-free, preferably certified seeds. Eradication of all crucifer weeds in or near seedbeds and fields and avoidance of growing crucifer's crops in infected field.
2. Crop rotation with non-host crop for at least 7 years.
3. Growing of crop in well drained and disease free soil.
4. Removal of diseased roots and crop debris from the field.
5. Carefully cleaning of machinery, wheelbarrows, tools, and footwear of soil to prevent the introduction ofclub root to disease-free field.
6. Resistant varieties are available for some crucifers but more work is required to evolve club root disease resistant varieties of cabbage, cauliflower and some other host. Numerous strains or races of the clubroot fungus are known, however a variety that is resistant in one county or locality may be susceptible in another.
7. Since the disease favour acidic soil amendment of soil is needed to increase the pH to 6.8 or 7 by applying lime @2.5t/ha or ground limestone (CaCO3), hydrated lime (Ca(OH)2), Penta Chloro Nitro Benzene (PCNB)
8. Biocontrol agents, such as *Bacillus subtilis* and *Gliocladium catenulatum*, reduced clubroot severity by more than 80% (Peng *et al.,* 2011). *Trichoderma* spp. and *Pseudomonas flouresence* have also been found effective for the management of club root.
9. Drenching of the seed beds with methyl bromide or vapam or aldrin gives good result.

Downy Mildew of Cabbage

Economic Importance

Downy mildew is an economically important foliar diseases causing considerable seedling loss in seed bed especially during cooler months (Timila, 2005). The fungus infects a wide range of plants, including broccoli, cauliflower, cabbage, mustard, radish and turnip. It is considered a serious disease of cabbage in India.

Symptoms

The disease is very serious both in nursery and in field planting.In the seedling stage there is development of the discoloured spots on the surface of the cotyledons, which may turn yellow, shrivels and die. Late infection of the host appear first as small light green yellow discoloured lesions or spots on the upper surface of the true leaves, later showing on the under surface.

The lesions turns yellow when it enlarges and during high humid condition a greyish white moldy growthi.e the mycelium is developed on the under surface of the leaves (Spencer, 1981 and Amin *et al.,* 2013). In severe cases the entire leaf is covered by the downy growth and the leaf may become papery, shrivels and dies. Cabbage heads also develops extensive greyish black discoloration which also penetrate the stem parenchyma (Gardener, 1920). The symptoms becomes apparent only near harvest.

Causal Organism

It is caused by an obligate pseudo fungi *Peronospora parasitica.* Asexual spores are produced on sporangiophores which arise from the aggregation of intercellular mycelium beneath the host epidermis on the under surface of leaf, stem and inflorescence. The sprorangiophores are dichotomously branched with slender curved tip at acute angle. Sporangiophore measure around 115-410 µm (average 295µm) in length ended with sterigmata bearing single sporangia. The sporangia are ovioid, hyaline and measure 18 to 28 X 25-45 µm (average 22 X 35µm) (Garibaldi *et al.,* 2006).

Epidemiology

The disease favours when there is high humidity above 85%, fog, drizzling rains and heavy dews. Low temperature of 15°C, leaf wetness of 4-6 hours at 20°C and 6-8 hours at 15°C are required for the infection and development of downy mildew (Jackson, 2017b). It is more prevalent in winter. The pathogen can overwinter as oospore in the soil and on crop debris. Disease contaminated seeds may also carry the disease. They produce wind and water blown spores *i.e.* sporangia which results in secondary spread of the disease from plant to plant. Weeds or other wild cruciferous host grown near the field may also harbour the disease.

Disease Cycle

Downy mildew have been known to affect mainly the foliage of the host plant. They reproduce both sexually via its male and female organ (Oogonium and antheridium) and asexually through the production of spores *i.e.* sporangia. The disease may be initiated by special resting spores which remain viable from one season to the next in the soil, crop debris or on weeds causing infection of the young seedlings. The sporangia germinate by formationof germ tube that will penetrate the plant. Once an infection is present, spore formation can occur as soon as 4 days after initial infection. The time period for these spores to germinate is 7 to 10 days, the fungus grows through the seedlings, producing characteristic white tufts of fungus on the seed, leaves and stem. These fungal tufts carry many minute fungal seeds or spores which are easily dislodged and spread about by wind or water-splash,insect and even contaminated equipments. Those spores which land on susceptible hosts when conditions are favourable for germination, produce new infections. When new fungal tufts are produced, the disease cycle is repeated, and further spread occurs (Moorman, 2014).

Integrated Management

1. Removal of infected plant debris and eradication of cruciferous weeds that may harbour the fungus from any fields or gardens by burning or burning them. This reduces overwintering sites for the spores and helps to lessen the chance of infection in the next year.

2. Use of overhead irrigation should be avoided during cool weather to evade periods of high humidity. Irrigation should be done early in the day so that leaves are wet onlyfor a few hours.

3. Proper spacing of the plants to ensure good air circulation so that when they are irrigated leaf surface dries fast.

4. During cool weather condition optimal for downy mildew, susceptible plants should be scouted to catch early infection so that treatment can be given as soon as possible and to avoid further infection.

5. Practice crop rotation with plants that are less susceptible to reduce the chance of soil-borne resting spore infection.

6. Spring-planted, summer-harvested crops have fewer problems than fall-harvested.

7. Maintenance of balance program of nutrition because a potash deficiency will increase the susceptibility of seedlings to downy mildew (Minchinton *et al.,* 1997).

8. Spraying the seedlings in the nursery beds with Copper Oxychloride (0.3%) is effective in controlling the disease. This spray can be applied every 7 to 10 days until the plants are transplanted in the field. The crop is sprayed with Copper Oxychloride (0.5%) for the field grown crops.
9. Fungicides like mancozeb, azoxystrobin, cyazofamid, fluopicolide, mafenoxam, and phosphouous acid have also been found effective. Often several applications are needed for visible results. Apply at 7-10 interval until the disease control.

Damping Off and Wire Stem of Cabbage

Economic Importance

Damping off is a very common disease occurring in the seed beds of the vegetables, it occurs not only in the nursery beds of orchards or open environment but also in the plastic houses. The wire-stem disease of cabbage seedlings, caused by the *Rhizoctonia* has been known for a long time and has been studied critically by Gratz 1925. He observed two other phase of *Rhizoctonia* injury to cabbage, known as bottom rot and head rot causing infection in older plants in the field and storage. Damping off and wire stem may cause losses in stand, yield, or quality, depending on the time of the season in which they occur (Hansen, 2009).

Symptoms: There are two disease symptoms:

Pre-emergence Damping off

It consists of a decay of the germinating seed or death of the seedling before it can germinate through the soil. Therefore patches are observed in the nursery beds due to poor germination and stand which are often recognized to inferior quality of the seedor the untreated seeds. It also causes seed decay.

Post-emergence Damping-off

- ***Wire Stem:*** It occurs after the seedlings have emerged from the soil and cause wire stem symptoms mostly caused by *Rhizoctonia solani*. The stem above and below the soil line shrivels and darkens, becoming tough and woody or wiry at the point of the lesion later the plant bents or twisted without breaking. Plants wilt, collapse and die. Diseased plants transplanted to the field grow poorly, are stunted, and may eventually die, especially if there is inadequate moisture shortly after transplanting. Even if the plants are able to produce new roots above the affected stem after transplanting, they are generally stunted, unthrifty, and they produce small heads.

- ***Head Rot:*** During the period between head formation and maturity a firm to slimy dark decay at the base of outer leaves and in cabbage heads develops. The outer leaves of the head wilt, become pale, and turn brown to black near the main stem.As leaves are killed, they dry at the base and remain held in place by the margin of the blade that folds over the top of the head subsequently leaves will die and drop off, exposing the stem beneath the head.Over the whole head surface, brown fungus mycelia are found on decayed tissues. The head remains conspicuous, upright, dark colored, and studded with tiny brown resting fungal bodies (sclerotia). Secondary rot bacteria usually invade the diseased tissue and turn the head into a slimy foul-smelling mass(Wellman,1932).
- ***Bottom Rot:*** The disease occurs in mid-season as a carry-over from wire stem seedlings and from new infections that occur when outer leaves come in contact with moist infested soil. The disease appears on the lower side of the head leaves that are in contact with the soil. The midrib produces lesions that are sunken, black, and sharply elliptical with the long side of the lesion parallel to the side of the midrib. Lesions may droop, dry out and become papery brown in appearance if the weather becomes dry, but remain attached. Later the Infected leaves are shed and only a naked stalk, capped with a small head.

Causal Organism

The disease is caused by two fungal pathogens i.e *Pythium* spp. (*Pythium debaryanum* and *Pythium aphanidermatum*) and *Rhizoctonia solani*. The sexual stage of *R. solani* is now known as *Thanatephorus cucumeris* (Ceresini,1999). *Pythium* are oomycetes saprobic soil fungi where the hyphae grow inter-or intracellularly in the host tissue. Haustoria are not produced. The sprorangia are globose or oval arising apically on the unspecialized hyphae and the tip of hyphae give rise to the sporangium. *Rhizoctonia solani* is a basidiomycete fungus that does not produce any asexual spores (called conidia) and only occasionally will the fungus produce sexual spores (basidiospores) so they are considered as 'Mycelia sterilla'. *R. solani* reproduces asexually and exists mainly as vegetative mycelium and/or sclerotia. The basidia are clavate,barrel shaped,which bear four sterigmata. Basidiospores are ellipsoid, hylaline, trunicate and 9-14x6-8 μ in size.

Epidemiology

They prefer warm wet heavy soil with acidic pH combined with windy condition, insect damage with low light and presence of weed. They prefer temperatures ranging between 8° to 32°C. Cabbage may become infected at temperatures ranging between 11° to 32°C with an optimum between 25° to 26°C.

Disease Cycle

The fungi survives indefinitely in the soil, passing through unfavorable conditions primarily as small, hard, chocolate-brown, kernel-like bodies called sclerotia or as mycelium in case of *Rhizoctonia* and as resting spore i.e. oospore incase of *Pythium*. They are spread through rain,irrigation water, tools etc. The sclerotia and mycelium of *Rhizoctonia* germinate in damp weather by forming mycelium that penetrate roots and leaves of susceptible through natural openings, wounds, or intact tissues and proceeds to invade other tissues regardless of external moisture. After the fungus has depleted nutrients in the host's tissue, or as other environmental conditions become unfavourable, the mycelium produces sclerotia to complete the life cycle. The fungus infects the seed without exhibiting symptoms. It remains in the seed in a dormant state until the seed begins to germinate then starts attacks the seed which serves as primary inoculum of the pathogen. Incase of *Pythium* oospore upon germination may produce more hyphae, or develop a zoosporangium, which produces motile zoospores that swim and infect plants. Thezoospores from the zoosporangium will reach the plant root surface and form cysts that will germinate, infect and invade the plant root. The hyphae during its growth tend to release certain enzymes which will eventually destroy the root tissue of the host plant and absorb nutrients.

Integrated Management

1. Growing of seedlings in well drained raised beds with adequate space to allow maximum air movement.
2. Excessive irrigation should be avoided to reduce humidity around the plants.
3. Crop rotation with grain crops and harvesting of cabbage heads early since cabbage leaves become more susceptible the longer they stand in the field.
4. Disease free seed should be used for planting and the seedlings showing symptoms of 'wirestem' disease should be discarded.
5. Avoid excessive amounts of nitrogen fertilizer.
6. Seed treatment with antagonistic fungi like *Trichoderma viride* @4g/kg seed can be used as biocontrol agent.
7. Treating seed beds with 4% formaldehyde solution or Bordeaux mixture and cover with polythene.
8. Seed treatment with Thiram or captaf or mancozeb (2-3 g/kg of seed) and soil treatment with Brassicol (PCNB) in 2g/litre water has also been found effective.

Leaf Spot and Blight of Cabbage

Economic Importance

Alternaria black leaf spot and blight disease is one of the most common and destructive foliar diseases of cabbage (Meah *et al.,* 2002). *Alternaria brassicae* and *Alternaria brassicicola* are the two major pathogen on Brassicaceae occurring throughout the world (Reis and Boiteux, 2010). Both the fungi have been frequently isolated and prevalent on the *Brassica oleracea* complex. Whereas *A. brassicae*is more prevalent on the *B.rapa* complex and weed species. These disease causes considerable reduction in quality and quantity of yield by reducing the photosynthetic potential, accelerating of senescence, defoliation, premature pod shattering and shrivelling of seeds (Dharmendra *et al.,* 2014).

Symptoms

It is a destructive disease of seed crop and appears on leaves and stem of cabbage seedlings off (Gorny *et al.,* 2010). The disease in more susceptible on older leaves where it first appears as a minute yellow speck on older leaves and stem. In later stage the spots will enlarge and darken to circular areas, turn from tan to dark brown spots with alternating light and dark concentric ring surrounded by yellow halos giving the spot a bull's eye appearance . Violets to tan spots also develops on infected cabbage. Older spot are black, brown in color and may fall of giving a shot hole appearance and the entire plant defoliates. In later stage it spreads to all aerial parts of the plant and under humid condition, lesions produces spores (French and Schultz 2011). Spots produced by *A. brassicae* are grey in colour while *A. brassicicola* produces black sooty velvety spots (Meena *et al.,*2010). *A. brassicicola* is a necrotrophic fungal pathogen and more virulent and common than *Alternaria brassicae* (Humpherson,1985).

Causal Organism

A complex two *Alternaria* species *i.e A. brassicicola* and *A. brassicae* has been known to cause leaf spot and blight of cabbage. The mycelium of *A. brassicae* is septate, brown to brownish grey in colour. The conidiophores arises singly or in groups of 2-12, pale to mid olivaceous brown, unbrancehed, septate arise in fascicles measuring 14-74×4-8 μm. Conidia are brownish black, obclavate borne singly or in chains of 2-4, muriform with long beak and with conidial size of 148- 184×17-24 μm with 10-11 transverse and 0-6 longitudinal septa. Mycelium of *A. brassicicola* is septate, olive grey to greyish black in colour. The conidiophore are septate, branched, olivaceous in colour measuring around 35-45μm in length and 5-8 μm in width. Conidia are dark in color, cylindrical to oblong, muriform without beak measuring 44-55 μm in length and 11-16 μm in width having 5-8 transverse and 0-4 transverse septa.

Epidemiology and Disease Cycle

They are saprophyte and can survive outside the host so the primary inoculum survive on diseased crop debris or volunteer weeds. The fungi may be seed-borne, soil-borne, and wind-borne. Both the pathogens produce many large spores that can be blown, splashed or carried by tools, human and animals throughout the fields. Seeds may be contaminated by surface-borne spores or internally infected by the fungus. Spores on the seed surface can remain viable for up to 2 years, and if the contamination is internal can remain viable for more than 12 years. Environmental factors viz., temperature, humidity, rainfall and sunshine are the most critical factors in the development of the disease. The disease favours moist and warm weather condition with a temperature between 24-27°C. Spore produced at night and released during the day.Prolonged periods of overcast weather with long dew periods and frequent rainfall are favourable for continuous spore production and disease development. Under favourable condition spores are produced within a week and disperse during the warmest, driest part of the day and spores germinate when the leaf surfaces are wet i.e. when rain or dew that persists for more than 9 hours (Sherf and MacNab, 1986).

Integrated Management

1. Planting high quality certified disease free seeds.
2. Practice crop rotation for three years or more with non-brassica crops.
3. Removal of crop residues after harvest and irradiation of brassica weeds, cull piles to reduce pathogen survival.
4. Regular checking for signs of disease, discard and rouging out any infected plants.
5. Maintainance of spacing by planting wide rows parallel to the prevailing wind direction to promote air movement in the canopy and avoid overhead irrigation.
6. Seed treatment with hot water at 50°C for 25-30minutes.
7. *Alternaria* leaf spot resistant varieties like Green Voyager, Cabbage No-139, Ankur Manas and Cabbage No-118 can be used (Dilip, 2015).
8. Bacterial isolate *Bacillus subtilis* @10^8 cfu\ml were proved to be highly effective in reducing *A. brassicicola* spore germination and disease parameters *in vivo*, followed by *Trichoderma viride* and *T. harzianum* @10^6cfu\ml. (Sabry *et al.,* 2015). Seed treatment with *T. harzianum* + foliar spray of *Pseudomonas fluorescence*, seed treatment and foliar spray with *T. harzianum* reduced the incidence of *A. brassicae* (Rai *et al.,* 2014).

9. Seed treatment with Iprodione, captan or thiram @4g/kg seeds.
10. Spraying of the plant at full bloom, pod set and pre- harvest stage with Captan (0.2%) or Copper Oxychloride (0.5%) for the control of disease.

Black Leg of Cabbage

Economic Importance

Black leg is a widespread diseases of cruciferous plants like cabbage (red, green, savoy, and white), broccoli, Brussels sprout, cauliflower, Chinese cabbage, kale, kohlrabi, mustards (black, leaf, and white), a few strains of radish, rape, some varieties of rutabaga, and some turnips. Under certain conditions it can cause serious damage. The first description of the fungus appeared in Germany in 1791 and it was described as a disease of cultivated brassicas in France in 1849 (Nattrass, 1951).

Symptoms: This disease cause infection in seedlings,stems,roots, leaves and head however it generally does not reduce seed crop yields; but a low levels of seed infection combined with favourable weather results in spread of the disease in seedbeds which can lead to severe losses after transplanting.

Seedling: Pale grey lesions appears on the cotyledons causing the seedlings to die early and the fungus tends to produce numerous conidia on the hypocotyl, cotyledon and first true leaves prematurely killed seedlings.

Stem: Symptoms on the stem appears as an elongated light brown sunken lesions with purplish margin and the centre of such lesions turn to an ash colored grey. Pycnidia which are tiny, black, fungus fruiting bodies are soon formed in the diseased area. The lesions gradually enlarges until it girdled the stem and turn black. Such kind of affected plants wilt suddenly and die.

Root: Root system becomes blackened and largely destroyed due to black leg. New roots produced above the diseased area may keep the plant alive. Affected plants keep growing until a fair sized head has formed, then topple over when the decaying roots can no longer support the head weight. Pycnidia are common on the surface of all decayed tissues.

Leaves: Circular, light brown to greyish spots form on the leaves and develop ash greywith a large number of black pycnidia are scattered. The affected leaves remain attached to the stem which helps to distinguish black leg from Blackrot disease, which causes the leaves to fall off.

Heads: Sunken black lesions will develop around the base of the head only after harvest which where late infected.

Causal Organism

Theses is caused by the fungus *Phoma lingam* (sexual stage: *Leptosphaeria maculan*) known as "Canker" in Britain and as "Black Leg" in the United States. The fungus produces round, dark brown to black pycnidia measuring aroung 185-384×220-510 μ in size. The pycnidia occurred individually at first, forming stromata later on. Pycnospores were single-celled, hyaline, short cylindrical, mostly strait but some slightly bent ranging around 1.25-2.36 ×2.60-5.70 μ (Mitrovi and Marinkovi, 2007).

Epidemiology

The fungus can survive on the infected plant debris for atleast three years producing both the sexual and asexual spore types. *Phoma lingam* can also be seedborne surviving many year in infested seeds. They are dessiminated by wind upto several miles, rain or irrigation splash, contaminated tools and require cool, windy and wet condition with a temperature ranging between 8-16°C.

Disease Cycle

The fungi reproduce both sexually and asexually. The seed pods may become infected when crucifer plants are carried through the winter for seed production the following growing season and the fungus penetrates into the seed coat, where it remains dormant. The sexual stage fruiting body i.e. Pseudothecia surviving on plant debris, fall planted brassica crops and winter weed releases ascospores which are later blown by wind. The asexual stage are produced on infected plants or crop residues, producing pycnidiospores or conidia that are dispersed by splashing water. Spores are released from these fruiting bodies on infected plants or residues on the soil surface.

Integrated Management

1. Use of disease free seeds, removal plant debris and deep ploughing along with eradication of disease susceptible weeds.
2. Avoid planting in or adjacent to a site where black leg has occurred within the last 3-4 years.
3. Minimum of 5 year Crop rotation with non-cruciferous plants since the fungus remains in the soil for at least 3 years.
4. Control of run-off water from infected areas to clean areas to avoid spreading of the fungus.
5 Rogueing out diseased plants from nursery and field.
6. Plant resistant/tolerant varieties e.g. Gloria F1,Pusa Drumhead.
7. Hot water treatment of seeds at 50°C for 30 mins is recommended.

8. Seed treatment with Captan or Thiram 4g/kg of seed, followed by seed treatment with *Trichoderma viride* 4g/kg of seed.
9. Spray seedlings with Copper oxychloride based products (Cupracaffaro 50g/20L of Water).

Yellows or *Fusarium* Wilt

Economic Importance

It has many names like *Fusarium* wilt ofcabbage, *Fusarium* yellows, brassica *Fusarium* wilt (Jackson, 2017a). These disease threatened the entire cabbage industry in the early 1900's. Five strains of the fungus have been known: Strain 1 infects cabbage, Brussels sprout, cauliflower, collard, and kale; strain 2 infects radish; strains 3 and 4 are found on flowering stock; and strain 5 occurs on cabbage.

Symptoms

The disease affects both the seedling in nursery and 2-4 weeks old transplanted plants.The first sign of yellows is a lifeless, yellowish green colour overall on one side of the plant. Young plants becomes stunted, turn yellow, and die. The leaves and stem develop a one-sided warping or curling. The lower part of the leaf blade adjoining the petiole or midrib wilts and dies.

The lower leaves turn yellow and later the yellow leaves turn brown and the affected tissue becomes dry, brittle and leaves drop prematurely and growth is retarded. The vascular vessels in the stems, petioles and leave veins turn yellow to dark brown, resembling symptoms of Blackrot where in this case turns black. Very susceptible plant may die within a few weeks.

Causal Organism

Fusarium yellows is a warm-weather disease caused by *Fusarium oxysporum f.* sp. *conglutinans*.The fungus produce both microconidia and macroconidia where the microconidia are formed in aggregations on short conidiophores, and macroconidia with septa formed in aggregations on short conidiophores. The microconidia are hyaline, elliptic, single-celled measuring 5 to 15μm X 1.5 to 3μm in size on the other hand macroconidia are hyaline, sickle-shaped, and 1-7 septate ranging from 22 to 45 μm X 2 to 5 μm in size (Enya *et al.*, 2008).

Epidemiology

The disease is more serious in summer than winter so it develops more rapidly at a temperature ranging from 24°C to 29°C, and lesser below 15°C. Potassium deficiency also intensifies the syndrome. It is a soil borne pathogen and disseminated from infected soil and plants on drainage water, farm machinery, tools.

Disease Cycle

The pathogen enters the plants through young roots or wounds in older roots at transplanting stage and gradually moves upward into the stem and leaves via the vascular system i.e. xylem but does not invade other tissues until part or all of plant dies. Later in severe case spores inside and outside of the infected stem and get disseminated from seedbed to field or from one field to another. Some highly susceptible varieties may die within two weeks and other less susceptible may decline gradually throughout the season. They can produce a thick walled Chlamydospore that can survive long periods in the soil and resistant to unfavourable environmental conditions (Mcavoy, 2012).

Integrated Management

1. Cleaning fields by washing soil from machinery, tools and shoe.
2. Inspection of seedlings frequently and destroying those that are infected.
3. Selection well-drained fields and planting on raised beds. Running-off of water from infested fields should be prevented.
4. Very early cabbage generally avoids the disease because of cool soil temperatures.
5. Diseased plant parts and crop debris left after harvest should be removed and burned.
6. Deep summer ploughing to expose the spore to sunlight and killing it.
7. Crop rotation for atleast 4-5years involving lettuce, capsicums, tomatoes or a root crop prevent build-up of the soil borne fungus in the soil.
8. Growing of varieties that have resistance or tolerance like Harvester queen, hybrid 15, Guardian, green express (Sherf, 1979).
9. Seed treatment with *Trichoderma viride* @4g/kg seeds or *Bacillus subtilis* @1X10^6µm/ml prior to sowing.
10. Seed treatment with captan or thiram @4g/kg seed 24hours prior to sowing.

Sclerotinia rot/ White Mould

Economic importance

The disease cause losses of cabbage in field,storage, under transit and also in market during cool and moist conditions all over the world. It has a wide host range of over 360 species plants and in cruciferae family alone it attacks on 18 genera and 32 species (Dillard, 1987).

Symptoms

The first initial symptom of white mold appears on the stem at the soil lineand leaves at their base as tan, water soaked, circular lesions that enlarges and develop a white cottony fungal mycelium growth that also spread to other plant tissues. Infection of roots may lead to girdling of the stem, and under severe condition plants may wilt, topple, and die. On the head, a soft, water-soaked lesion may affect the entire head and produces large, black, seedlike structures called sclerotia on the diseased tissues.

Causal Organism

The fungus Sclerotinia sclerotiorumis the causal organism of White mold of cabbage. This fungus produces only sexual spores. The dormant resting body sclerotia will germinate to produce fruiting bodies called apothecia, which are small cup-like structure about 5-15mm in diameter. The apothecium is lined with asci, whereas cospores are present.

Epidemiology

The development and occurrence of the disease depends upon many factors such as soil inoculum,soil moisture,rainfall, cultivar susceptibility, excessive nitrogen fertilization, continuous crops in monoculture, row width and plant density. The disease favours cool and damp condition so it will be more severe on soil with high inoculum and moisture. Temperature between 18-23°C and high humidity are required for its growth. (Tu,1997).

Disease Cycle

Infection of the plants begins with the germination of sclerotia from the soil where the mycelium starts to infect the roots, crown and leaves in contact with the soil. Under favourable condition the soilborn esclerotia located within 2 cm of the soil surface produce apothecia and ascospores. These ascospores are forcibly discharged and moved by air currents to the susceptible tissue and most likely initiate a focus of infection in the field. A highly susceptible weak parts such as flower petals or senescing leaves, is typically the first tissue to become infected and subsequently penetrate the healthy plant parts. Later sclerotiaare formed either on the plant surface or within stems and other plant parts. The slerotia will fall to the soil after the death of the plants or its part where they will survive for many years.The disease has only one cycle per year with no secondary infection cycle.

Integrated Management

1. Fields with a history of white mold should be planted with non-susceptible crops such as grains (corn, rye, wheat, etc.).

2. Promotion of air circulation by increasing row spacing.
3. Soil solorization during the month of May and June to kill the dormant fungi.
4. Irrigate early in the morning and stop all irrigation by midday which allows the foliage to dry (less than 16 hours of leaf wetness) reducing disease incidence.
5. Mechanical injuries to cabbage heads during harvesting operations should be avoid.
6. The fungus *Coniothyrium minitans* is the most widely available and tested biological control organism for *S. sclerotiarum* 1947 (Campbell, 1947). Bacteria such as *Pseudomonas fluorescence*, *P. putida* and *Bacillus subtilis* and *Trichoderma harzianum* have direct and indirect impact on the survival of sclerotia (Peltier *et al.,* 2012).
7. Pre sowing soil application of carbendazim (5%) granule.

Future Approaches

Cabbage is an important part of the human diet along with a good source of chemo protective phytochemicals and rich in nutrients. Since it is a leafy vegetable once it gets infected by any pathogen losses incurred by it is quite devastating and reduce the profit of the commercial cabbage growers in India and worldwide. Cabbage has been infected particularly by fungal pathogens and a few bacterial and viral pathogens. The pathogens causing clubroot, fusarium wilt, damping off and wire stem are soilborne so it is difficult to control, especially the clubroot organism is remarkably difficult to eradicate from infested soils since they remain dormant in soil for many years. Since majority of the diseases cause significant yield loss we need to give more emphasis on its management which are safe to the consumers and give significant good control measures. Integrated diseases management is the best method which combines the use of all methods in a coordinated way such as crop rotation to reduce survival of foliar pathogens, field sanitation, removal of alternate hosts, scouting to detect disease outbreaks, use of resistant varieties, seed treatment with hot water or fungicides, and judicious application of protectant fungicides. But we need further study on the disease life cycle and epidemiology to target the stage at which the management is most effective and to find out which condition favours best for the control of any disease. Progressions in fields of biotechnology like genomics and bioinformatics can help in development of biotic stress tolerant crops. Priority can be given on study of resistant genes to develop disease resistance cabbage. More Research is also needed on gene editing and gene silencing such as RNAi,host induced gene silencing that targets the disease causing genes and silencing them. Further research on different races of the pathogen especially downy mildew and club root is needed and develop resistance to these and other pathogens. Lastly as the climate is changing day by day due to global warming new disease will start cropping up which might cause unpredicted losses or epidemic, so we should remain alert and study the diseases more thoroughly whether they are evolving or changing from its normal life cycle.

Conclusion

Many varieties of cabbage are susceptible to diseases caused by fungi and bacteria that injure the plant and reduce the yield. Blackrot and club root are most important diseases and management of these diseases remains more challenging.There are other certain minor diseases caused by virus and fungi that has not been discussed here as it of minor importance and does not occur locally causing less significant loss. If fungicide are used they should be used in an integrated disease management strategy that takes advantage of cultural and biological control. Fungicide should be used less and applied in combination with both systemic and contact fungicide. As using only systemic fungicide increases the chances of development of fungicide resistance.

References

Amin, N., Ali, B., Muzaffar, S., Majeed, I. and Sayeed, N. (2013). Symptoms, pathogen biology and control of downy mildew of Brassica–A mini review. *J. Biosphere*, **2**(1): 6-9.

Bhattacharya, I., Dutta, S., Mondal, S., and Mondal, B. (2014). Clubroot disease on Brassica crops in India. *Canadian J. Pl.Pathol..*, **36**(1): 154-160.

Campbell, W. A. (1947). A new species of *Coniothyrium* parasitic on sclerotia. *Mycologia,* **39**: 190 –195.

Celetti, M. and Kristen, C. (2002). Blackrot of Crucifer Crops. Ministry of Agriculture, Food and Rural Affairs. Ontario State. http://www. omafra.gov.on.ca/english/crops/facts/02-025.htm. Accessed on 2nd Dec 2019.

Ceresini, P. (1999). Rhizoctonia solani. N C State University.https://projects. ncsu.edu/cals/course/pp728/Rhizoctonia/Rhizoctonia.html. Accessed on 1st Dec 2019.

Dharmendra, K. M., Neelam, Y. K. Bharati, A., Kumar, K., Kumar, S., Kalpana, C., Gireesh, K., Chanda, S. K., Singh, R. K. and Mishraand, K. A. (2014). Alternaria blight of oilseed Brassicas: a comprehensive review. *African J. Microbiol. Res.*, **8**(30): 2816-2829.

Dilip, K.T. (2015). Studies on Alternaria leaf spot of cabbage. M.sc thesis. Department of plant pathology, Faculty of Agriculture, Dr. Balasaheb Sawant Konkan Krishi Vidyapeeth, Dapoli, Maharastra.

Dillard, H. R. (1987). Sclerotinia Rot of Cabbage. Vegetable crops Cornell cooperative extension. Diseases of crucifers Sclerotinia Rot of Cabbage Fact Sheet. http://hdl.handle.net/1813/43297. Accessed on 7th Dec 2019.

Enya, J., Togawa, M., Takeuchi, T., Yoshida, S., Tsushima, S., Arie, T. and Sakai, T. (2008). Biological and phylogenetic characterization of *Fusarium oxysporum* complex, which causes yellows on *Brassica spp*., and proposal of *F. oxysporum f. sp. rapae*, a novel forma specialis pathogenic on *B. rapa* in Japan. *Phytopathol.*, **98**(4): 475-483.

Flynn, P. (1999). Blackrot of Cabbage.https://hortnews.extension.iastate.edu/1999/8-13-1999/blackrotcab.html. Accessed on 29th Nov 2019.

French, R. D. and Schultz, D. (2011). Alternaria Leaf Spot of Cabbage. Texas Agri Life Extension Service; The Texas A&M System. http://sickcrops.tamu.edu. Accessed on 4th Dec 2019.

Gahatraj, S., Shrestha, S. M., Devkota, T. R. and Rai, H. H. (2019). A review on clubroot of crucifers: symptoms, life-cycle of pathogen, factors affecting severity, and management strategies. *Archives Agri. Environ. Sci.*, **4**(3): 342-349.

Gardner, M.W. (1920). Peronospora in turnip roots. *Phytopathol.,* **10**: 321-322.

Garibaldi, A., Minuto, A., and Gullino, M. L. (2006). First Report of Downy Mildew caused by *Peronospora parasitica* on *Iberissemper virens* in Italy. *Pl.Dis.*, **90**(9): 1261-1261.

Gorny, A., Kreis, R. and Dillard, H. (2010). Alternaria Leaf Spot of Cabbage. Cornell University College of Agriculture and Life Sciences New York State Agricultural Experiment Station. http://web.pppmb.cals.cornell.edu/dillard/pdf/alternaria.pdf. Accessed on 4th Dec 2019.

Gratz, L. O. (1925). Wire Stem of cabbage. N. Y. Cornell Agr. Expt. Sta. Mem. illus.pp 60-85.

Hamilton, H. A. and Crête, R. (1978). Influence of soil moisture, soil pH, and liming sources on the incidence of clubroot, the germination and growth of cabbage produced in mineral and organic soils under controlled conditions. *Canadian J. Pl. Sci.*, **58**(1): 45-53.

Hansen, M. A. (2009). Wire Stem and Bottom Rot of Cabbage. Extension Plant Pathologist, Department of Plant Pathology, Physiology and Weed Science, Virginia Techhttps://vtechworks.lib.vt.edu/bitstream/handle/10919/55185/450713.pdf?sequence=.1. Accessed on 5th Dec 2019.

Henry, L.D. and Devasahayam, T. H. L. (2011). Crop Disease: Identification, Treatment and Management. New India Publishing Agency. pp 371-373.

Humpherson, J. F. M. (1985). The incidence of *Alternaria spp.* and *Leptosphaeria* in commercial brassica seed in the United Kingdom. *Plant Pathol.*,**34**: 385-390.

Jackson, G. (2017a). Pacific Pests and Pathogens - Fact Sheets Cabbage Fusarium wilt (132) http://www.pestnet.org/fact_sheets/cabbage_fusarium_wilt_ 132.htm. Accessed on 5th Dec 2019.

Jackson, G. (2017b). Pacific Pests and Pathogens - Fact SheetsCabbage downy mildew (192). http://www.pestnet.org/fact_sheets/cabbage_downy_mildew_ 192.htm. Accessed on 30th Nov 2019.

Jackson, G. (2017c). Pacific Pests and Pathogens - Fact Sheets. Cabbage club root (283). http://www.pestnet.org/fact_sheets/cabbage_club_root_283.pdf. Accessed on 29th Nov 2019.

Luna, L. C., Mariaono, R. L. R., Souto-Mairo, M. A. (2002). Production of a biocontrol agent for crucifers' Blackrot disease. *Braz. J. Chem. Eng.,* **19**(2): 133–140.

Mcavoy, G. (2012). Pest Of The Month: Fusarium Yellows of Cabbage.https://www.growingproduce.com/vegetables/pest-of-the-month-fusarium-yellows-of-cabbage/.Accessed on 5th Dec 2019.

Meah, M.B., Hau, B. and Siddique, M. K. (2002). Relationship between disease parameters of alternaria blight (*Alternaria brassicae*) and yield of mustard. *J. Plant Dis. Pl. Prot.*, **3**: 243-251.

Meena, A. K., Godara, S. L. and Gangopadhyay, S. (2010). Efficacy of fungicides and plant extracts against Alternaria blight of cluster bean. *J. Mycol Pl. Pathol.*, **40**(2): 272-275.

Meenu, G., Vikram, A., and Bharat, N. (2013). Blackrot-A devastating disease of crucifers: a review. *Agric. Rev.*, **34**(4): 269-278.

Minchinton, E., Pierce, P., Mebalds, M., Hepworth, G. and Cass, A. (1997). Downy Mildew of Brassicas. Agriculture Victoria, Institute for Horticultural Development, Knoxfield, Victoria. https://www.ausvegvic.com.au/pdf/r%26d _NY9406_Downy_Mildew_ seedling_ factsheet. pdf. Accessed on 1st Dec 2019.

Mitrovi, P. and Marinkovi, R. (2007). *Phoma lingam*–a rapeseed parasite in Serbia. In *Proc. of the 12th Intern. Rapeseed Congress Wuhan China*. pp. 217-219.

Moorman, G.W. (2014). Downy Mildew. Penn State Extension. https://extension.psu.edu/ downy-mildew. Accessed on 27th Nov 2019.

Nattrass, R. M. (1951). The "Canker" or "Black Leg" Disease of Cabbages and Cauliflowers. *The East African Agric. J.*, **16**(4):178-180.

Peltier, A. J., Bradley, C. A., Chilvers, M. I., Malvick, D. K., Mueller, D. S., Wise, K. A. and Esker, P. D. (2012). Biology, yield loss and control of Sclerotinia stem rot of soybean. *J. Integr. Pest Manag.*, **3**(2): B1-B7.

Peng, G., McGregor, L., Lahlali, R., Gossen, B. D., Hwang, S. F., Adhikari, K. K. and McDonald, M. R. (2011). Potential biological control of clubroot on canola and crucifer vegetable crops. *Pl.Pathol.*, **60**(3): 566-574.

Pfeufer, E. and Davis, D. (2019). Blackrot of Crucifers. College of Agriculture, Food & Environment. https://plantpathology.ca.uky.edu/files/ppfs-vg-01.pdf. Accessed on 28th Nov 2019.

Rai, D., Mishra, A. K. and Pandey, A. (2014). Integrated management of Alternaria blight in mustard.*Bioscan*. **9**(3): 1247-1249.

Reis, A. and Boiteux, L. S. (2010). *Alternaria* species infecting Brassicaceae. *J. Pl.Pathol.,* **92**(3): 661-668.

Sabry, S., Ali, A. Z., Abdel- Kader, D. A. and Abou-Zaid, M. I. (2015). Control of cabbage alternaria leaf spot disease caused by *Altrenaria brassicicola.Zagazig J. Pl.Pathol.*, ***42***(5).

Sherf, A. (1979). Vegetable Crops. Fusarium Yellows of Cabbage & Related Crops. Cornel University, Vegetable MD online. http://vegetable mdonline. ppath. cornell. edu/factsheets/ Crucifers_Fusarium.htm. Accessed on 7th Dec 2019.

Sherf, A. F., and MacNab, A. A. (1986). Vegetable diseases and their control. John Wiley & Sons. New York, USA. pp 728.

Singh, R. S. (2009). *Plant disease*. 9th edition. Oxford and IBH Publishing. pp.201.

Smart, C. D. and Lang, H. W. (2013). Managing Blackrot of Cabbage and other Crucifer Crops in Organic Farming Systems.https://eorganic.org/node/4957. Accessed on 28th Nov 2019.

Spencer, D.M. (1981). Plant pathology. Academic Press, New York, pp. 31-43.

Timila, R. D. (2005). Management of Downy Mildew Disease of Cauliflower Seedlings in Seed Bed. *Nepal J. Science Technol.*, **6**(1).

Tommerup, I. C., and Ingram, D. S. (1971). The lifecycle of *Plasmodiophora brassicae* Woron. in Brassica tissue cultures and in intact roots. *New Phytol.*, **70**(2): 327-332.

Tu, J. C. (1997). An integrated control of white mold (*Sclerotinia sclerotiorum*) of beans, with emphasis on recent advances in biological control. *Botanical Bulletin of Academia Sinica*, 38.

Vauterin, L., Hoste, B., Kersters, K., and Swings, J. (1995). Re-classification of *xanthomonas*. *Int. J. Syst. Evol. Microbiol.*, **45**(3): 472-489.

Vicente, J. G. and Holub, E. B. (2013). *Xanthomonas campestris* pv. *campestris* (cause of Blackrot of crucifers) in the genomic era is still a worldwide threat to brassica crops. *Mol. Plant Pathol.*, **14**(1): 2-18.

Wellman, F. L. (1932). Rhizoctonia bottom rot and head rot of cabbage. *J. agric. Res.*, **45**: 461-469.

Zitter, T. A. (1985). Clubroot of Crucifers. Vegetable crops Cornell cooperative extension.http://vegetablemdonline.ppath.cornell.edu/factsheets/Crucifers_ Clubroot.htm.Accessed on 10th December 2019.

7

Diseases of Knol Khol (*Brassica oaeelorcepa* var. *Gongylodes* L.) and Their Integrated Management

***Pranjal Kr. Kaman*[1] and *Pranab Dutta*[2]**

[1]*Department of Plant Pathology, Assam Agricultural University, Jorhat-785 013 Assam*

[2]*School of Crop Protection,College of Post-Graduate Studies in Agricultural Sciences Central Agricultural University, Umiam, Meghalaya-793 103*

Cruciferous vegetables are important Kharif vegetable crops, which are grown both for table and seed purposes. This vegetable group constitutes crops like cauliflower, cabbage, radish, turnip and knol-khol. These crops are grown throughout the country and are attacked by number of diseases which not only reduce the quantity but also quality of the produce. Important diseases of Knol Khol are described below:

Major Fungal Diseases of Knol-Khol

Alternaria leaf spot

Economic Importance

It is of concern in north and central India where appreciable seed losses in terms of size, colour and germinability are reported (Tripathi *et al*., 1987), Similarly, losses up to 30 % in cauliflower by *Alternaria brassicae* in Columbia and South America have also been recorded (Tamayo *et al*., 2001).

Symptoms

Initally the disease appear as small brpwn spots on the leaf lamina or on margin of the leaves which later increases rapidly in size and mostly become circular to irregular in shape and turn light to dark brown in colour with slight distinct zonations with or without halo (Awasthi and Kolte, 1989). Kadain and Sahran, 1983 described the symptoms as concentric rings on the leaf. The leaf spot later enlarge, becomes nearly circular, 4-12 mm in diameter olivaceous colour with zonations, and profuse sooty sporulations in case the infection is caused by *Alternaria brassicola* (Singh, 1998).

Causal Organisms: *Alternaria brassicola*

Disease Cycle

Epidemiology

Sporulation occurs between the temperatures of 8 to 24^0 C where mature spores occur after 24 to 14 hours, respectively. Optimum temperatures are between 16 and 24 °C where sporulation time ranges from 12 to 14 hours. *A. brassicicola* sporulates in a temperature range of 8 to 30^{0}C, where mature spores occur after 43 and 14 hours respectively. Optimum temperatures are between 18 and 30^{0}C where the average sporulation time is 13 hours. Moisture in the presence of rain, dew, or high humidity is essential for infection, and a minimum of 9-18 hours is required for both species (Humperson-Jones and Phelps, 1989). Continuous moisture of 24 hours or longer practically guarantees infection (Chupp and Sherf, 1960; Rangel, 1945). Relative humidity of 91.5% (at 20 °C) or higher will result in the production of large numbers of mature spores in 24 hours (Humpherson-Jones and Phelps, 1989).

Infected seeds, with spores on the seed coat or mycelium under the seed coat, are likely the main source of transport for these pathogens. Spores are disseminated by wind, water, tools and animals. The fungus can survive in susceptible weeds or perennial crops (Chupp and Sherf, 1960; Rangel, 1945; Maude and Humpherson-Jones, 1980a, b).

Infected crops left on the ground after harvest also serve as a source of infection for *A. brassicae* and *A. brassicicola.* In one study, infected leaves of oilseed rape and cabbage placed outdoors on soil, produced viable spores for as long as leaf tissues remained intact. For oilseed rape this was up to 8 weeks and for cabbage up to 12 weeks (Humpherson-Jones, 1989). This type of spread is likely to occur in seedling beds as well, and seedlings from infected seed beds can carry the inoculum to the field (Rangel, 1945).

Integrated Management

Control: Rotation with non cruciferous crops and eradication of cruciferous weed hosts can help control these pathogens. Since spores can survive on leaf tissue for 8 to 12 weeks and stem tissue for up to 23 weeks, fields that are replanted soon after harvest often coincide with a large amount of inoculum which is likely to effect the crop's emergence and early growth stages (Humpherson-Jones, 1989).

Biological Control: Studies with the actinomycete fungus, *Streptomyces arabicus*, indicated an antifungal effect on *A. brassicicola* in both laboratory and field studies (Sharma *et al*., 1984; Sharma *et al*., 1985). In Finland, surface treatment with powdered *S. griseoviridis* (at 15 mg/g seed) has been shown to provide control against *A. brassicicola.* Seed treatment with Org-Trichojal (*Trichoderma asperellum*) @ 5ml/litre found to be effective in controlling the disease. Bacterial biocontrol agents like *Pesudomonas fluorescence* is also found to be effective in managing the disease.

Chemical Control: Seed treatment with Agrosan @ 2-3gm/kg of seed is effective in controlling the disease upto a certain extent. Foliar spray with Mancozeb @0.2% or Copper Oxychloride @ 0.3% twice at fornightly interval is found effective in field conditions.

Clubroot

Economic Importance

Russian scientist Mikhail Woronin eventually identified the cause of clubroot as a "plasmodiophorous organism" in 1875, and gave it the name *Plasmodiophora brassicae*. Clubroot can be a reoccurring problem for years because it is easily spread from plant to plant and is able to infect 300 species of cruciferous plants, making this disease a recurring problem even with crop rotation (http://allotment.org.uk/vegetable)

Symptoms: Developing plants may not show any symptoms but as the plants get older they will start to show symptoms of chlorosis or yellowing, wilting during hot days, and exhibit stunted growth. Below ground, the roots experience cell proliferation due to increased auxin or growth hormone production from the plant as well as the pathogen (Agrios, 2005) This causes the formation of galls that can grow big enough to restrict the xylem tissue inhibiting efficient water uptake by the plant. Galls appear like clubs or spindles on the roots. Eventually the roots will rot and the plant will die.

Causal Organism: *Plasmodiophora brassicola*

Disease Cycle

Epidemiology

Resting spores of *P. brassicae* can be disseminated through the transport of infested soil by means of tools, equipment, animals, and humans. Clean fields can be infested with diseased transplant material, and runoff from infected fields. Datnof *et al.* (1984) demonstrated that resting spores can be present in pond water sediment, especially those fed by runoff, where that water is used for irrigation. Pumping water from ponds with resting spores present is likely to spread the pathogen into pathogen-free seedbeds and fields. Cool, wet and acidic soils provide the most favorable environment for the pathogen. In acid soils, the temperature range for infection is between 50 and 95 ºF, the optimum being between 68 and 77 ºF. In alkaline soils, the temperature range is lower (Sherf and MacNab, 1986).

Integrated Management

Cultural Control

A number of cultural practices should be followed to help curb *P. brassicae* infestation which includes buying or using disease free transplants, using well drained and pathogen free soil, eliminating nearby crucifer weeds, incorporating a 7-year rotation of non-cruciferous crops, adjusting soil pH to 7.2 or higher, and using resistant crop varieties. Raising soil pH can provide good control of *P. brassicae* because spores germinate poorly or not at all and secondary zoospores are not produced at all, thus no clubs develop. Soil pH should be adjusted at least 6 weeks before planting the crop.

Biological Control

So far no biocontrol agents have been found to effective against clubroot diseases. But Lin *et al*., (1990) reported that S-H mixture a biological control agent, effective in both greenhouse and field tests. S-H mixture is believed to inhibit zoosporangium formation, and it probably raises the pH of the soil a little since the mixture has a pH value of 8.0.

Chemical Control

Fumigation using metam sodium in a field containing diseased cabbages is yet another way to decrease the buildup of the pathogen. Treat the soil of seed bed areas with chloropicrin, methyl bromide or vapam two weeks before planting. Drenching soil with a solution of Brassicol (Pentachloronitrobenzene)is an effetive, non phytotoxic, and economically feasible control measure for use in large fields. Datnoff *et al*., (1987) reported that chlorinating irrigation water significantly reduces clubroot incidence.

Damping-off (Wirestem)

Economic Importance: It is a disease of nursery beds and young seedlings resulting in reduced seed germination and poor stand of seedlings. It is reported yield loss 25-75%.

Symptoms

Pre-emergence Damping off: Seedlings disintegrate before they come out of soil surface. This is known as pre-emergence damping-off which results in poor field emergence / poor seed germination.

Post-emergence Damping off : It is characterised by development of disease after seedlings have emerged out of soil surface but before the stems are lignified. Lesion formation at collar region Infected areas appear brown and water soaked. Plants shrivel and collapse as a result of softening of tissues. Infected stems become hard, thin (Wire stem symptoms) and infected seedling stopple.

Causal Organisms

Pythium aphanidermatum, P. debaryanum, P. ultimum, Fusarium and *Rhizoctonia*

Disease Cycle

Epidemiology

The severity of the disease depends on the amount of pathogen present in the soil and enviromental conditions. Cool cloudy weather, high humidity, wet soils and overcrowding especially favour the development of damping off. Seedling stage upto 20 days is prone to attacked by *Rhizoctonia solani* as compared to 35-50 days old seedlings (Chauhan *et al.*, 2000)

Integrated Approaches

Cultural: Soil used for preparing raised beds should be well- drained. Excessive irrigation should be avoided to reduce humidity around the plants. The seedlings in the seedbed should be adequately spaced to allow maximum air movement. While transplanting, the seedlings showing symptoms of 'wirestem' disease should be discarded.

Biological: Seed treatment with antagonist fungal culture of Org Trichojal (*Trichoderma viride*)@ 5ml/l is effective.

Chemical: Seed treatment with Thiram @2-3 g/kg of seed is found to effective in managing the disease . Soil around the affected seedling should be drenched with Dithane M 45 (0.2%) or Bavistin (0.1%) to control the spread of the disease.Soil drenching with 1% Bordeaux mixture or Copper Oxychloride@ 0.3% prevent the damping off.

Powdery Mildew

Economy Importance

It is a disease of little economic importance in cruciferous crops, however it is is prevalent on the crop in certain semi arid areas or in more humid areas after dry spells (Walker, 1952).

Symptoms: Whitish or dirty grey, tiny powdery growth on foliage, stems and young growing parts. The superficial growth ultimately covers the entire leaf area. The diseased areas turn brown and dry. This leads to premature defoliation anddeath. Fruits remain undeveloped and are deformed

Causal Organisms: *Erysiphie chichoracearum.*

Disease Cycle

Epidemiology

Growth and development of powdery mildew is affected by climatic conditions during the growing season. Free moisture is necessary for release of ascospores from cleisthothecia, Powdery mildew thrives well under dry conditions with moderate temperatures. Mild weather results increased in powdery mildew growth (Ypema and Gubler, 1999)

Integrated Management

Cultural: Eliminate weeds in and around the fields. Plant resistant varieties where available. Choose planting sites with good air movement and free of shade.

Biological: So, far only one biocontrol agents that is *Ampelomyces quisqualis* a bacterial biocontrol agents is found to effective in managing powdery mildew.

Chemical: Dust finely ground sulphur at 30 kg/ha or spray wettable sulphur @ 0.3%or Dinocap @ 0.1%, three to four times at 15 days intervals.

Downey Mildew

Economic Importance

In India the diseases has been reported by Thind (1942) and Lund and Wyatt (1978). The disease is particularly important on seedlings but can also cause poor growth and reduced yield and quality of produce at later plant stages.

Symptoms

The initial symptom of downy mildew infection is the appearance of purplish, irregular spots on the leaves. During cool, moist conditions, the spots will enlarge and become yellow-brown in color. Cool, wet conditions also promote the development of a white mold between the veins on the undersides of leaves. As affected areas enlarge, they turn tan and papery. In severe cases, the whole leaf will die. Later the infections appear first as small discolored spots and yellowing on the upper surface of the true leaves with sparse patchy growth of mycelium on the lower surface (Spencer, 1981).

However, Jenkin (1964) reported that infection and partial or complete, destruction of some leaves might be total expression of the disease in the field.

Causal Organisms: *Hyaloperonospora parasitica*

Disease Cycle

Epidemiology

Environmental factors like temperature, moisture, light and relative humidity play an important role in the development of downy mildew on Knol Khol and their effect has been worked ouyt by different workers from time to time. Felton and walker (1946) found that in the pre-penetration stage, conidial germination was most rapid at 8-10^{0}C, while penetration of the host by the infection hyphae and formation of haustoria was rapid was at 16^{0}C and 20-24^{0}C, respectively. Similar temperature ranges of 8-16^{0}C for disease development were reported by Jonnson (1966). However, Nakovo (1972) found that 15-20 ^{0}C was most favourable for the development of disease. Studies on the influence of light, temperature, suscept stage and suscept nutrition on oospore formation indicated that conditions favouring senescence of leaf favoured oospore formation when anthridial and oogonial strains were present (Dorothy, 1959).

Integrated Approaches

Cultural: All the weeds serving as alternate host to the fungus should be destroyed. The crop should be irrigated judiciously to avoid periods of high humidity. Avoidance of thick sowing is effective in preventing the diseases. Use seed treated with hot water or seedlings raised from such treatment.Plant disease-free seedlings.Maintain a balanced program of nutrition deficiency of potash will increase the susceptibility of seedlings to downy mildew.Plough in field crop debris immediately after harvest and use a crop rotation of two to three years if possible.

Biological: Till now no biocontrol agents have been found to effective against downey mildew.

Chemical: Spraying the seedlings in the nursery beds with Copper Oxychloride (0.3%) is effective in controlling the disease. The first spray should be given as soon as the seedlings appear. Subsequent sprayings are given at weekly intervals until the plants are transplanted in the field. For controlling the disease in the field, the crop is sprayed with Copper Oxychloride (0.5%).Spray metalaxyl @0.25% or Mancozeb@ 0.2% at 10 days interval is found to be effective in managing the diseases at field condition.

Fusarium Yellows

Economic Importance

Yellows or *Fusarium* wilt was first reported in 1899 on cabbage in New York State. The disease was first observed in Canada in Ontario in 1931. Losses may be significant on a wide range of Cruciferae during warm growing seasons or where resistant cultivars are not commercially available. During extensive surveys of diseases on vegetable crops in Ontario in 1967, the disease was observed in 5 of 24 fields of early heading cabbage, with a range of 10 to 95% of the plants affected (Armstrong, 1966).

Symptoms

The first and most apparent symptom is a dull green to yellowish green color of the leaves of plants in the seedbed or in the field within a month after transplanting. Young plants may be stunted, turn yellow, and die rapidly in warm soils. The lower leaves develop a one-sided warping or curling, with the yellow color being more intense on one side of the leaf midrib than on the other. The stem may be twisted toward one side. Yellowing, browning, and withering of the leaves progresses up the plant. Affected leaves drop prematurely and growth is retarded. Very susceptible plants may die within a few weeks, while others in the same field or garden may live for a month or more. A yellow to dark brown discoloration can be seen in the water-conducting vessels (xylem) in the stem, petioles, and veins of the leaves when they are cut through.

Causal Organism:*Fusarium oxysporum* f.spp *conglutinans*

Disease Cycle

The pathogen usually infects through seedling rootlets and through roots wounded during transplanting. The fungus moves directly to the water-conducting xylem in the stem and leaves. Part or all of the plant may die. The pathogen then produces conidia and chlamydospores, both inside and outside the affected tissue.

The browning of vessels and yellowing of the leaves in advance of the fungus is caused by a toxin, which is produced in the xylem and moves upward in the sap stream. The pathogen can survive in soil for a number of years in the absence of a host. Once the pathogen has become established, it can spread rapidly to other areas by wind-borne soil, surface drainage water and soil adhering to farm implements. Long distance spread is by infested seed or diseased seedlings.

Epidemiology

Yellows is favored by warm weather. It develops to its maximum at 27 to 29°C soil temperatures. It is inhibited above 32°C and does not develop well at soil temperatures below 16°C. Soil moisture and pH have little or no influence on disease development (Subramanian, 1970).

Integrated Approaches

Cultural: Susceptible crucifers should not be planted in areas that are likely to receive surface-drainage water from infested fields. Transplanting of seedlings in the soil should be done that has been disinfested by steam or a soil fumigant. Rotate the field or garden to non host crops such as lettuce, peppers, or tomatoes for several years to prevent buildup fo the *Fusarium* fungus in the soil. Early varieties planted early may escape the disease until they approach maturity. Only resistant cultivar should be used.

Chemical: Not suitable as it is economically not feasible.

Major Bacterial Diseases

Blackrot

Economic Importance

A crop loss up to 50 % in cabbage and 50 to 70% in cauliflower has been recorded (Jorgensen and Walter, 1955). Substantial crop losses have been reported from the rapid spread of the bacteria under favourable conditions, especially during seedling production by Berg *et al.* (2005).

Symptoms: Blackrot can appear on plants at any growth stage. On young plants, margins of cotyledons turn black and they may drop off. On true leaves, symptoms appear along leaf margins as yellow, V-shaped lesions, with the base of the V usually directed along a vein. As the lesion expands toward the base of the leaf, the tissue wilts and eventually becomes necrotic .The infection may move down the vascular tissue of the petiole and spread up or down the stem of the plant and into the roots. Systemic infection can produce scattered, yellow lesions on leaves anywhere on the plant.

The veins of infected leaves, petioles, stems and roots turn black as the pathogen multiplies, thus impairing the normal flow of water and nutrients. Severely affected plants may lose a few or many leaves. Blackrot infection is often followed by soft rot organisms that further reduce the quality and storage life of the product. The presence of black veins in yellow lesions along leaf margins is diagnostic of black rot. When affected stems and petioles are cut crosswise, the vascular ring may appear black.

Disease Cycle

Epidemiology

Black rot of crucifers occurs under warm and wet weather conditions. Meier (1934) found that the disease occurred only under the conditions of high atmospheric humidity. Walker (1934) attributed absence of disease from Skaget valley of Washington to low rainfall conditions. Disease has been reported to be severe and highly destructive under the conditions of warm wet weather at various places (Cass-Smith, 1943; Anonymous, 1953 and Knosel, 1960). Derieand Gabrielson (1988) reported warm summer weather to be associated with symptom expression of black rot in crucifers.

Integrated Approches

Cultural: Seed that has been certified as disease-free should be selected whenever possible. A hot-water treatment, if done carefully, is an effective method of eradicating black rot bacteria from seed. Seed of Knol khol should be treated for 15 minutes. Use clean, pathogen-free seed. Seed from Western States generally is free from this disease. Use a 2-year rotation out of crucifers in production fields. Control cruciferous weeds and insect pests

Biological Control: Strong antibacterial activity of the essential oil of *Anacardium occidentale* Linn. was recorded against *X. campestris* pv. *campestris* (Garg and Kasera, 1984). *Terminalia chebula* extract was found effective in reducing the disease incidence (Bora and Bhattacharyya, 2000). Seed treatment with biocontrol agents like *Trichoderma asperellem* @ 5 ml/l or *Pseudomonas fluorescence* @ 5ml/l is found to be effective in controlling the disease.

Chemical Control: Seed treatment with Mercuric chloride solution for 30 min or Agrimycin or Aureomycin 0.01%. Spray Agrimycin-100 or Streptocycline-50ppm at transplanting and during stem formation. Drenching seed bed with 5% formalin or any antibiotic solution in nursery beds.

Future Approches

The following future approaches is being formulated

In vitro studies have shown that there is a relationship between temperature and sporulation. If these relationships apply under field conditions, they may be of value in predicting sporulation and subsequent disease development of *A. brassicae* in a wide range of susceptible crops (Humpherson-Jones and Phelps, 1989). Identification of resistance genes from wild species of knol khol may be utilized in breeding programme for development resistant variety.

Development of a suitable variety through breeding approaches resistant to clubroot is the only future approaches that can be seen as chemicals are not economically feasible and causing toxic to environment and soil.

Future research, combining experimental and modeling approaches, should focus on a better understanding of the role of abiotic stresses in damping-off diseases. In addition, diagnoses of commercial fields with various levels of damping-off symptoms could also help analyze the effects of interactions between cropping practices and production situations on the biotic and abiotic drivers of damping-off. The developed models would thus significantly improve our understanding of the critical interactions between biotic and abiotic factors that affect damping-off diseases and would help design integrated management strategies of dumping-off diseases.

Conclusion

The great economic importance of Knol-khol diseases and increasing concerns in finding sustainable solutions to this problem imply that opportunities exist to develop Integrated Disease Management strategies. Achieving this outcome will require a greater understanding of the ecology, genetics, and pathogenicity of the microbes associated with the diseasem.

References

Anonymous. (1953). Plant diseases: bitter pit of apples, brown fleck or internal brown spot of potatoes, black rot of cabbage and cauliflower. *Agric. Gaz. N.S.W.* **64**: 215-218.

Armstrong, G.M., and J.K. Armstrong. (1966). Races of *Fusarium oxysporum* f. sp. *conglutinans*; race 4, new race; and a new host for race 1, Lychnis chalcedonica. *Phytopatho*. **56**:525-530.

Awasthi, R.P. and Kolte, S.J. (1989). Variability in *Alternaria brassicae* affecting rapeseed and mustard. *Indian Phytopath*.42: 275 -280.

Berg,T., Tesoriero L. and Hailstones. D.L. (2005). PCR-based detection of *Xanthomonas campestris* pathovars in *Brassica* seed. *Plant Pathology* 54(3): 416-427.

Booth, C. (1971). The Genus *Fusarium*. Commonw. Mycol. Inst., Kew, Surrey, England. pp 237. Bora, L.C. and Bhattacharyya A.K. (2000). Integrated management of black rot of cabbage caused by *Xanthomonas campestris* (Pammel) Dowson. *J. Agric. Sci. Soc. NE India* **13**: 229-233.

Cass-Smith, W.P. (1943). Black rot of cabbage, cauliflower and related plants. *J. Dep. Agric. W. Aust. Ser.* **20**: 298-302.

Chauhan, R.S., Maheswari, S.K. and Gandhi, S.K. (2000) Effect of soil type and plant age on stem rot of cauliflower. *Agri Sci. Digest*. **20**: 58-62.

Chupp, C., and Sherf, A.F. (1960). Vegetable diseases and their control. Pp. 267-269. The Ronald Press Company. New York. 693 pp.

Datnoff, L.E., T.K. Kroll, and G.H. Lacy. (1987). Efficacy of chlorine for decontaminating water infested with resting spores of *Plasmodiophora brassicae. Plant Disea* **71**: 734-736.

Datnoff, L.E., T.K. Kroll, and J.A. Fox. (1984). Occurrence and population of *Plasmodiophora brassicae* in sediments of irrigation water sources. *Plant Disea* **68**: 200-203.

Derie, M.L. and R.L. Gabrielson. (1988). Black rot of crucifers in cabbage seed field in Western Washington. *Plant Dis.* **72**: 453-456.

Dorothy, M. 1969. Other hosts for *Peronospora parasitica* from cabbage and radish. *Phytopath.*, **59**: 693-696.

Felton, M.W. and Walker , J.C. (1946) Downy Mildew Disease of Crucifers: Biology, Ecology and Disease Management. *J. Agri. Res.* **72**: 69-81.

Garg, S.C. and H.L Kasera. (1984). Antibacterial activity of the essential oil of *Anacardium occidentale* Linn. *Indian Perfumer* **28**(2): 95-97.

George N. Agrios (2005). Plant Pathology (5th ed.). Burlington, MA: Academic Press.

Humpherson-Jones, F.M. (1989). Survival of *Alternaria brassicae* and *Alternaria brassicicola* on crop debris of oilseed rape and cabbage. *Ann. Appl. Biol.* **115**: 45-50.

Humpherson-Jones, F.M., and K. Phelps. (1989). Climatic factors influencing spore produciton in *Alternaria brassicae* and *Alternaria brassicicola. Ann. Appl. Biol.* **114**: 449-458.

Jenkin, J.E.E. (1964). Downey Mildew an emerging diseases *Plant Patho.*13: 46-50.

Jorgensen, M.C. and Walter J.M. (1955). The 1953 outbreak of black rot at Ruskin. *Proc. Fla. Hort. Soc.* **67**: 109-111.

Kadian, A.K. and Saharan, G.S. (1983). Physiological specialization in *Alternaria brassicae. Crucifereae Newsletter* **8**: 32-33.

Knosel, D. (1960). An extraordinarily severe occurrence of vein-blackening of Filder cabbage in the year 1958. *Zbl. Bakt. Abt.* **113**: 212-214.

Lin, Y.S., S.K. Sun, S.T. Hsu, and W.H. Hsieh. (1990). Mechanisms involved in the control of soil-borne plant pathogens by S-H mixture. Pp. 249-259 In: Biological Control fo Soil-borne Plant Pathogens. D. Hornby, Ed. CAB International, Wallingford UK.

Lund, B.M. and Wyatt, G.M. (1978). Post harvest spoilage of cauliflower by downy mildew *Peronospora parasitica. Plant Path.* **27**: 143-144.

Maude, R.B., and F.M. Humpherson-Jones. (1980a). Studies on the seed-borne phases of dark leaf spot *(Alternaria brassicicola)* and grey leaf spot *(Alternaria brassicae)* of *brassicas. Ann. appl. Biol.* **95**: 331-319.

Maude, R.B., and F.M. Humpherson-Jones. (1980b). The effect of iprodione on the seed-borne phase of *Alternaria brassicicola. Ann. appl. Biol.* **95**: 321-327.

Meier, D. (1934). A cytological study of the early infection stage of black rot of cabbage. *Bull. Torrey Bot. Club* **61**: 173-190.

Nakovo, B. (1972). Scientific work. Higher Agricultural Inst. Vasil Kolarov **21**: 109-116.

Pagoch, K., Srivastava, J. N. and Singh, A. K. (2015). Damping-Off Disease of Seedlings in Solanaceous Vegetables: Current Status and Disease Management. In *Recent Advances in the Diagnosis and Management of Plant Diseases* (pp. 35-46). Springer, New Delhi.

Rangel, J.F. (1945). Two Alternaria diseases of cruciferous plants. *Phytopatho* **35**: 1002-1007.

Saharan, G. S., Mehta, N. and Meena, P. D. (2017). *Downy mildew disease of crucifers: biology, ecology and disease management*. Singapore: Springer.

Sharma, A.K., J.S. Gupta, and R.K. Maheshwari. (1984). The relationship of *Streptomyces arabicus* to *Alternaria brassicae* (Berk.) Sacc. and *Alternaria brassicicola* (Schew.) Wiltshire on the leaf surface of yellow sarson and taramira. Geobios New Reports **3:** 83-84.

Sharma, A.K., J.S. Gupta, and S.P. Singh. (1985). Effect of temperature on the antifungal activity of *Streptomyces arabicus* against *Alternaria brassicae* (Berk) Sacc. and *A. brassicicola* (Schew.) Wiltshire. *Geobios* **12**: 168-169.

Sherf, A.F. and A.A. MacNab. (1986). Vegetable diseases and their control, second edition. Pp. 256-260. John Wiley & Sons, Inc. New York. 728 pp.

Singh, R.S. (1998). Plant diseases, 7th Edn. Oxford & IBH Pub. Co. Pvt. Ltd. New Delhi.

Spencer, D.M. (1981). *Plant pathology*. Academic Press, New York, pp. 31-43.

Subramanian, C.V. (1970). *Fusarium oxysporum* f. sp. *conglutinans*. CMI Descriptions of Pathogenic Fungi and Bacteria, No. 213. Commonw. Mycol. Inst., Kew, Surrey, England. 1 p.

Tamayo, M.P.J. (2001) *Alternaria brassicae*, causal agent of head rot in Cauliflower, *ASCOLFI Inform* **27**: 10-11.

Thind, K.S.(1942). Downey Mildew in Brassicae. *J Indian Bot Soc* **21**: 197-215.

Tripathi, N. N., Sahran, G.S. and Kaushik, C.D. (1987). Magnitude of losses in yield and management of Alternaria leaf blight in Rapeseed and Mustard. *Har. Agri. Univ. J. Res.***17**: 14-18.

Walker, J.C. (1934). Production of cabbage seed free from *Phoma lingum* and *Bacterium campestre. Phytopath.***24**: 158-160.

Walker, J.C. (1952). Diseases of vegetable crops. Mc Graw Hill Book Co, Inc. New York, pp 552.

Ypema, H.L. and Gubler, W.D. (1999). Long term effect of temperature on powdery mildew of crucifers. *Plant Dis* **81**: 1187-11892.

8

Diseases of Broccoli (*Brassica oleracea* var. *itialica* L.) and Their Integrated Management

Diganggana Talukdar

College of Horticulture, Sikkim, Central Agricultural University, Imphal

Introduction

Broccoli (*Brassica oleracea* L) belonging to family Brassicaceae, is an herbaceous annual or biennial plant which is cultivated for its edible flower heads being are used as vegetable or salads. The broccoli plant has a broad green stalk, or stem, that give produces thick, leathery, oblong gray-blue to green in color leaves. The plant produces large branching green flower heads covered with numerous white or yellow flowers. Broccoli can be annual or biennial depending on the variety. Broccoli is a rich source of vitamins, minerals and antioxidants and has a reputation as super foods (Hatanaka and Ware, 2020).

It is admitted that they are susceptible to a numerous of serious diseases that leads to reduction in yield. Thus, it is the utmost requirement to control such biotic stressors in order to obtain desired quality and decent yields. The most important diseases and its control measures are being portrayed in this chapter for both home gardener and the commercial producers for cruciferous vegetables particularly broccoli (Mani *et al.*, 2020).

Alternaria Blight

Economic Importance

Among the different diseases caused by the genus *Alternaria*, the blight disease is one of the most dominant one that causes average yield loss in the range of 32-57% (Conn and Tewari, 1990). *Alternaria brassicae*, however, is the more dominant invasive species amongst all. *Alternaria* causes significant loss in viable seeds and produce. The resulting lesions caused by Alternaria greatly reduce available photosynthetic area, leading to wilt and ultimately plant death. Javed, *et al.*, (2018 reported that both *Alternaria brassicicola* and *A. brassicae* are the most obnoxious species that cause leaf blight in cultivated or wild members of cruciferous family. *A. brassicicola* is prominent for its ability to infect vegetable crops while *A. brassicae* dominates in causing diseases to oil seed crops (Ayuke, *et al.*, 2017; Manhas and Kaur, 2017). Both of these species can damage leaves, stem, pods, seeds and seedlings. The primary and secondary inoculums of these *Alternaria* species survives in crop debris up to several years and spread to other fields too. *Alternaria raphani* is another species that is reported to cause blight in broccoli & radish (Sue, *et al.,* 2005).

Symptoms

On seedlings, symptoms appear as small dark spots on the stems which eventually cause the seedling to collapse downward. The typical syymptoms on the foliage also appear as small dark spots which enlarge rapidly. The two *Alternaria* species, *A. brassicae* and *A. brassicicola,* cause similar symptoms like small, dark specks that develops first on leaves and later enlarge into circular, dark or tanned spots measuring 0.25-0.5 inch in diameter approximately. The spots caused by *A. brassicicola* tend to be darker than those caused by *A. brassicae* (Mani, *et al.*, 2020). If conditions are favorable, dark green spores of the pathogen will grow on the spots. Such growth causes the spots to have concentric rings in them. Old leaf spots become papery in texture and may tear. When the dry tissue cascades out, a shot hole appears on it. Within these infected areas, huge masses of dark spores are created. The pathogen may attack the heads of cauliflower and broccoli rendering them unmarketable. The disease is also a problem in storage of Brassica crops.

Causal Organisms

It is caused by two species namely *Alternaria brassiace* and *A. brassicicola* and it is given in details in below Table 1.

Table 1. Morphological structure of *Alternaria* spp. causing leaf of Rapeseed and Mustard

Fungal structure	*A. brassicae*	*A. brassicicola*
Mycelium	Septate, brownish grey	Septate, olive grey to greyish black
Conidiophore	Dark, septate, arise in fascicles, 14-74μx 4-8μ	Olivaceous, septate, branched, 35-45μ x 5-8μ
Conidia	Brownish black, obclavate, muriform, produced singly or inchains or 2-3	Dark cylindrica l to oblong, muriform, produced in chains of 8-10 spores
Spore body (μ)	96-114 x 17-24	96-114 x 17-24

Spore beak length (μ)	45-65	None
Transverse septa tion	10-11	5-8
Longitudinal septation	0-6	0-4
Infection	Penetrates leaf only through stomata	Penetrates leaf directly or through stomta

Source: Talukdar *et al*., (2015) and Meena, *et al* (2010)

Disease Cycle

Mani, *et al.,* (2020) reported that the pathogen can overwinter on old, infected plant debris and in seed. The fungus produces masses of spores which are easily disseminated by wind, rain, or equipment. Warm, moist weather conditions favor disease development. Alternaria leaf spot is usually not an economic concern on crops. It is a frequent problem on cabbage during cool, rainy months. This pathogen can also infect other crucifers like brussels sprouts, broccoli, and even cauliflower. Leafy crucifers like red mustard, Chinese cabbage, tat tsoi, and Mizuna mustard that are harvested for their leaves are also susceptible to *Alternaria brassicae* and can be seriously devastated by this pathogen. Disease is favored by moist conditions. Spores are spread by winds and splashing water. The fungus does not survive in soil, but is carried over in crucifer seed, on weed or volunteer hosts, or on un-decomposed crop residue.

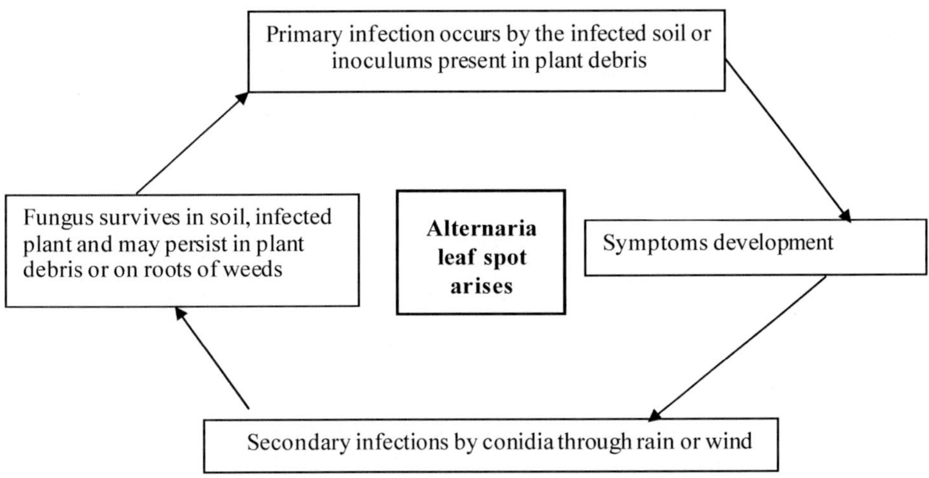

Disease cycle of *Alternaria* blight of Broccoli

Epidemiology

The infected seeds with spores on the seed coat or the presence of mycelium under the seed coat are the major source of transport of these pathogens. Wind, water, tools and animals are the ways for the dissemination of spores. The fungus can survive in susceptible weeds or perennial crops (Chupp and

Sherf, 1960; Humpherson-Jones, 1992). Furthermore, existence of infected crops present on the ground after harvest also serves as a source of infection for majority of the *Alternaria* species. In one study, infected leaves of cruciferous placed outdoors on soil produced viable spores for as long as leaf tissues remained intact. (Humpherson-Jones, 1992). This type of spread is likely to occur in seedling beds as well, and seedlings from infected seed beds can carry the inoculum to the field (Rangel, 1945).

Integrated Disease Management

- Promising plants, reported by Kabir, *et al.,* (2014) are Garlic, Neem (*Azadirachta indica*), Mahogany (*Swietania mahagoni*) and Koromcha (*Carissa carandas)* that has their ability against *Alternaria* blight of broccoli.
- Percent amount of reduction of plant infection, leaf infection, and leaf area diseased were highest in combined Garlic+Neem (*Azadirachta indica*)+Koromcha (*Carissa carandas)* and Rovral 50WP (Kabir *et al.*, 2014).
- Sharma, *et al.*, (2007) described that among all other botanicals, neem leaf extract showed highest efficacy to inhibit the radial growth of *A. solani* that ranged from 43.3 and 26.7% at 0.1% and 0.01%, respectively.
- Infected residues of susceptible varieties in field from a previous crop retained on the surface should be avoided.
- A balanced crop nutrition especially with potassium supliments should be supplied (Talukdar, 2015).
- Complete resistance against Alternaria blight is not available in the Brassica gene pool. Population improvement programme using the selected lines as base population is very likely to genetically enhance the level of resistance. Due to non-availability of clearly defined isolates, the utilization of multilocational breeding programme can effectively executed to address pathogenic variability besides efforts on biocontrol of the disease and epidemiological studies at different locations in field trials (Talukdar, 2015).
- Studies on variability at pathogenic and genetic level in *A. brassicae* could enable easier development of disease resistant material (Talukdar, 2015)

Downy Mildew

Economic Importance

Broccoli growers report that downy mildew is the only disease of economic importance in their greenhouses, occasionally resulting in 100% loss of yield (Hong *et al.*, 2008).

Symptoms

Infection usually takes place on the underside of lower leaves where the fungus produces a white fluffy growth. As the disease develops, yellow spots develop on the upper leaf surface which later turn tan in color. When the pathogen attacks cabbage heads, symptoms appear as small, dark sunken spots. Similar symptoms can be observed on curds of cauliflower and broccoli. Plants may become infected during any stage of growth, but usually the disease is more of a problem on early-seeded plant beds or on late-maturing crops. Susceptible hosts consist of cabbage, broccoli canola, brussels sprouts, kale, cauliflower, radish, rutabaga, horseradish, mustards, Chinese cabbage, ornamentals such as stock and wallflower, and many cruciferous weeds.

Causal Organisms

It is caused by the fungus *Peronospora parasitica.*

Disease Sycle

Mani, *et al.,* (2020) admitted that this fungus overwinters on old crop debris where it produces masses of spores which can be disseminated by wind and rain. Cool, moist weather conditions favor disease development. Downy mildew fungus becomes a great problem in early spring, late summer, and fall during damp, cool weather especially during overwintering in crop debris, cruciferous weeds, and even on crop seeds. Small yellow-brown spots appear in beginning on lower leaves, that eventually expands and develop grayish black lace-like markings on it. During moist weather, bluish-white downy mold is appears on the underside of these leaf spots.

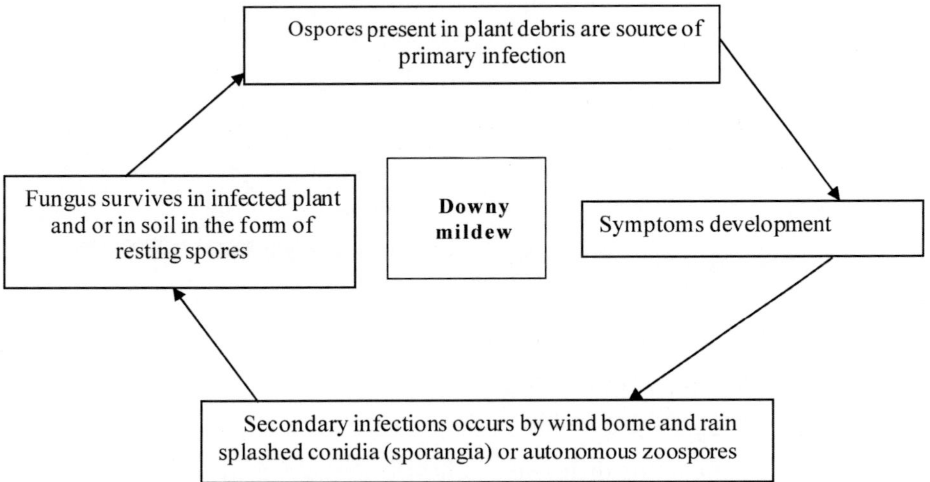

Disease cycle of Downy mildew of Broccoli

Profuse sporulation and rapid disease development occur at more than 98% relative humidity, when leaves are wet, and temperature ranging from 8-16°C. Symptoms may occur as pale brown or greyish discoloration on the curd (flower), or grey to black spots and streaks on the stems below the curd.

Epidemiology

Peronospora parasitica is obliged cool, moist weather for infection and disease development. The pathogen survives on weed hosts or as resilient oospores in crop residue. The spores are exclusively airborne in nature. This disease is most severe on young seedlings. The young plant may die if the cotyledons and the first true leaves are harshly infected.

Integrated Disease Management

Control of this disease is achieved by integrating management practices and fungicide protocols as reported by Minchinton (2017):

- Use seed treated with hot water or seedlings raised from such treatment.
- Plant disease-free seedlings.
- Do not water seedlings in the morning when spores are released.
- Keep seedlings as dry as possible. One heavy watering is preferable to a long light watering.
- Use a preventative fungicide spray program if weather conditions favour disease.
- Maintain a well ventilated environment - a lower relative humidity minimises spore production. This may mean fewer plants and trays per square meter.
- Maintain a well ventilated environment - a lower relative humidity minimises spore production. This may mean fewer plants and trays per square meter.
- Maintain a balanced program of nutrition - deficiency of potash will increase the susceptibility of seedlings to downy mildew.
- Remove any source of spores, such as heavily infected trays of seedlings, old infected seedlings, and weedy crucifer weeds.
- Disinfect glasshouses and igloos and rotate areas housing seedlings.
- Plough in field crop debris immediately after harvest and use a crop rotation of two to three years if possible.
- Alternate chemicals of different classes to avoid the fungus developing resistance to a particular group of fungicides.

Black Rot

Economic Importance

Black rot is a potentially lethal bacterial disease and must be considered the most important worldwide disease that affects cruciferous vegetables such as broccoli, Brussels sprouts, cabbage, cauliflower, kale, rutabaga and turnip, as well as cruciferous weeds such as shepherd's purse and wild mustard. Black rot occurs worldwide wherever cruciferous plants are grown and makes cruciferous vegetables unfit for the marketplace Pape, (2013). As with many bacterial diseases, this pathogen thrives in warm, humid climates and is most serious in tropical, subtropical, and humid continental regions.

Symptoms

This disease affects the plants at any stage of growth. The water pores around the leaf margins are the main entrance for the bacterium to enter. Eventually, the tissue turns yellow once the bacterium has entered the leaf. The infection progresses toward the centre of the leaf, forming a V-shaped pattern which is the most diagnostic symptoms (Mirik, *et al.,* 2008). The veins within the yellow-colored tissue turn black. The bacterium slowly moves down the leaf and then gets distributed throughout the entire plant. The infected leaves turn yellow in clolour and in due course fall upon the ground. Some plants might be entirely defoliated from top to bottom. When cut crosswise, the affected stem, disclose a black ring within the water-conducting tissue. The bacterium is carried both on and in seed. That the reason why, when infected seed is planted, the bacteria pass from seed parts into the leaves of the small seedling and eventually symptoms develop.

Causal Organisms

The disease is caused by the bacterium *Xanthomonas campestris* pv. *campestris*. It is non-spore forming, gram negative, rod shaped, motile, aerobic, oxidase-negative, catalase-positive, and amylolytic-positive (Mirik *et al.*, 2008).

Disease Cycle

The bacterium overwinters both in and on the seed and in the plant debris left in the field after harvesting. The bacterium is spread in a field by planting infected seeds, planting infected transplants, or planting in fields where the disease was a problem the previous year. Moreover, this bacterium is spread by rain-splashing or running water and vectors like insects.

Epidemiology

Black rot is a seed-borne disease. When the infected seeds are showed, it spreads over the plants. The disease is favoured by warm, humid conditions and can spread rapidly from rain dispersal and irrigation water (Vicente and Holub, 2012).

Integrated Disease Management

The integrated disease management is explained below as reported by (Vicente and Holub, 2012):

- Hot water seed treatment is an effective way to control this pathogenic spread.
- Fungicide treatment with copper-based products may help slow down the progress of the disease.
- Plant bed sanitation by taking care to avoid planting beds that have been used for Brassica crop production proves effective. Wait at least three years. Drainage water from old compost heaps and old cabbage fields can contaminate the soil. If disease-free soil is not available, soil fumigation may be another fruitful choice.
- Plant inspection and handling of plants thoroughly and looking for disease symptoms sound effective to avoid injury.
- A three-year crop rotation with unrelated crops is necessary since the bacterium can overwinter in the soil for two years.
- Spraying with copper bactericides in the fields may help reduce the spread of the bacteria.
- For the control of black rot, disease-free seed and transplants are essential. It is also important to use clean, disinfested containers and planting materials. Crucifer plantings should be located where air and soil drainage are good.
- Cultural practices such as fall tillage hasten the decomposition of infested crop residue and will significantly reduce pathogen populations in fields of previously infected crucifer crops. Rotating fields out of crucifer crops for 3 years is advisable for both disease management and soil improvement.

Black leg

Economic Importance

Although blackleg is an important disease of cabbage and some other brassica (crucifer) crops including broccoli, Brussels sprouts, cauliflower, rutabaga, and turnip, it is the most important disease of oilseed rape, including canola, with a potential to devastate the crop (Ash, 2000).

Symptoms

Plants may become infected in the seedbed or in the field at anytime during the growing season. The primary symptom is a circular depressed canker that develops at the base of the stem, enlarges, and eventually surrounds the entire stem. Yellow spots with gray centres appear on the leaves. Small dots appear on the infected tissue on both the stem and leaves. These black dots are nothing but fungal fruiting bodies called "Pycnidia" indicating the presence of the pathogen. The pathogen destroys the supportive stem tissue for which severely infected plants topple over.

Causal Organisms

This disease, caused by the fungus *Phoma lingam*, is a major concern in areas involved with *Brassica* crop production because the pathogen can infect a huge range of *Brassica* spp. world wide.

Disease Cycle

Of particular importance is the ability of this pathogen to be carried in and on seed. This is how the fungus is introduced into greenhouse and field plantings (Ash, 2000). The pathogen survives in crop debris if the plant parts are not fully decomposed. Cool, moist conditions enhance disease development. Spores are spread with splashing water and also can be blown long distances on wind currents. The fungus overwinters in the soil on old, infected plant debris for at least three seasons. When infected seed is planted, the fungus infects the seedling and produces spores which, in turn, are disseminated to other plants. Rain and surface drainage water spread the pathogen. The disease can spread very rapidly even though only a few plants may be infected initially.

Epidemiology

The disease is favored by wet, rainy weather.

Integrated Disease Management

The methods needed to control black leg (fungus) are the same as those described for black rot (bacteria) control. These control measures include-

- Planting western-grown, disease-free seed.
- Using a hot water seed treatment.
- Applying a fungicide treatment.

- Maintaining good sanitation of planting beds.
- Inspecting plants and handling plants carefully.
- Using crop rotation
- Planting resistant cultivars.
- Utilizing disease-free seed and transplants from decent sources is indispensable for black leg control.
- A combination of cultural and chemical control practices can prevent serious losses in fields with a history of black leg grown earlier.
- A 4 year rotation out of crucifers should be used because the pathogen may survive for several years in association with crop residue.
- Fall plowing will reduce the over-wintering pathogen population; avoiding poorly drained fields will reduce the likelihood of infection.
- A fungicide (Rovral) is labeled for treatment of young plants in fields prone to black leg infection.

Clubroot

Economic Importance

Clubroot, caused by the obligate parasite *Plasmodiophora brassicae*, is an economically important disease affecting plants in the family Brassicaceae worldwide (Dixon, 2009).

Symptoms

Distortion and thickening of the roots, particularly the tap root takes place. Plants gets wilted when temperature is very high and eventually it lacks vigor and it growth gets stunted. Infected roots are unable to effectively take up water and nutrients, and eventually the plant will die. Infected roots enlarge, become distorted, and resemble clubs, hence the name. Often disease development on the roots of affected plants can be extensive before above-ground portions show any symptoms. Leaves on infected plants turns yellow, wilt, and die.

Causal Organisms

This important *Brassica* crop disease is caused by the fungus *Plasmodiophora brassicae.*

Disease Cycle

The fungus enter into the plant by attachment with the root hairs. As the fungus begins to develop in the roots, it produces spores. These spores are released and can be disseminated by water and infested soil. Acid soils and cool wet weather favor pathogen development. The disease is not seed-borne. Club root infects all of the Brassica crops, as well as many weeds in the mustard family. The fungus remains viable in soil as thick-walled resting spores for 10 years or even longer. Acid soils above pH 7.0 with adequate moisture favor infections. The thick-walled resting spores germinate by releasing swimming zoospores. Such zoospores infect and colonize root hairs. A second type of zoospore also appears infecting the main roots. Second type of zoospore causes the galling and clubbing of roots due to colonization. In due course of time the roots get decayed and the fungus is dispersed from field to field by various ways like diseased transplants and movement of infested soil by tools and machinery and running of surface water.

Epidemiology

Colhoun, (2008), reported that this disease can occur in both acid and alkaline soils. A mean air temperature of 18–23°C and high soil moisture is quite favourable for this disease to proliferate. The disease occurs in both naturally and artificially contaminated soils with pH of 7.8 at planting time. In alkaline soils infection is favoured by both high moisture and high temperature.

Integrated Disease Management

IDM explained by Colhoun (2008) is as follows:

- Location of the seedbed is very important in club root control. To avoid local and wide-spread distribution of the pathogen, plant in disease-free soil.
- Hydrated lime incorporated into the soil to raise the pH to 7.2 reduces club root; however, on muck soils the application of lime is of little value because of the high soil buffering capacity.
- Club root can be reduced by application of PCNB fungicides.
- A modification of field soil acidity to pH 7.2-7.5 is fruitful for reducing the disease incidence.
- It is necessary to maintain clean fields.
- Although crop rotation with non host crops does not provide effective control, however, a 2-year rotation away from crucifer crops and into cereals might be helpful to some extent.
- Spread of the pathogen can be minimized by using pathogen-free transplants. It is preferable to use transplants that are produced in soil-less rooting mixes in trays.

White Rust

Economic Importance

White rust or white blister caused by the oomycete *Albugo candida* (Pers.) (O.) Kunze is an important disease of crucifer (*Brassicaceae*) crops and weeds worldwide. It is common in Hawai'i and can severely affect the production of cabbage and watercress on large farms and in residential gardens Patnude and Nelson (2013). *Albugo candida* has a wide host range that includes over 29 genera of crucifers (Jacobson, 1998; Farr, *et al.*, 1989). *Albugo candida* infects both wild and domesticated species of crucifers which includes broccoli, field mustard, Indian mustard, brown mustard, rutabaga, cauliflower, the cabbages and radish (Ferreria and Boley, 1991).

Symptoms

Both leaves and floral parts are affected. Underneath the plant epidermis a distinctive white, raised pustules to formed. These blister like pustules sometimes result in twisted, deformed growth of the stem, leaves, or flowers. On maturity, the epidermis covering the pustule gets ruptured releasing powdery white sporangia. This sporangia is then carried by winds or splashing water onto neighbouring host plants. Severely infected leaves can wither and die.

Causal Organisms

A fungus *Albugo candida*

Disease Cycle

Patnude and Nelson (2013) reported its disease cycle in details. The pathogen has both sexual sexual and asexual stages. The asexual stage starts with the dispersal of sporangia by wind or rain splash. These propagules while landing on a susceptible host, germinates and penetrates the epidermal layer of leaves which forms a mycelial wall between plant cells in due course of time. The invading hyphae develop special organs called haustoria within the epidermal cells and absorb nutrients from the host cells. Intercellular hyphae which produces sporangiophores, erupts through the plant tissue. These sporangiophres appear as white blisters or pustules and release sporangia. Under optimum conditions, between 16 and 25° C (60–77° F) and high relative humidity, the sporangia germinates and produces biflagellate, motile, infective zoospores. If moisture is available on leaf surfaces, the zoospores form germ tubes that enter leaves through the stomata causing systemic infections affecting vascular system of a host and the pathogen can move to plant inflorescences. In sexual reproduction, the pathogen produces oospores which are quite resistant and hard in texture. These spores survive for extended periods of time in soils or infested plant debris.

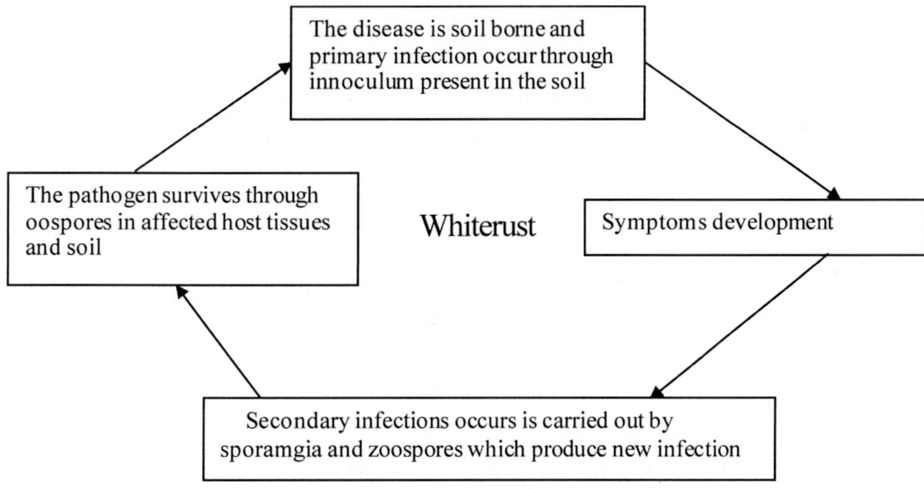

Disease cycle of White rust of Broccoli

Epidemiology

The white rust pathogen infects cruciferous groups of plants including arugula, bok choy, broccoli raab (rappini), brussels sprouts, cabbage, cauliflower, Chinese cabbage, collards, Japanese or Mizuna-type mustards, radish, tahtsai, and turnip. Economic damage is only found on the crucifer until and unless the leaves are marketed. The special spore called sporangia germinates during high moisture and cool temperature by producing several smaller motile spores called zoospores that swim and enter susceptible young tissues. As *A. candida* is reliant on cool and wet environments, the disease pre-dominant during winter and early spring months. Besides sporangia, *A. candida* also produces a second type of spore called oospores that can resist dry conditions which enable the fungus to survive in a dormant state in soil or crop residue. The white rust pathogen has multiple forms of distinct races that is adjustable with different climatic conditions both in macro and micro climates.

Integrated Disease Management

Patnude and Nelson (2013), reported about IDM and explained that managing white rust poses challenges, as pathogen spores are dispersed readily by wind and splashing water and may survive for years in soils. The pathogen may be seed borne and thereby introduced into a field unwittingly. The following ways are been used-

- Use pathogen-free seed stock.
- Practice sanitation by removing infected plant tissue or debris from growing sites and destroy it by burning or dumping underground.

- Time of planting matters a lot. Since the disease is more prevalent in areas of high humidity or in moderately wet climates, planting crops in drier seasons can reduce infections and the dispersal of the pathogen.
- Polycrop or intercrop crucifers with non-susceptible hosts. Managing white rust poses
- Use pathogen-free seed stock.
- Grow resistant plant varieties for example, 'Hirayama' (white rust resistant) and 'Waianae Strain' (slightly tolerant to white rust).
- Modify the environment. Knowledge of the epidemiology of diseases caused by Albugo species can help with their management. As spores of A. candida require free water to germinate, and not just high humidity, minimizing the occurrence and duration of leaf wetness, as by using a drip-irrigation system, can reduce infections.
- Time of planting. Since the disease is more prevalent in areas of high humidity or in moderately wet climates, planting crops in drier seasons can reduce infections and the dispersal of the pathogen.
- Polycrop or intercrop crucifers with non-susceptible hosts.

Wire Stem

Economic Importance

Rhizoctonia species is ubiquitous in nature which is prevalent across all areas in the world where its host crops are located. The severity of infection can vary and are devastating to the vegetable growers. Some of these consequences are major yield losses (from 25% to 100%), increased soil tare (because the soil sticks to the fungus' mycelium), and poor industrial quality of the crops based on increased levels of sodium, potassium, and nitrogen. Due to the number of hosts that the pathogen attacks, these consequences are numerous and detrimental to a variety of crops.

Symptoms

On Brassica crops, *Rhizoctonia* causes two types of disease symptoms: damping off (or wire stem) and bottom rot. Damping off or wire stem occurs on very young transplants affecting the hypocotyl region or lower stem tissue that remains in contact with the soil. This results in browning and cracking of the epidermis and the formation of lesions. As infection progresses, the outer stem decays, leaving only the fibrous inner xylem intact and hence the name wire stem. Affected plants wilt, turn purple, and remain stunted. Seedlings may break off at the soil line. The pathogen can usually be identified by its coarse mycelia that often cause soil particles to adhere to and dangle from diseased stems. The leaves develop dark brown, oval lesions when it comes in contact with soil. Secondary infection also takes place by decaying microbes that makes the lesions more soft and watery. Infected leaves may wilt, exposing the head of the broccoli.

Causal Organisms

This disease is caused by a soil borne fungi *Rhizoctonia solani*. This particular disease affects the nursery beddings with pre and post emergence damping off diseases in seedlings (Keinath, 2019).

Disease Cycle

When wirestem is severe, the entire outer layer of the taproot rots away, leaving only the tough water-conducting tube in the center of the root. Lateral roots also rot. Secondary above ground symptoms associated with lesions that girdle the stem include wilting, stunting, and a blue color to the youngest leaves. Moist soil may stick to the lesions, held there by the fungal pathogen. Wirestem can occur on greenhouse-grown transplants if potting mix becomes contaminated or is mixed with infested native soil. Wirestem can kill plants outright. Just as important, surviving plants may be stunted permanently and never produce a harvestable yield. Because of stand loss and stunted plants, wirestem reduces the number and weight of marketable-sized heads of cabbage and broccoli or collard plants per acre (Keinath, 2019).

Epidemiology

Being a common soil inhabitant, *Rhizoctonia solani* survives for long period of time in soil or crop residue or as sclerotia. Wet but warm soils favours wire stem development. Susceptibility in seedlings decreases as plants mature and get older (Keinath, 2019).

Integrated Disease Management

Prepare good quality seedbeds before planting. If possible, plant when soils are warm because seeds germinate faster and seedlings are more vigorous. Avoid excessively wet soils during early stages of seedling growth. If transplants are used, planting too deep must be avoided. No other control measures are recommended.

Bacterial Leaf Spot

Economic Importance

Bacterial leaf spot is distributed worldwide. *Pseudomonas syringae* has impacted many crop and orchard industries with its various pathovars leading to catastrophic catastrophic losses of crops.

Symptoms

Bacterial leaf spot infections start as small, dark specks on leaves. As disease develops, numerous water soaked leaf spots appear. Leaf spots remain small (0.125 inch or 3 mm in diameter). Older leaf spots turn tan and may or may not have a purple border around them. Leaf spots are visible from both top and bottom sides of leaves. Symptoms on transplants may resemble downy mildew symptoms.

Causal Organisms

Pseudomonas syringae pv. *maculicola,* is a rod-shaped, gram negative bacterium with polar flagella. This bacterium infects a wide range of species and has as over 50 different pathovars (Arnold and Preston, 2019).

Disease Cycle

Jeong *et al.,* (2016) reported the disease cycle of this fungus. *Pseudomonas syringae* overwinters on infected plant tissues developing s of necrosis or gummosis (sap oozing from wounds on the tree) in the regions. This pathogen can also overwinter in healthy looking plant tissues. During spring season, water from rain washes the bacteria onto leaves or blossoms. Then the bacteria starts to grow and even survive throughout the summer. This phase of *P. syringae's* life cycle is called epiphyte where it will multiply and spread without causing any disease or ailments. Once the bacteria enters the plant through a leaf's stomata or necrotic spots or wounds the disease start to appear. The pathogen will then penetrate into the intercellular space causing the leaf spots and cankers. *P. syringae* can also survive in temperatures slightly below freezing.

Epidemiology

Disease by *P. syringae* tends to be favored by wet and cool temperature. Optimum temperatures range from 12–25 °C. This bacterium is be seed-borne, and is dispersed from one plant to another by rain splash. These bacteria can also survive as saprophytes in the phyllospheric region of the plants whenever conditions are unfavorable for disease. But, instances are found that these saprotrophic strains of *P. syringae* can be utilized as biocontrol agents against postharvest rots (Hirano and Upper, 2000).

Integrated Disease Management

Planting of clean disease-free seeds and transplants is necessary. Crop rotation away from fields where the disease has recently occurred might reduce the inoculums in soil or infected debris. Furrow or drip irrigation compared to sprinkler reduces the disease spread.

Generalised Control Measures

- For producing transplants or direct seedlings of broccoli, be sure to use resistant cultivars as much as possible. Consult your seed catalogues, seed company representatives, and crop specialists for further information on available resistant cultivars.
- Use only new vigorous seed with high percent germination. Old, improperly stored seeds germinate more slowly, producing weak plants which are more susceptible to disease.
- Use seed which is certified free from black-rot bacteria and Phoma, and has been treated with hot water and fungicides to control seed-borne diseases. Even though hot-water seed treatment may reduce germination, simply conduct your own germination test. Calculate the percentage germination after seed treatment, and sow extra seed to supply enough plants for your needs.
- Sow seed in soil which has been fumigated (seedbeds) or sterilized (greenhouse). If fumigation is uneconomical, sow seedbeds in land which has grown no crucifers for at least 2 years.
- Use only new or sterilized transplant flats and greenhouse equipment. Be sure seeds are not too dense, and provide optimum conditions of ventilation, watering, fertility, temperature, and light for growth (see previous section on damping-off).
- Locate field-growth away from existing crucifer crops to avoid introduction of disease.
- Maintain weed-free seedbeds, cold frames, and greenhouses.
- Avoid run-off from land currently or previously growing a cruciferous crop, either directly, or indirectly in irrigation water.
- Inspect the seedlings regularly, removing and destroying localized infections "hot spots" early.
- Avoid dipping plants in water or trimming them before transplanting in the field, as this can easily spread bacteria and fungi.

Future Approaches

In order to have environmentally friendly and sustainable agricultural practices especially for the protection of Broccoli cultivation, Agro-ecosystem Analysis based Integrated Pest Management must be followed along with the regular IPM practices which has been mentioned in the chapter above. Huge scope still remains to harness technically proper agro- ecosystem Analysis based IPM system. Ecological Besides Engineering for disease and pest management, a new paradigm, is gaining recognition as a strategy for promoting Bio-intensive Integrated Pest and disease management strategies in recent times which still needs to execute on the field trials during production of diseases free broccoli produce.

Conclusion

Broccoli is one of the most important crops as it is bestowed with plethora of health benefits. The production and productivity of broccoli is at stake due to various diseases causing huge yield loss. This chapter depicts explanation of the most important diseases of this noble vegetable and its management. Since there is a rising public concern about the potential adverse effects of chemical pesticides on the human health, environment and biodiversity, integrated management strategies are been elucidated in this chapter for controlling the diseases. The intensity of the negative externalities of diseases through cannot be eliminated altogether, but can be minimized through development, dissemination and promotion of sustainable biointensive approaches for disease management. This approaches for the promotion of plant health will enable the farmers to take informed decisions based on experiential learning and it will also enable the farmers to take judicious approached for the uses of the chemicals in management of diseases with the highest utilization of the other non-conventional techniques like cultural, biological and botanical and tolerant or resistant varieties.

References

Arnold, D.L. and Preston, G.M. (2019). "Pseudomonas syringae: enterprising epiphyte and stealthy parasite". *Microbiology*. **165** (3): 251–253.

Ash, G. (2000). Black leg of oilseed rape. 2000. *The Plant Health Instructor*.DOI: 10.1094/PHI-I-2000-1111-01.

Ayuke, F.O., Lagerlof, J. and Jorge, G. (2017). Effects of biocontrol bacteria and. earthworms on the severity of *Alternaria brassicae* disease and the growth of oilseed rape plants (*Brassica napus*). *Appl. Soil. Ecol.*, **117**: 63-69.

Christine D. Smart (2019). Managing Black Rot of Cabbage and other Crucifer Crops in Organic Farming Systems. *eOrganic.* https://eorganic.org/node/4957.

Chupp, C., and Sherf, A.F. (1960). Vegetable diseases and their control. The Ronald Press Company. New York. pp 693.

Colhoun, J. (2008). A Study of the Epidemiology of Club-root Disease of Brassicae. Annals of Applied Biology, **40**(2):262 – 283.

Conn, K.L. and Tewari, J.P. (1990). Survey of *Alternaria* blackspot and Sclerotinia stem rot in central Alberta in 1989. *Can. Plant Dis. Survey*, **70**: 66-67.

Dixon, G. R. (2009). The Occurrence and Economic Impact of Plasmodio-phorabrassicae and Clubroot Disease. *Journal of Plant Growth Regulation*. **28**(3):194-202.

Ferreira, S.A., and Boley, R.A. (1991). *Albugo candida. Crop Knowledge Master*. University of Hawai'i: Department of Plant Pathology, CTAHR.

Hatanaka, M. and Ware, M. (2020). Health benefits of Broccoli. *Medical news today*. https:/ www.medicalnewstoday.com.

Hirano, S. S., Upper, C. D. (2000). "Bacteria in the Leaf Ecosystem with Emphasis on Pseudomonas syringae-a Pathogen, Ice Nucleus, and Epiphyte". *Microbiology and Molecular Biology Reviews*. **64** (3): 624–53.

Hong, S. Y., Jee, H. J., Choi, Y. J , Shin, H. D. (2008) First confirmed report of downy mildew caused by *Hyaloperonospora parasitica* on broccoli in Korea. *Plant Pathology*, **57**: 1777.

Humpherson-Jones, F. M. (1992). Epidemiology and control of dark leaf spot of brassicas. In : Alternaria Biology. *Plant Dis. and Metabolites*: pp. 267-288.

Javaid A , Shahzad, G. I. R, Akhtar, N. and Ahmed, D. (2018). Alternaria leaf spot disease of broccoli in Pakistan and management of the pathogen by leaf extract of *Syzygium cumini*. *Pak. J. Bot*., 50(4): 1607-1614.

Jeong, R. D., Chu, E.H., Lee, G.W., Park, J..M. and Park, H. J. (2016). "Effect of gamma irradiation on Pseudomonas syringaepv. tomato DC3000 – short communication". *Plant Protection Science*. **52** (2): 107–112.

Kabir. M.H., Rashid, M.M., Bhuiyan, M.R,. Mian, M.S, Ashrafuzzaman, M., Rafii, M.Y. and. Latif M.A. (2014). Integrated Management of Alternaria Blight of Broccoli. *J. Pure App. Microbiol*, **8**(1): 2-8.

Keinath, A.P. (2019). Identifying and Managing Wirestem on Vegetable Brassica (Cole) Crops. Clemson (SC): Clemson Cooperative Extension, Land-Grant Press by Clemson Extension. LGP 1029. http://lgpress.clemson.edu/publication/identifying-and-managing-wirestem-on-vegetable-brassica-cole-crops/.

Manhas, R.K. and T. Kaur. 2017. Biocontrol potential of *Streptomyces hydrogenans* strain DH16 toward *Alternaria brassicicola* to control damping off and black leaf spot of *Raphanus sativus*. Front. *Plant Sci*., 7: Article Number: 1869.

Mani, A., Dutta, P., Chatterjee, S. (2020). Diseases in Brassica vegetable crops and their Integrated Disease Management (IDM). *Agriculture & food: e- Newsletter*, **2**(1): 2581 – 8317.

Meena, P.D, Awasthi, R.P., Chattopadhyay, C., Kolte, S.J. and Kumar, A. (2010). Alternaria blight: a chronic disease in rapeseed-mustard. *J. of Oilseed Brassica,* **1**(1): 1-11.

Minchinton, E. (2017). Downy mildew of brassica (http://agriculture.vic.gov.au/agriculture/ pests-diseases-and-weeds/plant-diseases/ vegetable/downy-mildew-of-brassicas.

Mirik, M., Selcuk, F., Aysan,Y., Sahin, F. (2008). First Outbreak of Bacterial Black Rot on Cabbage, Broccoli, and Brussels Sprouts Caused by *Xanthomonas campestris pv. campestris* in the Mediterranean Region of Turkey. *Plant Dis.*, **92**(1):176.

Pape, A.(2013). Black Rot of Crucifers. Wisconsn horticulture, http: //hort.extension. wisc.edu /articles/ black-rot-crucifers/.

Patnude,E. and Nelson, S. (2013). White Rust of Cruciferous Vegetables in Hawai'i. *Plant Disease* PD-94.

Rangel, J.F.(1945). Two *Alternaria* diseases of cruciferous plants. *Phytopathol.*, **35**: 1002-1007.

Sharma, A., Dass, A. and Pau, M.S. (2007). Antifungal effect of neem extract on some common phytopathogenic fungi. *Adv. Plant Sci.*, **20**(2): 357-358.

Su, X.J., Yu H., Zhou, T., Li, X..Z. and Gong, J. (2005). First Report of *Alternaria raphanin* causing black patches on Chinese radish during postharvest storage in Canada. *Plant Dis.*, **89**: 1015-1015.

Talukdar, D. (2015). Alternaria diseases of oilseed crops of India and its management strategies. In: Crop Diseases in India : Management strategies (Eds. Dutta, P and Kaushik, H), Aavishkar publishers , Jaipur-302003, ISBN-978-81-7910-4-92-7.

Vicente, J.G. and Holub, E.B. (2012). *Xanthomonas campestris* pv. *campestris* (cause of black rot of crucifers) in the genomic era is still a worldwide threat to brassica crops. https://bsppjournals.onlinelibrary.wiley.com/doi/10.1111/j.1364-3703.2012.00833.x.

9

Diseases of Radish (*Raphanus sativus* L.) and Their Integrated Management

Amar Bahadur[1] **and** ***Pranab Dutta***[2]

[1]*College of Agriculture, Tripura Lembucherra, Agartala-799210 (Tripura)*
[2]*School of Crop Protection, College of Post Graduate Studies in Agricultural Sciences Central Agricultural University, Umiam, Meghalaya-793103*

Radish *(Raphanus sativus)* is a cool weather crop and its origin is in Mediterranean, rapid growing annual or biennial plant belongs to family Brassicaceae, grown for its edible taproot. Its edible roots are having different color from as white to red. Radish is generally grown as an annual plant, surviving only one growing season. It is crop of tropical and temperate region. It is growing as herb in West Bengal, Bihar, Uttar Pradesh, Karnataka, Punjab and Assam areas are major radish growing states. Radish is good source of vitamin B6, Calcium, copper, magnesium and riboflavin. Also it is rich in ascorbic acid, folic acid and potassium.

Alternaria Blight

Economic Importance

Alternaria leaf spot diseases occur in every part of the world, in both vegetable crops and seed crops and most important disease of Brassica species, often results in severe yield losses. It is considered as a number one disease of radish affecting both fresh vegetable and seed production. The disease is more severe when the plants are left for seed production in the field and causes significant reduction in seed yield. Alternaria black spot of radish; black pod blotch of radish. *A. japonica* is a seed-borne pathogen of plants in the Brassicaceae. A. japonica is known to affect most parts of infected plants depending upon growth stage. There are reports of it on the hypocotyls and cotyledons of seedlings and on the leaves, petioles, stem, inflorescence, fruits (siliquae) and seeds of adult plants. *A. japonica* occurs in all parts of radish (*Raphanus sativus*) seed. Its presence in the embryo can result in seed death, but in milder infections it will affect seedlings after germination, causing seedling blight both before and after emergence. Yoshii (1941) reported *A. japonica* to occur on radish (Raphanus sativus), cabbage and cauliflower (*Brassica oleracea*) and Chinese cabbage (Brassica chinensis) (Brassica rapa Chinensis Group). Leaf lesions are small (1-10 mm diameter), black to grey, dry, with a raised margin, and they are often surrounded by a translucent yellow halo (Atkinson, 1950).

Symptoms

Symptoms of the disease first appear on the leaves of seed stem in the form of small, yellowish, slightly raised lesions. The pathogen affects on leaves, stem, pods and seeds. Infection is rapidly spread in rainy season. Fungus spread on pods and seeds loose viability.

Alternaria blight causes dark yellow, dark brown or black circular spots on leaves as characteristic ring spots sometimes have a yellow halo. The leaf spots enlarge and are brown in colour, with concentric rings of sporulating fungal growth. The center of the ring often dries out and drops, leaving the leaves with a shot-hole appearance, spots coalesce to form large necrotic patches; leaf drop may occur.

Lesions appear later on the stems and seed pods. The fungus penetrates in pod tissues, ultimately infecting the seeds. The infected seed fails to germinate. In seed crops, the seed capsules are damaged; seriously infected seed capsules ripen prematurely and lose their seed.

Causal Organism

Disease caused by *Alternaria raphani and A. japonica*

Disease Cycle

These pathogens are seed-borne and occur on both the outside and the inside of the seeds. Infected seed can have a lower germination capacity and produce seedlings with infected cotyledons. Infection occurs via airborne spores in crop residues, adjacent crops and weeds. Conidia of the fungus are disseminated primarily by wind and rain splash (Verma and Saharan, 1994; Tewari and Buchwaldt, 2007). The fungus may also be moved in infested debris and by any natural dispersal of infected seeds. The role of the chlamydospores in survival in the debris or in soil is undetermined. The fungus was found in all parts of radish seeds, with conidia present on the seed coat. The amount of inoculum decreased with depth in the seed (Vannacci and Pecchia, 1988). Vannacci and Harman (1987) also found that the fungus was present on seeds both internally and externally. Under temperate conditions, the pathogen survives in the form of mycelia or conidia on the previous year's crop debris and as chlamydospores or microsclerotia at cooler temperatures (Tsuneda & Skoropad, 1977; Humperson-Jones & Maude 1982; Humpherson-Jones & Phelps, 1989).

Epidemiology

Pathogens resides in alternative hosts such as susceptible weeds or perennial crops during the off season (Chupp & Sherf, 1960; Maude & Humpherson-Jones, 1980). Disease emergence favors warm, wet weather. Sporulation requires a high level of humidity for at least 12 hours of leaf wetness period. The disease cycle lasts just five to seven days and epidemics can develop quickly during hot, wet weather. The spores landing on the host surface as conidia adhere to the surface and penetrate the host mainly through the stomata after forming a germ tube, which then grows into the host epidermal cells through the formation of an appressorium, often triggered by host signaling mechanisms (Cho *et al.*, 2006). The infection results in the appearance of severe disease symptoms, including numerous black spots covering the pods, leading to losses in seed yield of up to 60% (Kolte, 2002; Meena *et al.*, 2010). An epidemic, very large numbers of leaf spots can develop, causing the more mature leaves to die off. During the growing season, the disease spreads across the crop and infection can occur on the stems.

Management

Plant only certified and use of disease-free seed; rotate crops to non-brassica species; hot water treatment of the seed kills the fungus, regular spraying with Difolatan (0.3%) or Dithane M 45 (0.2%) controls the disease effectively. Irrigate in the morning to allow the foliage to dry. If infestation is observed, spraying of Mancozeb@2gm/lit of water or Carbendazim@1gm/lit of water should be done to control this disease.

Downy Mildew

Economic Importance

This fast growing fungus affects brassicas, cucumbers, melons, peas. It is most often a problem in humid areas, especially in cool and wet weather.

Symptoms

Symptoms appearance on the leaves of some yellow, irregular spots that further become brown. Fungus enters the plant through wounds and natural openings and first appears on older leaves, small angular lesions on upper surface of leaves which enlarge into orange or yellow necrotic patches; white fluffy growth on undersides of leaves. Small, angular lesions develop on leaves and inflorescences. These lesions enlarge and become irregular, yellow to orange necrotic patches, with dense sporulation on leaf undersides. Heavy sporulation gives leaf undersides a gray to purple, downy appearance. On seedlings, cotyledons and hypocotyls may become infected and seedling loss can occur. Downy mildew also attacks the taproots of radish and infected organs develop a black, epidermal blotch and an internal discoloration.

Causal Organism

Downy mildew disease is caused by the fungus *Hyaloperonospora parasitica*

Diseases Cycle

Downy mildew overwinters on winter-sown host crops or cruciferous weeds. Infection of leaves and inflorescences results from sporangia produced on living hosts. Secondary sporangia are spread by wind and splashing water. Oospores, if produced, survive in crop residues and in the soil. The pathogen is favored by cool, moist conditions.

Epidemiology

Mildew spreads very easy with the wind, and the high humidity favor expanding. Disease emergence favors cool damp weather. The disease is favoured by cool temperatures with an optimum range of 8-16°C. The vegetative spores need water to germinate and can infect seedlings within three hours of contact with a leaf. Deficiencies of potash have been shown to increase the susceptibility of downy mildew.

Management

Removal of crop debris and weed hosts may reduce inoculums, rotate with non-brassicas crops; it is possible to control downy mildew with the application of fungicide Blitox 50 (0.3%), Daconil (0.1) at 8-10 day interval. Spores overwinter on crop debris, so clean up the beds in fall. The spores can travel long distances on the wind, especially in moist air. Spray with fungicides in the vegetation period as Aliette (fosetyl-Al).

White Rust

Symptoms

It is common diseases of radish crop of India. Disease attacks the leaves and flowering shoots. Affected flowering shoots get deformed and bear only malformed flowers. White powdery substance in patches is observed on the under surface of the leaves. White rust appears as white pustules on foliage and flowers. The leaves which are severely attacked dry and fall. The disease manifests through the appearance on the leaves of some yellow spots, and on the lower part. White, raised spore masses appear on the underside of leaves. The host epidermis is ruptured and powdery masses of spores are produced. White pustules appear on cotyledons, leaves, stems and flowers which coalesce to form large areas of infection.

Causal Organism

White rust caused by fungus *Albugo candida.*

Epidemiology

Fungus can survive for long periods of time in dry conditions; disease spread by wind With mild temperature and high humidity, the disease spreads quickly. Extreme hot or cold temperatures and dry weather will slow the disease progress.

Management

Crops rotation; use only disease-free seed; Use fungicide if the disease progresses such as. Dithane Z 78 (0.2%), chlorothalonil (0.1%) effectively controls the disease. The following practices should control white rust: destroy crop residue, and cultivate fields prior to planting to destroy all volunteer plants. Collecting and destroying the plant residues after harvesting. Seed treatments with Ridomil, Bravo. Clean cultivation and use of resistant varieties help to prevent the disease.

Clubroot

Symptoms

It is easy to recognize disease due to the symptoms that appear on the leaves and root. Aerial symptoms are represented by plant dwarfism, yellowing of the leaves and their wilt. On the root the disease manifests through the appearance of some swellings, extensive gall formation and dysfunctional radicular system. The disease can appear from the seedling phase.

Causal Organism

Clubroot is caused by the fungus *Plasmodiophora brassicae.*

Management

Fungus can survive in soil for periods of 10 years; can be spread by movement of contaminated soil and irrigation water to uninfected areas. Cultivate crop selected alkaline lands. Elimination of the pathogen is economically unfeasible; rotating crops generally does not provide effective control; plant only certified seed and avoid field grown transplants unless produced in a fumigated bed; applying lime to the soil can reduce fungus sporulation.

Damping-off (Wirestem)

Symptoms

Damping off (wirestem) is a common fungus found in the soil in areas of high humidity. Radishes are prone to seed rot or the collapse of seedlings when afflicted with damping off. Death of seedlings after germination; brown-red or black rot girdling stem; seedling may remain upright but stem is constricted and twisted (wirestem).

Causal Organism

Disease caused by *Rhizoctonia solani*

Management

Use pathogen-free seed or transplants that have been produced in sterilized soil; apply bio-control (*Trchoderma* sp.) for seed treatment to kill off any fungi; shallow plant seeds or delay planting until soil warms. Don't plant seeds in cold, moist soil and be sure the soil is well-draining.

Cauliflower Mosaic Virus

This disease attacks all the plants in the Brassicaceae family. The plant leaves shows discoloration, nerves become transparent and the tissue between the nerves becomes yellow. The affected seedlings remain small, and the mature plant has low production. The virus is transmitted from year to year through infested seeds and in plant to plant the virus is transmitted through insects.

Management

Use healthy and certified seeds. Removal and destruction of plants which show symptoms of the disease. Use of insecticide to manage the insects.

Future Approaches

Radish is widely used as root vegetable belongs to the family of Brassica. It's one of the nutritious root vegetables use in both raw as salads and recipes. Fresh roots are a good source of vitamin-C. Today it is widely cultivating the crops throughout the world. Radish sharp, pungent flavor of radish comes from "isothiocyanate" like other cruciferous family vegetables, contains *isothiocyanate* antioxidant compound called sulforaphane. Use for treatment against prostate, breast, colon and ovarian cancers by its cancer-cell growth inhibition, and cytotoxic effects on cancer cells. During crop production crop suffering several diseases so need to apply management strategies for their production and prevent quality.

Conclusion

Radish is important crop that have medicinal property, during cultivation several major diseases are attacking and hampered crop production but few are minor disease are also important due to changing climate, it may become epidemic, hence the grower should know their diseases and management practices for their production in future.

References

Atkinson, R.G. (1950). Studies on the parasitism and variation of *Alternaria raphani. Canadian Journal of Research, Section* C, **28:**288-317.

Chupp C, Sherf AF. (1960). Vegetable diseases and their control. New York: The Ronald Press Communication. 693.

Cho Y, Davis JW, Kim KH, Wang J, Sun QH, Cramer CB, Lawrence CB. (2006). A high throughput targeted gene disruption method for *Alternaria brassicicola* functional genomics using linear minimal element (LME) constructs. *Molecular Plant-Microbe Interaction* **19:** 7-15.

Humpherson-Jones FM, Maude RB. (1982). Studies on the epidemiology of *Alternaria brassicicola* in *Brassica oleracea* seed production crops. *Annals of Applied Biology* **100**(1): 61-71.

Humpherson-Jones FM, Phelps K. (1989). Climatic factors influencing spore production in Alternaria brassicae and A. brassicicola. *Annals of Applied Biology* **114**: 449-458.

Kolte SJ. (2002). Rai M, Singh H, Hegde DM, eds. Diseases and their management in oilseed crops, new paradigm in oilseeds and oils: research and development needs. Hyderabad: *Indian Society of Oilseed Research*. 244-252.

Maude RB, Humpherson-Jones FM. (1980). Studies on the seed-borne phases of dark leaf spot (*Alternaria brassicicola*) and grey leaf spot (*Alternaria brassicae*) of brassicas. *Annals of Biology* **95**: 331-319.

Meena PD, Awasthi RP, Chattopadhyay C, Kolte SJ, Kumar A. (2010). Alternaria blight: a chronic disease in rapeseed-mustard. *Journal of Oilseed Brassica* **1:** 1-11.

Tewari, J.P. and Buchwaldt, L. (2007). Alternaria diseases. In: Rimmer SR, Shattuck VI, Buchwaldt L, eds. Compendium of Brassica Diseases. St. Paul, Minnesota, USA: American Phytopathological Society, **15**-18.

Tsuneda A, Skoropad WP. 1977. Formation of microsclerotia and chlamydospores from conidia of *Alternaria brassicae. Canadian Journal of Botany* 55:1276-1281

Vannacci, G., Pecchia, S. (1988). Location and transmission of seed-borne Alternaria raphani Groves and Skolko in *Raphanus sativus* L.: a case study. *Archiv für Phytopathologie und Pflanzenschutz.* **24** (4): 305-315.

Vannacci, G. and Harman, G.E. (1987). Biocontrol of seed-borne *Alternaria raphani* and *A. brassicicola. Canadian Journal of Microbiology,* **33:** 850-856.

Verma, P.R. and Saharan, G.S. (1994). Saskatoon Research Station Technical Bulletin 1994-6E. Monograph on Alternaria diseases of crucifers. Saskatoon, Canada: Agriculture Canada Research Station., 160 pp.

Yoshii, H. (1941). On the black spot of Asiatic crucifers and the black mould of cabbage. *Journal of Plant Protection.* 14-18.

10

Diseases of Elephant Foot Yam (*Amorphophallus paeoniifolius*) (Dennst. Nicolson) and Their Integrated Management

***Gunadhya Kumar Upamanya*[1], *Pranab Dutta*[2] and *Ananya Dutta*[3]**

[1]*SCS, College of Agriculture, Dhuburi, Assam Agricultural University, Assam*

[2]*School of Crop Protection, College of Post Graduate Studies in Agricultural Sciences Central Agricultural University, Umiam, Meghalaya*

[3]*Department of Nematology, College of Agriculture, Assam Agricultural University (AAU), Jorhat, Assam*

Introduction

Elephant foot yam (*Amorphophallus paeoniifolius* (Dennst.) Nicolson) is an important edible tropical tuber crop since time immemorial in the tropics and sub-tropics (Behera *et al*., 2009) and regarded as "King of tuber crops" due to its higher biological efficiency as a food producer (Mukhopadhyay and Sen, 1999; Nath *et al*., 2007; Nedunchezhiyan, 2014). It belongs to the family Araceae. Tender stem and leaves of this tuber crop are used for vegetable purpose. The corms are rich in nutrition and contain 18.0% starch, 1-5% protein and up to 2% fat. Leaves contain 2-3% protein, 3% carbohydrates and 4-7 % crude fibre. It has great scope for commercial exploitation as a medicinal crop in pharmacological industry and widely recommended as a remedy in all three of the major Indian medicinal systems: Ayurveda, Siddha and Unani. The corm is prescribed for bronchitis, asthma, abdominal pain, emesis, dysentery, enlargement of spleen, piles, elephantiasis, diseases due to vitiated blood, and rheumatic swellings. Pharmacological studies have shown a variety of effects, specifically antiprotease activity, analgesic activity, and cytotoxic activity. It has also been found to be a potential inhibitor of bacteria activity when used with antibiotics. In addition, *Amorphophallus* corm is a good sources of phyto-aestrogens and are effective alternate or complementary to conventional hormone replacement therapy for symptoms associated with menopause and chronic degenerative diseases in women. It is a crop of South-east Asian origin, and grown widely in Philippines, Malaysia, Indonesia and South eastern Asian countries. In India, the crop is popular in the state of West Bengal, Andhra Pradesh, Kerala, Uttar Pradesh, Bihar, Madhya Pradesh, Maharashtra, Orissa, Tamil Nadu and Karnataka. West Bengal is the highest producer of elephant foot yam (293.84 thousand MT in the year 2015-16) followed by Andhra Pradesh (207.52 thousand MT) and Kerala (110.37 thousand MT) (Source: National Horticultural Board). However, the crop is gaining popularity dueto its shade tolerance, easiness in cultivation, high productivity, less incidence of pests and diseases, steady demand and reasonably good price.

Major Diseases of Elephant Foot Yam

Very few diseases have been reported in elephant foot yam out of which the diseases such as collar rot caused by the fungus *Sclerotium rolfsii,* dry rot disease caused by *Rhizoctonia solani* and *Fusarium solani* and mosaic disease caused by virus are worth mention. Some of the important diseases are discussed below.

Collar Rot

The disease is more destructive during rainy season followed by warm dry weather. Injury to collar region during intercultural operation, poor drainage and water logging acts as predisposing factors for infection. Soft and tender pseudostems are more vulnerable to this disease.

Symptoms

The infection begins on the succulent stem as a dark brown lesion just below the soil line. Soon at first the lower leaves and the upper leaves turn yellow or wilt or die back from the tips downward. The disease spreads to the entire pseudostem and cause complete yellowing of the plant. The fungus moves even more rapidly downward into the roots and finally destroys the root system. The white mycelium is always present in or on the infected tissues and from these it grows over the soil to adjacent plants, starting new infections. The pathogen grows into the cortex and girdles the plant and in severe case, the plant collapses leading to complete crop loss. On all infested tissues, and even on the nearby soil, the fungus produces numerous small roundish scleritia of uniform size that are white when immature, becoming dark brown to black on maturity.

The Pathogen

The pathogen, *Sclerotium rolfsii* is distributed in tropical and sub-tropical regions of the world where high temperatures prevail. The fungus has a wide host range of 500 species in about 100 families including groundnut, green bean, lima bean, onion, garden bean, pepper, potato, sweet potato, tomato and water melon (Aycock, 1966). It produces abundant, white, fluffy, branched mycelium that forms numerous sclerotia but is usually sterile.

Development of Disease

The fungus overwinters mainly as sclerotia. It is spread by moving water, infested soil and contaminated tools. The fungus attacks tissues directly. However, the mass of mycelium produces and secretes oxalic acid, pectinolytic, cellulolytic and other enzymes which kills and disintegrates tissues before it actually penetrates the host. Once established in the plants, the fungus advances and produces mycelium and sclerotia quite rapidly especially at high temperature and high moisture.

Management

The control of *Sclerotium* diseases is difficult. Crop rotation, cultural practices such as deep ploughing to burry fungal sclerotia in surface debris, application of calcium compounds to the soil before planting or in the furrow during planting provide only partial control. Use of disease free planting material planting material reduce the incidence of the disease. In recent years some control of *S. rolfsii* has been obtained by soil solarization and by use of antagonistic microorganism like *Trichoderma harzianum.* Soil drenching with the fungicide like 0.2% Captan or 0.1% Brassicol is also useful.

Leaf Blight

The wide spread cultivation of acridity free and high yielding cultivars of elephant foot yam (*Amorphophallus paeoniifolius* Dennst) has led to severe problem of leaf blight disease, throughout the country. The difficulty to control with chemical means after its outbreak, particularly in rainy season, leaves us with the only option to use resistant cultivars. This disease appears in severe form wherever elephant foot yam is grown with specific perspectiveto Northern and Eastern parts of the country.

Symptoms

Leaf blight of elephant foot yam are generallyobserved in the older leaves during August to September. Small water-soaked lesions develop on the leaflets. Thesespots enlarge and coalesce and give a blighted appearance inlater stages. The spots are surrounded by brown, green and yellow shades. A whitish growth of sporangia around the edge of the lesions during periods of high relative humidity is the prominent sign of infection by *P. colocasiae*. The intensity of the attack depends on the weather conditions prevalent during the growth period of the host, being much greater in years of abnormal humidity and is greater in shady sheltered places than in open field.

The Pathogen

Phytophthora colocasie Racib is the causal agent of the disease. The mycelium of the fungus is coenocytic, the sporangiophores are up to 50 micron long, unbranched, very slender and extremely narrow at the tip. They bear single pear shaped or elongated sporangia which measure 38-60 x 18-26 microns. As many as 12 reniform biflagellate planospores are produced in each sporangium. Direct germination of the sporangia is also known. Chlamydospores are produced which are thick walled, round and hyaline. The antheriudium is amphigynous. The oospores are spherical, 20-28 μm in diameter and lie free in the oogonium. The oogonia are yellowish and almost spherical. The antheridia persist at the base of the oogonia for a considerable period after the oospores are formed.

Development of the Disease

The disease perpetuate through the infected corms and oospores persisting in the soil. The fungus has poor competitive ability and is not soil borne in free state. Thus the infected corms are the major source of primary inoculums. The mycelium enters the host and intercellular in freely attacked tissues but intracellular in advanced infection.

The maximum sporangial production occurs at 21^{0}C when the relative humidity is 100 per cent. Relative humidity lower than 90 per cent totally inhibits sporulation. Viability of the planospores is lost rapidly at relative humidity less than 90 per cent. Germination of sporangia by planospore production is optimum at 20-21^{0}C. The planospores germinate within 30 minutes after release. They are capable of causing secondary infection

.

Management

Infected leaves and corms are important source of perennation and spread of the disease. In early stages of attack, spotted leaves should be removed and destroyed. The healthy corms should be selected for planting. Crop rotation reduces the intensity of the disease. Planting in shady localities should be avoided and proper drainage must be ensured. Treatment of corm with cowdung slurry mixed with *Trichoderma* is useful. Spraying of the crop with 0.3% Metalaxyl at 15-21 days interval has been found effective. Spraying with Carboxin 75% WP @ 1 gm/lit or Hexaconazole 5 % EC @ 1.5 ml/lit or Propiconazole 25% EC @ 0.75 ml/lit of water may be recommended.

Anthracnose

Symptoms

Infections on leavesare circular to crescent shape and tan to dark brownspots. The spots frequently coalesced into extended patches blighting theleaves and gradually the entire foliage. Brown lesions containing dark spore producingbodies on leaves, and later lesions were enlarged and coalescedto kill the entire leaf.

The Pathogen

The fungus, *Colletotrichum siamense* Prihastuti, L. Cai & K.D. Hyde is the cause of the disease. The shape and size of fifty conidia were determinedby microscopy. Single spore colonies grown on PDA at 27 ±1°C were atfirst white and later became pale brownish to pinkish, reverse paleyellowish to pinkish colonies. Mycelia were greyish-white, dense, cottony,with visible conidial masses. Acervuli brown to dark brown, conspicuousfor their brown setae. Conidia 8-17.2 × 3-4 µm, one-celled, smooth-walled, guttulate, hyaline, fusiform with obtuse to slightly rounded ends, sometimesoblong (Prihastuti *et al.*, 2009).

Management

Use of disease free planting material, crop rotation, removal and burning of infected parts and spraying with fungicide like Carbendazim 50% WP is effective.

Pythium Root Rot

Pythium sp., one of the most common and most important causes of seed rot, seedling damping off and root rot of all types of plant and also of soft rots of fleshy fruits in contact with the soil. Roy and Hong (2007) first reported the occurrence of the root rot disease of elephant foot yam in India from West Bengal.

Symptoms

Symptomatic plants ranged from chlorotic and stunted to completely blighted as the disease progressed. Necrotic root symptoms began at the tip, but progressed quickly eventually killing the whole root. The cortex of severely affected roots could be easily sloughed off, leaving only parts of the vascular system intact.

The Pathogen

The organism responsible for the disease was identified as *Pythium helicoides* (Roy and Hong, 2007). Sporangia of the fungusis ovoid to globose, papillate, caducous and non-proliferating (41.46 µm Â± 2.9 x 37.64 µm Â± 2.0). Isolates were homothallic. Oogonia are spherical, smooth walled with elongate diclinous antheridia and aplerotic oospores (39.16 µm Â±2.26 diameter). Hyphal swellings is absent.

Development of Disease

Root of the plant can be attacked in any stage of the plant growth. The oomycete enters root tips and proliferates, causing a rapid collapse and death of the rootlet. Invasion of older roots is usually limited to the cortex. Relatively young and fleshy roots may be invaded and form lesions several centimetres long. The disease and losses caused by *Pythium* infections are more severe when the soil is kept wet for prolonged periods, when the temperature is unfavourable (usually too low) for host plant, when there is an excess of N in the soil and when the crop is planted in the same field for several consecutive years.

Management

Certain cultural practices are sometimes helpful in reducing the amount of infection. Such practices include providing good drainage and good air circulation among plants, planting when temperature favourable for fast plant growth, avoiding application of excessive amounts of nitrate forms of nitrogen fertilizer and practicing crop rotation. In the field corm treatment with one or more effective chemical is the most important disease preventive measure.

Amorphophallus Mosaic Disease

The mosaic caused by the Dasheen mosaic virus (DsMV) are economically important. In addition to Elephant foot yam, DsMV is known to infect a wide variety of cultivated aroids and ornamental plants worldwide. It causes serious damage to the ornamentals like Caladium, Dieffenbachia and Zantedeschia, and is ubiquitous in commercial plantings of the tropical root crops, of the genera *Colocasia, Xanthosoma* and *Amorphophallus*. Natural infection of DsMV is mainly restricted to the plants of aroid family (Brunt *et al.* 1996).

Symptom

The symptoms include mosaic mottling, puckering, stunting and filiformy/ shoestring of leaves and distortion of leaf lamina. The disease cause more proliferation of lateral buds and poor growth of roots. The infected plant produce smaller corm.

The Pathogen

Dasheen mosaic virus (DsMV, a flexuous Potyvirus particles) has been reported to be the causal agent of the disease.

Development of Disease

The disease is transmitted by insect vectors viz; *Myzus persicae*, *Aphis gossypii*, *A. craccivora* and *Pentalonia nigronervosa*. (Brunt *et al.*, 1996). Hosts of DsMV include important root and tuber crops such as Alocasia, Colocasia and Xanthosoma, and the ornamentals Caladium, Dieffenbachia and Philodendron. Most of the commercially cultivated hosts are vegetatively propagated and the infection primarily spreads by infected planting material. Although the virus is known to be transmitted by aphids, the major cause of spread of the Elephant foot yam mosaic in India seems to be through the use of virus-infected cormas planting material.

Management

Production of virus-free planting material is essential for effective management of the disease. Proper identification and accurate detection ofthe virus in planting material is needed for successful production of virus-free planting materials. Systemic insecticides like Acephate 75% WP @ 0.75 g /lit or Dinotefuron 35% SL @ 0.5 gm/litre of water have been recommended for spraying to prevent the secondary spread.

Nematode Diseases

Nematodes like the root knot nematode, *Meloidogyne incognita* and the root lesion nematode, *Pratylenchus* spp. are the two most important nematode pests causing economic damage to the elephant foot yam in India.

Causal Organism

Meloidogyne incognita, *Tylenchorhynchus indicus*, *Pratylenchus* spp.

Symptom

The Juvenile of *Meloidogyne incognita*, enters the tubers through roots and produce typical galls on the roots. In corms and cormels, the infestations lead to irregular wart like projections, which become severe on advanced stages. Under severe infestations, the tuber tissues get dissolved and rot. In later stages, the infection leads to premature multiple sprouting in the field. Severe weight loss of the infected corms occurs due to continuous feeding of tuber tissue by the nematode and also by the moisture evaporation from the wound caused by the nematode infestation. As the damaged tissues dry up, it gives a general dry rot appearance and a number of fungi invade the dead tuber tissue, subsequently.

Pratylenchus spp., popularly known as lesiuon nematode, initially produces lesion on the root, cormels and corms. As the nematode multiply, the lesions spread on the entire surface. Cracking of tuber surface is also noticed. Besides, *Tylenchorynchus indicus* causes severe damage in the elephant foot yam tubers under storage conditions.

Management

No single management approaches is found to be effective for the disease caused by nematodes. So, integrated approaches should be followed for the management of nematode diseases of elephant foot yam.

1. Avoid mechanical injuries of the tubers occurred during harvesting and subsequent transportation of the tubers to market or retailers and pre-harvest infection as the infestation of the roots and tubers by root knot nematode (*Meloidogyne incognita*) also act as a predisposing factor for fungal infection.
2. The infected tissues need to be removed with a sharp knife in such a way that no infected portion is left on the tubers. While removing the infected portion, even the healthy tissues adjourning the infected portions should be removed.
3. Reducing or eliminating overhead irrigation can prevent dispersal of the nematode through water splashing.One should avoid planting infected plants, and it is recommended that new plants is to be planted in an isolated area to monitor the plant for the development of symptoms before transplanting the plant near established plants. All symptomatic plants should be destroyed immediately. Dead plant material should also be handled with caution.
4. Add organic matter to the soil. This will make the water and nutrients more available to the plants and healthier plants will tolerate more nematode damage.

Conclusion

There is a great demand of organically produced tuber vegetables among Asians and Africans living in Europe, USA and Middle East. Tropical tuber crops in general and edible aroids like elephant foot yam in particular, respond well to organic manures. Moreover, there is not much serious pests and diseases of elephant foot yam. Hence, there is ample scope for organic production in these crops. However, more researches are necessary to develop organic disease management practices against the important diseases of this crop.

References

Aycock, R. (1966). Stem rots and other diseases caused by *Sclerotium rolfsii*. North Carolina Agricultural Experiment Station, Tech. Bull. No. 174, 202p.

Behera, K.K., Mahayana, T., Sahoo, S. and Prusti, A. (2009). Biochemical quantification of protein, fat, starch, crude fibre, ash and dry matter content in different collection of greater yam (*Dioscorea alata* L.) found in Orissa. *Nature Sci*., **7**: 24-32.

Brunt, A.A., Crabtree, K., Dallwitz, M.J., Gibbs, A.J., Watson, L. (1996). Viruses of plants. Wallingford, UK: CAB International. p. 1484.

Mukhopadhyay, S.K. and Sen, H. (1999). Effect of *Azotobacter*on corm yield of elephant foot yam. *J. Root Crops*, **25**: 159-62.

Nath, R., Kundu, C.K., Majumder, A., Gunri, S., Biswas, T., Islam, S. K. J., Chattopadhyay, A. and Sen, H. (2007). Seed corm production of elephant footyam [*Amorphophallus paeoniifolius* (Dennst.) Nicholson] through mini corm setts in rainfed laterite ecosystem of Eastern India. *J. Root Crops*, **33**: 30-37.

Nedunchezhiyan, M. (2014). Production potential of intercropping spices in elephant foot yam.(*Amorphophallus paeoniifolius*). *Indian J.Agron.*, **59** : 596-601.

Prihastuti, H. Cai, L. Chen, H. McKenzie, EHC, Hyde, K.D, (2009). Characterization of *Colletotrichum* species associated with coffee berries in northern Thailand. *Fungal Diversity* **39**: 89-109.

Roy, G and Hong, C.X. (2007). The first finding of Pythium root rot and leaf blight of elephant foot yam (*Amorphophallus paeonifolius*) in India. *New Disease Reports* **15**: 48.

11

Fungal and Viral Diseases of Taro (*Colocasia esculenta* L.) and Their Integrated Management

Pranab Dutta[1] **and** ***Munni Das***[2]

[1]*School of Crop Protection, College of Post Graduate Studies in Agricultural Sciences Central Agricultural University, Umiam, Meghalaya*

[2]*Department of Food Science and Technology, Assam Agricultural University Jorhat-785013, Assam*

Introduction

Colocasia esculenta (L.) Schott, also known as taro, belongs to the family Arecaceae. Taro is believed to have originated in Southern or South-East Asia, and to have been dispersed to Oceania through the Island of New Guinea to many centuries ago but cultivated in Asia for more than 10,000 years. Taro is considered as an important vegetable grown throughout India. Taro is herbaceous perennial plants with a large corm on or just below the ground surface and can be grown in the ground or in large containers. They can be grown in almost any temperate zone as long as the summer is warm.

Taro root contains very significant amount of dietary fibre and carbohydrates, as well as high levels of vitamin A, C, E, B6 and folate, as well as magnesium, iron, zinc, phosphorus, potassium, manganese and copper. Thogh negligible, the plant is also known to provide some protein in our diet. As Taro is a great source of fibre and other nutrients, it offers a variety of potential health benefits, including improved blood sugar management, gut and heart health.

Globally, taro is grown in an area of 1.6 m ha producing 11.66 mt with an average productivity of 7.25 t ha^{-1} (FAO, 2009). The plant can be grown in wide range of climatic condition but during growth period taro is prone to attack by at least twenty- three (23) pathogens, of which, only a few causes serious reduction in the potential yield of the crop. Some of the mentionable diseases that commonly occurs and causes yield loss of the plants are as follows:

1. Taro Leaf Blight
2. *Pythium* Root Rot
3. *Sclerotium* Rot
4. *Phyllosticta* spot
5. Dasheen Mosaic Disease
6. Alomae and Bobone
7. Bacterial soft rot

In this chapter, the aetiology of these diseases, symptoms, epidemiology and management approaches are discussed briefly.

Taro Leaf Blight

Taro Leaf Blight is a fungal disease which is a highly infectious. Taro Leaf Blight causes varying losses in corm yield depending on reactivity of the cultivars towards the causal agent of Taro Leaf Blight infection and damage.

Causal organism: Taro leaf blight is caused by *Phytophthora colocasiae*

Epidemiology

Mycelium, zoospore, sporangia, oospore are the inoculum of the fungi belongs to Pythiacae family.Sporangia and zoospores of the pathogen are spread by rain splash and wind- blown rain. They germinate on leaves and petioles or are washed into the soil, where they are able to infect taro corms. Corms left in field after harvest, also serve as inoculums sources for the newly planted taro crops.

Symptoms

a) Symptom of the disease is characterised by the formation of large brown lesions on the leaves of infected taro plants. Lesions are the result of oomycetes leaching nutrients out of the leaves via haustoria to create white powdery rings of sporangia.
b) Initially symptom develops as small, dark brown flecks or light brown spots on the upper leaf surface.
c) The spots enlarge rapidly, becoming circular, zonate and purplish brown to brown.
d) Spots have a dry gray appearance and hard globules of plant exudates occur on the lower leaf surface.

Management

The diseases caused by Phytophthora are very difficult to control. An integrated apaches can help to minimize the diseases successfully.

a) Maintain field sanitation as it decrease inoculums levels early in the season.

b) Adopt wide spacing for cultivation of taro as fungus can be eliminated to a marked degree by increasing the interplant- spacing distance.

c) Protectant chemical sprays containing copper, manganese or zinc should be applied at recommended dose at an interval of 7 days in the favourable environmental condition. Use sticker along with the fungicide suspension during rainy days.

i) Disease can be controlled by spraying with 4-4-50 Bordeaux at 10 to 17 days intervals.

ii) Use of resistant varieties offer the best long term solution to manage Taro leaf blight disease.

Pythium Root Rot

Pythium root rot is probably the most widely distributed disease of the crop. The normally firm flesh of the corm is converted into a soft, mushy, often evil-smelling mass, unfit for human consumption. *Pythium* species are soil borne pathogens and affect the roots and underground parts of many plants. The plants become stunted, with leaf stalk shortened and leaf blades curled and crinkled.

Causal organism: *Pythium myriotylum* is the causal agent of the disease.

Epidemiology

Root rot caused by *P. myriotylum* is most severe in high soil moisture and high temperature (25- 30°C). Under unfavourable conditions, *Pythium* produces resistant spores "oospores", that remain alive but inactive in the soil for many years. They germinate when conditions are suitable and produce swimming oospores which infect the roots of susceptible plants.

Symptoms

a) Infection causes, rotting of corms and drying up of the outer older leaves.

b) With the progress of disease, the number of leaves declines, young leaves are shorter and smaller than usual.

c) Plant can be pulled easily from the soil.

d) Firm flesh of the corm is converted into a soft, mushy, often evil- smelling mass.

Management

Pythium is a soil borne plant pathogen and can survive many years by producing Oospores, so the integrated approaches are best to curb the disease caused by *Pythium* spp.

a) Sites should be chosen carefully.

b) Flooded area should be avoided.

c) Heavy clay soils should be avoided.

d) Soil types where the disease is rare should be used.

e) Fungicides are not recommended for the control of this disease. If a fungicide was required, phosphorus acid could be used.

Sclerotium Rot

Upland taro may be attacked by *Sclerotium* rot. Almost all varieties of taro are prone to this disease. The disease is generally seen on over mature plants or plants suffering from inadequate irrigation.

Causal Organism

This disease is also caused by soil borne sclerotia producing plant pathogen *Sclerotium rolfsii.*

Epidemiology

Sclerotium rolfsii, persists in the soil as mycelium and commonly as sclerotia. Sclerotia are mustard seed like light brown to dark brown coloured. Sclerotia germinate in the presence of moisture to produce mycelium. As it is a necrotrophic pathogen it causes death of plant cell in advancement of their penetration and causes rotting young and old roots and over mature corms.

Symptoms

a) Affected plants are usually stunted.

b) On uprooting, the lower portion or the corm is seen to be partially rotten away.

c) Corms are covered with numerous, small, almost spherical, lemon- yellow to dark brown bodies, called sclerotia.

d) The rotted tissue is ocherous to brown in colour, soft but not watery.

Management

a) Ploughing infested field to a depth of 4 to 8 inches reduces infection in the next season.

b) Infected plants and the surrounding soil should be removed and burned.

c) The taro should be harvested before it becomes over mature.

d) Soil drenching with Carbendazim based fungicides @ 0.2%. Apply 1-3 litre of fungicide solution per plants.

e) Apply *Trichderma harzianum* based liquid biofertilizer @ 0.5% surrounding the plants at the time of planting.

f) Soil application of *Org-Trichojal* enriched biopesticides @100 g per plants.

Phyllosticta Spot

This leaf spot occurs exclusively on upland taro, occasionally becoming serious. Losses are due to destruction of parts of the leaves. The disease is influenced by temperature but more particularly by rainfall.

Causal Organism

Phyllosticta colocasiophila is the fungal pathogen that causes the disease.

Phyllosticta colocasiophila produces spores in specialised fruiting bodies called pycnidia. Spores ooze from the pycnidia in wet weather and are splashed from plant to plant, where they germinate to reproduce the disease. The mycelium of the parasite probably also lives in the soil on decaying taro material.

Epidemiology

The disease is influenced by temperature but more particularly by rainfall. Cloudy, rainy weather, accompanied by cool winds, is conductive to infection. Taro plants probably lose only three to five leaves per season from this disease.

Symptoms

a) The pathogen produces spots on both surfaces of the leaf.

b) The spots on the leaf vary from a quarter of an inch to one inch or more, and are oval or irregular in shape.

c) The young spots are buff- coloured and older spots are dark brown in colour.

d) The diseased areas become rotten in the later stage.

Management

a) The collection and burning of diseased leaves seem to be of practical value.

b) *Phyllosticta* spot can be controlled by covering the leaves with a fungicide, which will prevent germination of spores landing on the foliage.

c) Spray 1% *Bordeaux* mixture at the initiation of the disease.

Dasheen Mosaic Disease

Dasheen mosaic virus, a flexuous rod 0f 750 nm, was initially described in 1970 as a polyvirus infecting members of the Arecaceae. It is a stylet borne virus carried by aphids. Plants generally become asymptomic three to four months after initial symptom expression.

Causal Organism: Dasheen Mosaic Polyvirus (DsMV)

The disease is transmitted byAphids or green flies, spread Dasheen Mosaic Virus (DsMV). Spread occurs when winged adults fly from diseased to healthy plants. The aphids suck up the viruses as they feed on infected plants, and the viruses stick to their mouthparts. The aphids then fly or are blown by wind to healthy plants. As the aphids feed again, the viruses enter the plant.

DsMV is also Spread in Other Ways

a) When infected plants are taken to new garden.

b) From mother plants to suckers.

c) When healthy leaves comes in contact with sap of infected plants.

Symptoms

a) Light green streaks appear along the main leaf veins or between them.

b) Streaks from different patterns,- some are light green, feather- like patches between the veins, others are over the main veins throughout the leaf.

c) Leaves of taro may be distorted where DsMV patterns occur, especially at the leaf margins.

Management

a) Plants with severe symptoms should be removed and destroyed.

b) If aphids are present, a rice bag should be put over the plant before pulling out and tip the plant into a fire to destroy the insects.

c) Spray *Metarhizium anisopliae* based bioformulation like Org-Metajal @ 0.5% as and when first pospulation of aphid or green fly is observed in the field.

Alomae and Bobone

Gollifer and Brown (1972) for the first time described two virus diseases from the Solomon Islands. Alomae, a disease apparently caused by two Bacilliform viruses, results in the death of susceptible cultivars. Early symptoms of Alomae are usually a conspicuous feathery mosaic of the leaves. The main difference between these two diseases are that, alomae is a lethal disease that affects nearly all varieties of taro whereas bobone does not kill taro; it reduces the yield by about 25 percent. Bobone only occurs in a few varieties of taro.

Causal Organism

Alomae and Bobone are diseases of taro caused by virus called rhabdovirus.The large virus particle present in plants with Alomae and Bobone is spread by planthopper *Tarophagus Proserpina,* and the small particle by the mealy bug *Planoccoccus citri.* When infected corms are planted, diseased shoots develop. Multiplication of the virus with reappearance of symptoms is often triggered when these taro are replanted.

Symptoms

Bobone

a) The petiole of the affected leaf is shorter than normal with irregularly shaped outgrowth, usually at the base.

b) The leaf blade hangs downward, is thick, extremely puckered and the veins are enlarged.

Alomae

a) Petioles are short, enations are present.

b) The main veins are often enlarged.

c) Leaves are small, remain rolled and slightly twisted.

d) Leaf production ceases and the plant rots and dies.

Management

a) **Destruction of diseased plants**

  i) For both bobone and alomae, removing infected plants can give good control.

  ii) Infected plants should be pulled out and destroyed, burnt or deeply buried.

b) **Cultural methods**

  i) New plants should not be situated next to infected plants; otherwise infestations of plant hoppers will be too high for rouging to be effective.

  ii) All debris should be collected after harvest, burnt and buried.

c) **Resistant cultivars:** Only the lower yielding "female" taro are immune to alomae.

d) **Chemical control:** Insecticides can be used against the insects that spread the diseases.

e) **Biological control:** *Cyrtorhinus fulvus,* an egg- predator of *Tarophagus* species, can be introduced to areas where alomae and bobone are endemic.

Bacterial Soft Rot

Bacterial soft rot is a destructive disease of fruits, vegetables and ornamentals found worldwide, and affects genera from nearly all the plant families. The gram negative plant pathologic bacteria causing soft rot mainly attack the fleshy storage organs of their hosts.

Causal Agent: *Erwinia carotovora* is the bacteria that causes the disease.

Epidemiology

Growth of the bacteria is possible between 32-90^{0}F, with the most ideal conditions between 70- 80^{0}F. Disease spread can be caused by simple physical interaction between infected and healthy tissues during storage or transit. The disease can also be spread by insects.

Symptoms

a) *Erwinia* causes infection in wet soil condition throughout the growing season.

b) Initially starts infection on the tender surface later in advance stages, it causesrots are often strong smelling, white soft rots.

c) Rot begins on the tuber surface and progress inward.

Management

Cultural Control

a) Harvest the corms leaving the "tops" and suckers attached, but remove the leaves, trimming the leaf stalks to about 10 cm.

b) Remove large soil clods, but do not wash the corms.

c) Maintain good drainage facility.

d) Use healthy resistant varieties.

Chemical Control

Other than the use of bleach, chemicals cannot be recommended for the control of rots.

Conclusion

Beside the aforementioned diseases of taro, a number of minor diseases has been reported which are nowadays may become a emerging diseases of the crop. So, during the formulation of integrated management strategies proper care must be taken to curb the multiple problems with least cost and less damage to the environment and with sustainable production.

References

Babu, B., Hedge, V., Makeshkumar, T. And Jeeva, L. M. (2011). Detection and identification of *Dasheen mosaic virus* infecting *Colocasia esculenta* in India. *Indian J. Virol.* **22(1)**: 59-62. Doi: 10.1007/S13337-011-0030-7.

Banjaw,D.T. (2017). Review of Taro (*Colocasia esculenta*) genetics and breeding. *J. Hortic.* **4(1)**: 1-4. Doi: 10.4172/2376-0354.1000196.

Das, S.R. (1997). Field efficacy of fungicides for the control of leaf blight disease of taro. *Journal of Mycology and Plant Pathology* **27(3)**: 337-338.

Dervis, S., Soylu, S. And Serce, C.U. (2014). Corm and root rot of Colocasia esculenta caused by Ovatisporangium vexans and R. Solani. Romanian Biotechnological letters. 19: 9868-9874. http://www.fao.org/3/ac450e/ac450e03.htm.

Hu, J.S., Meleisea, S., Wang, M. et al. (1995). Dasheen mosaic potyvirus in Hawaiian taro. *Australian Plant Pathology.* **24**: 112-117.

Khatoon, A., Mohapatra, A. And Satapathy, K. B. (2016). Fungi associated with storage rot of *Colocasia esculenta* L. Tubers in Bhubaneswar city, Odisha. *British Microbiology Research Journal.* **12(3)**: 1-5. Doi: 109734/BMRJ/2016/23137.

Mishra, R. S., Sharma, K. and Mishra, A. K. (2008). *Phytophthora* leaf blight of taro *(Colocasia esculenta)*- a review. *The Asian and Aus. Jour. of Plant Sc. and Biotech.* **2(2)**: 1-60.

Mishra, S. R., SK, Maheswari., Sriram, S., Sharma, K. and AK, Sahu. (2017). Integrated management of *Phytophthora* leaf blight disease of taro *(Colocasia esculenta* (L.) Schott). *Journal of Root Crops.* **33**: 144-146.

Nelson, C. S. (2008). Dasheen mosaic of edible and ornamental aroids. *Plant disease.* **44**: 1-8.

Nishimura, N. and Kudo, K. (1994). *Fusarium oxysporum* f. sp. *Colocasiae* n. f. sp. Causing dry rot of taro *(C. Esculernta). Ann. Phytopath. Soc. Japan.* **60**: 448-453.

Ooka, J. J. (1990). Taro diseases. *University of Hawaii*. p. 51-59.

Tomlinson, D. L. (1987). A bacterial leaf disease of taro *(C. esculenta)* caused by *Xanthomonas campestris* in Papua New Guinea, *Tropical Pest Management,* **33:4**, 353-355, Doi: 10.1080/ 09670878709371181.

Widodo, W. and Supramana, S. (2011). *Fusarium* sp. Associated with corm rot of taro in Borgor. *Microbiology Indonesia.* **5**. 132-138. Doi: 10.5454/mi.5.3.6.

Yusop, M. S. M., Saad, M. F. M., Talip, N., Baharum, N. S. and Bunawan, H. (2019). A review on viruses infecting taro *(C. esculenta* (L.) Schott). *Pathogens* **8 56**: 1-13. Doi: 10.3390/ pathogens 8020056.

12

Diseases of Bean (*Phaseolus vulgaris* L.) and Their Integrated Management

Tulika Medhi

Sugarcane Research Station, Assam Agricultural University, Assam, India

Introduction

India is the second largest producer of vegetable in the world after China. Every year diseases cause economic losses (8-23%) and also post-harvest losses (20-25 %) in vegetable crops. Bean (*Phaseolus vulgaris* L.) is a nutritionally important and remunerative crop grown throughout the world and is consumed both as a vegetable, shelled green beans as well as dry beans. It is known commonly known as french bean, common bean, kidney bean, field bean, snapbean, dry bean etc on the basis of its variety and edible purpose. The crop is infected by a number of diseases. Common bacterial blight, Anthracnose, White mold, Bean Common Mosaic, Bean yellow mosaic, Web Blight, Halo blight, Dry root rot and wilt, Bean Rust are some of the major diseases.

Bacterial Blight of Bean

Three types of bacterial blight are found to occur in beans *viz*. Common bacterial blight or Bacterial blight, Halo blight and Bacterial brown spot. These three diseases affect leaves, pods, stems and seeds in similar manner such that it is usually impossible to distinguish the diseases from one another based on the visual symptoms. While common blight is usually more prevalent in warmer weather, the other two occur more in cool weather.

Common Bacterial Blight/Bacterial Blight

Common bacterial blight (CBB) affects foliage and pods and is a major problem of beans worldwide. The disease is particularly severe in warm, humid climates with high levels of rainfall and causes losses in both yield and seed quality. The disease is reported to occur across the world in many countries including USA, many states of Australia, European countries like Italy, France, Netherlands, Greece, Latin American countries Chile, Brazil, Asian countries like Bangladesh, India, China, Japan and North and South Korea as well as African countries. The disease is of major economic importance in most lowland tropical and subtropical countries resulting in yield losses and reduction in seed quality through discolouration of infected seed. When young pods are infected, they drop off soon but infection on older pods results in deformity and blemishes and consequent reduction in their market value. Infected young seedlings do not develop properly and remain stunted and weak. In severe cases the whole plant may be completely destroyed. Yield losses in beans due to common bacterial blight are estimated to range between 10-45 per cent in susceptible genotypes. Common bacterial blight was the most economically important bacterial disease in the USA, causing an estimated loss of 4 million US dollars in 1976.

Causal Organism: *Xanthomonas axonopodis* pv. *phaseoli* (Smith) Vauterin *et al.* (*Xap*) 1995

Synonym: *Xanthomonas campestris* pv. *phaseoli* (Smith 1897) Dye 1978b

Host Range

Common bean (*Phaseolus vulgaris*), Tepary bean (*P. acutitfolius*), soybean (*Glycine max*), *Lablab bean* (*Dolichos lablab*), Common Lupine (*Lupinus polyphyllus*), Georgia velvet bean (*Stizolobium deeringianum*), Fuzzy bean (*Strophostyles helvola*), Moth bean (*Vigna aconitifolia*), Azuki bean (*V. angularis*), Black gram (*V. mungo*), Mung bean (*V.radiata*) and Cowpea (*V. unguiculata*).

Symptoms

Common bacterial blight of bean shows typical leaf spot and leaf blight symptoms and affects leaves, stems, pods as well as seeds.

Symptoms on Leaves

The symptoms on leaves appear first as flaccid, small water-soaked spots, usually more evident on the underside of the leaves as well as leaflets. These small, translucent and water-soaked lesions often enlarge to 10mm/1 inch or more and form dark brown lesions along the edge of the leaflets. Lesions coalesce into irregularly-shaped areas and later develop dry brown centers. A narrow bright lemon-yellow margin of tissue often encircles these dried and necrotic lesions. The whole leaflet and large portions of the foliage can be infected. Lesions may be found in both interveinal areas and along leaf margins. In severe infections, leaves remaining attached to plants, give a burnt appearance to foliage and premature defoliation may result. However, in some cases, dead leaves may remain attached to the plant up to maturity time.

Symptom on Stems

Lesions on stems have a propensity to be more restricted and sunken. With continued development of the disease, the vascular system may also turn brown forming surface cankers on the stem. On the other hand, stem girdle or joint rot may develop at the cotyledonary stage, above the cotyledonary node especially in plants that develop from infected seed. Severe infections from seed may cause stem girdling which causes the stems to break.

Symptoms on Pods

Lesions on pods also tend to be restricted and sunken. Pod symptoms can appear with little foliar damage. Symptoms on infected pods develop as large, water-soaked spots which are slightly circular and sunken. Lesions vary in shape and size depending on pod age. As symptoms progress, pod lesions may become distinctly brick-red in color and more deeply pitted . Under high humid conditions, pod lesions become covered with yellow coloured bacterial exudates or ooze that dry to a yellowish, crusty mass. Pod lesions sometimes are surrounded by a reddish margin.

Symptoms on Seeds

Infections of the seeds are usually not deep, but primarily in the surface to subsurface regions of the cotyledons. Usually, the seeds produced in lightly infected pods may not show any signs of the disease and develop normally or in some cases only small, shiny, yellow discolorations are found on the seeds. These yellow discolorations may be very small spots and follow the veins of the seed coat or envelop the entire seed. Infected seeds provide the primary inoculum which causes the characteristic spots on stems, leaves, and cotyledons of seedlings. Early infection during pod and seed development may cause the seeds to rot or shrivel and wrinkle, and the bacteria may cause yellow to orange discoloration under the seed coat of infected seeds. However, some seed may appear healthy.

If sown, such seed show signs of poor germination and vigor after emergence. Seedlings that develop from infected seed may incur damage to the growing tip and either dies or remain stunted. It has also been reported that sometimes seeds have no visible symptoms and can germinate vigorously, while supporting large symptomless epiphytic communities of the pathogen. The disease also reduces seed size and quality.

Disease Cycle

Contaminated or infected seeds are the major source of bacteria introduced into new fields of bean crops. Other than that, common bacterial blight can also arise when the crop is planted in fields with previous occurrence of the disease, and when climate is consistently hot, wet and humid. The bacteria may survive in infected crop debris which has not been destroyed by decay in the soil from season to season. It can also survive and multiply as an epiphyte or resident on the shoot surfaces of weed hosts, predominantly in members of the legume family. Alternative weed hosts, without showing symptoms, may also serve as sources of inoculum for the pathogen.

The bacterium can be sheltered both within and on the seed coat. Bacteria infect the seed by passing through the vascular system of the pedicel of the diseased pod and then to the funiculus, through the raphe leading into the seed coat or by growing from an external infection of the pod through the parenchyma, and into the conducting tissues. The pathogen either remains in the seed coat or passes to the cotyledon when the seed germinates. There are no reports of direct penetration through the seed coat. When the bacteria enter through the funiculus, only the hilum may become discolored. Bacteria are carried on and within the infected seed and contaminate the surface of the expanding cotyledons when the seed germinates causing primary infection. They progress through the parenchyma and into the vascular tissues, migrate through the large xylem elements, break out into parenchyma from place to place, and give rise to leaf lesions and stem cankers. The bacteria may enter the stem through the stomata of the hypocotyls and epicotyls and reach vascular elements from infected leaves or cotyledons. The bacteria that exit through the stomata provide inocula for secondary spread enter and bean plants through openings, such as stomata in leaves and other plant organs and through hydathodes at margins of leaves. Continued spread of secondary inoculum from primary lesions is brought about by rainfall, irrigation water, implements, and insects like whiteflies, leafminers, and beetles. Moisture and temperature plays an important role for infection as well as disease development and progression.

Epidemiology

Common bacterial blight is favored by warm temperatures. It causes greater damage to plants at 28-32°C than at temperature lower than 16°C. Symptoms usually appear on middle-aged to older leaves, even at low humidity. The time between initial infection and production of inocula for secondary spread is 10-14 days. High temperature, rainfall, and humidity favour rapid disease progress in the field. Severe disease outbreaks often occur within 7-10 days after a period of humid, rainy weather. Violent storms with hail and high winds cause plant wounding which enable pathogens to enter and infect plant tissues. When crop canopies become saturated by rain, mist or dew, large drops may form on the leaves. These large raindrops are the most efficient in the dispersal of inocula by rainsplash. However, it is reported that rainsplash only accounts for short distance dispersal- from leaf to leaf of the same plant or neighbouring plant. Rain-generated aerosols may have greater potential for transporting bacteria over modest distances.

The pathogen is also spread by windblown rain, soil and plant debris, contact between wet plant leaves, irrigation water, animals, and insects like leafminers, grasshopper, beetle and whiteflies especially when foliage is wet. Leaf-chewing insects are more efficient disseminators of the bacterium than sucking insects. Planting non-certified and contaminated seed also contributes to serious disease outbreaks due to the seed-borne nature of these pathogens.

Management of Bacterial Leaf Blight

High disease incidence of common bacterial leaf blight and severity result from a combination of susceptible genotypes, contaminated or infected seed and adequate rain and wind to spread inocula over a extensive area. An integrated approach consisting of the following strategies is need of the hour to effectively manage the disease.

Cultural Management

Crop rotation, use of pathogen-free seed, choice of production site, use of clean seed and field hygiene are some approaches of cultural control. A two year rotation with non-legumes is recommended. Seeds are to be produced in a region free of the pathogen or where environmental conditions are unfavourable for disease development. An ideal production site should have less than 300mm annual rainfall, daily relative humidity less than 60% and mean daily temperature less than 25^{0}C with a gravitational irrigation facility.

Infected bean residues should be destroyed by burning or burying by deep ploughing. However, in minimum or zero tillage operations, deep ploughing may not be a feasible. Diseased plants should be rogued or basal leaves should be removed to reduce sources of inocula for the secondary spread of the pathogen. Rogued plants should be buried, burnt or composted. Suspending field operations when leaves are wet reduces chances of field spread.

Different cultivars should be grown in alternating seasons to reduce large acreage of susceptible plants and lower pathogen population build up.

Chemical Control

Seed treatment or foliar applications before moderate or severe infection can be done using various chemicals. Copper based chemicals like copper sulphate, copper hydroxide, copper oxychloride and potassium methydithiocarbamate can control foliage infection.

Streptomycin and kasugamycin have been used to control external contaminant bacteria. Antibiotics like streptomycin should, however, not be foliarly applied as resistant bacterial mutants may be induced. Seed dressing with fungicides like quintozene, thiram and/or carboxin before planting is practiced in Zimbabwe to control contaminant pathogens.

An integrated management strategy of seeds treatment with streptomycin (50,000ppm), garlic extract and moringa leaf extract (10-1 dilution) combined with foliar spray of Copper hydroxide (77% WP) is reported to effectively manage the disease in Ethiopia.

Use of Resistant Cultivars

Using disease-free seed, chemicals and crop rotation short-term control is possible but long term control depends on the development of disease resistant cultivars. Breeding for resistance is the most effective control measure under the smallholder farming sector where farmers retain seed for the subsequent cropping cycles. In Zimbabwe, the cultivar Mkuzi is tolerant to the disease.

Integrated Disease Management

Integrated disease management uses all types of management to keep disease pressure below the economic threshold level. Integration of recommended strategies *viz.* crop rotation, sanitation, use of treated or healthy seeds, resistant or tolerant varieties, stress and wound avoidance and proper bactericide scheduling must be practiced to minimize the impact of the disease on bean. Use of resistant varieties supplemented with proper cultural practices and chemical seed treatment could be the best alternative options in managing common bacterial blight of bean and avoiding yield losses. Integrating of resistant varieties with chemical seed treatment and cultural practice were highly significant in reducing common bacterial blight development and increased seed yield and yield component of a bean in Ethiopia.

Halo Blight

Halo blight, which is considered to be a low-temperature disease, is regarded as a major problem throughout the world and is most destructive where moderate temperatures exist during bean production. It causes most devastation in developing countries. Halo blight is considered to be such a serious disease in United States, that the law of Idaho State demands that any crops found to contain the bacterium causing halo blight must be destroyed. Reductions in total yield have been reported upto 43% and further losses occur owing to the poor quality of infected pods. Yield loss potential is greatest at temperatures of 18 to 20°C.

Causal Organism

Pseudomonas savastonoi pv. *phaseolicola* (Burkholder,1926) Gardan *et al.*,1992.

Synonym

Pseudomonas syringae pv. *phaseolicola* (Burkholder, 1926) Young *et al.*,1978.

Host Range

Primary hosts are common bean (*Phaseolus vulgaris* L.), lima beans (*P. lunatus*). Others are scarlet runner beans (*P. polyanthus*), tepary bean (*P. acutitfolius*), soybean (*Glycine max*), Azuki bean, Black gram(*Vigna mungo*)), mung bean (*V. radiata*), purple bushbean (*Macroptilium atropurpureum*), kudzu (*Pueraria montana*).

Symptoms

The disease is frequently detected early in the season during early to mid-vegetative stages of plant growth. Typically, the pathogen most commonly infects young foliage and symptoms are rarely seen on older leaves. Pods and seeds are also infected.

Symptoms on Leaves

The first symptoms of infection are small, water-soaked lesions resembling pinpricks on the underside of leaves. The small lesions rapidly turn dry and reddish-brown, and are visible on both sides of the leaf. A broad yellow-green halo then develops around necrotic spots, effectively distinguishing this disease from the narrow, bright-yellow border characteristic of common bacterial blight. Further, the necrotic spots generally remain very small, unlike that of common bacterial blight. Yellow-green chlorosis becomes more pronounced at temperatures of 18 to 23°C due to the production of phaseolotoxin by the pathogen. When temperatures are above 23°C, toxin production often decreases and chlorotic symptoms become less noticeable. In severe cases, a general systemic chlorosis may develop in infected plants. Symptoms can spread throughout the plant, leading to yellowing and death of new foliage.

Symptoms on Pods and Seeds

Symptoms on pods begin as tiny, water-soaked spots or streaks along pod sutures that steadily enlarge to form dark, sunken lesions. If lesions expand into pod sutures or penetrate young pod walls, seeds can be discolored, shriveled, and small. The halo blight pathogen can be present in bean crops and on bean seed even without visible symptoms.

Bacterial ooze may emerge from the lesions of stems, pods, or leaves 7 to 10 days after infection giving lesions a greasy, water-soaked appearance. Bacterial ooze in centre of spots is cream to silver coloured which is distinctive and different from the yellow exudate found in symptoms of common bacterial blight pathogen.

Disease Cycle

Infected seeds serve as source of primary inoculum as pathogen survives in seeds from the previous year. The pathogen can also overwinter in previously infected bean debris, contaminated seeds, weed hosts or volunteer beans. The bacterium can enter the plants either through plant injuries or the natural openings in plants such as stomata and hydathodes. After 7 to 10 days of infection, bacteria ooze from lesions which serve as secondary inocula. The bacterium can spread from plant to plant and to nearby fields in windblown rain or irrigation water and soil and plant debris. During periods of high humidity or when the foliage is wet the pathogen can be dispersed by contact between wet leaves, rainfall, irrigation or man and animals moving through infested fields.

Epidemiology

Halo blight disease outbreaks are most serious when temperatures are moderately cool (below 26°C) and humidity is greater than 95 percent for 24 hours or longer. The optimal temperature for the bacterium to thrive is 20-23°C. High relative humidity or free moisture facilitates entry of the pathogen into the plants through wounds or stomata and hydathodes. Moist environments also allow the spread of the disease. Above 28°C, usually symptoms do not develop even though some water soaked spots might be present. Windblown rain splashes allow the disease to spread. Contact of diseased plant from a human or animal can spread the disease to a whole new environment and introduce the pathogen to new hosts.

Management

Management of halo blight is very similar to the control for common bacterial blight. Several management strategies can be followed:

- Certified disease-free seed produced in arid regions unfavorable for development of bacterial diseases should be planted.
- Sprinkler irrigation should be avoided since this can provide the needed moisture and humidity for halo blight. Instead furrow irrigation is recommended.
- In fields that have been infected by halo blight, infested debris should be deep ploughed and a 2 to 3-year rotation with a non-legume crop should be practiced.
- Once plants are infected, it is difficult to eradicate the bacterial pathogen. Copper-based bactericides should be applied before appearance of symptoms to reduce growth of bacteria on foliage and spread of bacteria to healthy foliage and pods. This should be repeated at 7 to14-day intervals.

White Mold

White mold is among the most devastating and widely distributed fungal diseases of both green and dry common bean. It is endemic in the cool humid highlands with wet years. It affects the plant in all stages of growth, i.e. seedling, matured and harvesting stages. Under favourable weather conditions, crop losses due to white mold disease outbreaks go upto 100% in seed yield and on susceptible common bean cultivars. Amount of losses varies widely from year to year and from location to location depending upon factors such as crop species/cultivars, locations and environmental conditions. The disease was reported to reduce seed yield by an average of 12 kg per hectare in Pinto (a variety of dry bean) in North America and by 23 kg per hectare in Navy bean in rain-fed production systems in North Dakota for every 1% increase in white mold incidence.

Causal Organism: *Sclerotinia sclerotiorum* (Lib.) De bary

Host Range: *S. sclerotiorum* is omnipresent and has a very wide host range. It is pathogenic to more than 500 plant species various developmental stages of plants. The pathogen poses a hazard to many dicotyledonous crops such as rapeseed and mustard, sunflower, soybean, chick-pea, peanut, dry pea, lentils and various vegetables and ornamentals. It is also reported to infect mono-cotyledonous species such as onion, tulip etc.

Symptoms

White mold disease makes its first appearance in the field at about 45-55 days after sowing. The disease may occur on all above ground plant parts including stems, leaves, petioles and reproductive organs.

The first field symptom of the disease appears as irregular, water soaked spots on basal portion of the stems near the soil line. Eventually, symptoms appear on leaves, branches and pods. Lesions on pods, leaves, branches, and stems are initially small, circular, dark green, and water soaked but rapidly increase in size, may become slimy, followed by a soft watery rot of affected plant parts and may eventually encompass and kill the entire organ. Under moist conditions, these lesions may also develop a white, cottony growth of external mycelium. Affected plant tissue dries quickly and bleaches to pale tan or almost white colour that is contrasting to the normal light tan color of senescent tissues. Entire branches, stems or the entire plant may be killed. The epidermis easily sloughs off when the stem or pod is rubbed. Colonies of white mycelium (immature sclerotia) develop into hard, black sclerotia in and on infected tissue which is the best diagnostic feature of the disease. Sclerotia, resembling mouse droppings are most commonly found on the outer surface of the diseased tissue, but sometimes inside of soft host tissues or cavities such as floral receptacles, fruits and the pith of stems.

Disease Cycle

Sclerotinia sclerotiorum overwinters in the soil as sclerotia that can persist there upto 5-8 years. Under cool (10-20°C), moist conditions, sclerotia in the top 2 to 4 inches of soil germinate. *S. sclerotiorum* is capable of reproducing both asexually (myceliogenic germination) and sexually (carpogenic germination). When germination is myceliogenic, the hyphae produced by sclerotia can directly infect plants, while in carpogenic germination, sclerotia produce apothecia and subsequently ascospores. Apothecia develop rapidly on sclerotia located at the surface or near the surface of the soil.

The apothecia eject ascospores that are then dispersed by wind throughout a field. Prolonged periods of leaf wetness (16 to 48 hours) with temperatures in the range of 12 to 24°C promote ascospore germination and infection. Sporulating apothecia can persist 5 to 10 days. Ascospores require a nutrient source in order to initiate infection; in fruiting crops, this source is usually the senescing tissue of a flower or leaf. The fungus then invades healthy stem or fruit tissue. Disease development occurs most rapidly at 20-25°C, especially in the presence of moisture. As the disease progress, sclerotia are formed either on the plant surface or within stems and other plant parts. As the plant or plant part dies, the sclerotia fall to the soil where they can survive for multiple years.

Epidemiology

Infection of the fungus may either be via germinating sclerotia or ascospores. Sclerotia should be conditioned at low temperatures for some time to overcome dormancy and germinate carpogenically. The optimum temperature range for germination of ascospores is 10 to 20°C. The congenial temperature for disease development lies between 20-25°C and relative humidity of 95-100%. White mold is a monocyclic disease, having only a primary cycle of infection, and once the host tissue is diseased, it is not transmittable within the same season.

Severity of the disease in a field arises from the sclerotia that are overwintered within the field soil. Under conditions favourable for sclerotia germination, presumably all sclerotia in the top 2-4 inches of soil will germinate. During carpogenic germination of sclerotia, ascospores are released into the air typically at ground level under closed canopies, and consequently, sharp foci of infection are often observed within the field. However, occasionally infections initiated by wind-disseminated ascospores from another location have been observed. *S. sclerotiorum* infection and mycelium development is maximized in the presence of free water on the plant surface. Introduction of the pathogen from outside sources also may occur by contaminated machinery, mass movement of water, and by wind-blown crop debris. Internally infected seeds with mycelium of *Sclerotinia*, or contaminated with sclerotia, several broadleaf weed hosts also may contribute to a persistence of sclerotia within a field.

Management

Diseases caused by *S. sclerotiorum* are difficult to control because of the long term persistence of sclerotia in the soil and the production of air-borne ascospores. Management of *S. sclerotiorum* are done at several stages of crop development. Successful disease control usually requires integration and implementation of multiple methods.

Cultural Control

White mold is more severe where the plant canopy is dense. Hence, reduction in canopy density can be achieved by increasing row width, plant spacing, cultivar selection, and application of balanced nitrogen. Excessive irrigation after petal fall should be avoided and irrigation should be timed so as to allow plants to dry before nightfall. Crop rotation is of limited value because of the wide host range of the pathogen. If feasible, rotation should be practiced with non-host crops like grasses and cereals for upto 8 years. Broad leaved weed species should be removed so as to eliminate potential hosts. Field flooding during warm temperatures destroys sclerotia.

Chemical Control

Fungicide should be applied prior to disease development for maximum benefit. In general, fungicide application should begin when 100 percent of the plants have their first open blossom. Satisfactory results have been obtained when only 50 percent of the plants have their first open blossom or when plants have pods up to 1 inch long. Fungicides like Thiophanate-methyl, Procymidone, Pencycuron and Tolclofos-methyl can be applied.

Biological Control

Several fungi have been shown to be parasites of sclerotia of *S. sclerotiorum*. One of these organisms, *Coniothyrium minitans*, has been released as a commercial product for suppression of this disease. The product should be incorporated into the soil and is best applied 3-4 months before crop planting. In practice, dried spores of this parasite are sprayed onto pathogen-infested crop debris either at the end of a season, or onto the soil surface before planting. This treatment is most useful when combined with other components of an integrated disease management program. Several species of *Trichoderma* are also found to be effective in reducing disease and increasing yield. The most promising method seems to be the application of the antagonistic fungi on organic carriers which extend their persistence in the soil.

Host Resistance

White mold resistance from small-seeded white bean has been successfully transferred into pinto bean. Similarly, white mold resistance from *P. coccineus* and *P. costaricensis* has been introgressed into common bean.

Integrated Disease Management

- Soil application of Contans (*Coniothyrium minitans*) and Sumisclex (Pocymidone) mixture was found to be highly effective in suppressing the white rot disease incidence that produced 100% survival of plants.
- Thorough an ploughing of field upto a depth of 24cm followed by flooding for 4 weeks and subsequent application of Trichoderma harzianum was successful in reducing the disease by almost 90%.

Bean Anthracnose

Anthracnose of bean is one of the most important diseases of legume crops. It distributed worldwide, but is more serious and causes more loss in temperate and subtropical zones than in tropics. Yield losses can reach 100% when contaminated seeds are planted and environmental conditions remain favourable for disease development for long periods. Marketing losses are attributed to seed spots and blemishes, which lower their quality rating and saleability.

Common bean and tepary beans are very susceptible to this disease. Scarlet runner bean, lima bean, and mung bean are somewhat susceptible.

Host Range

Lima bean (*Phaseolus lunatus* L.), scarlet runner beans (*P. coccieus*), tepary beans (*P. acutitolius* var. *latifolius* L.), Mung bean (*Vigna radiata*), cow pea (*V. unguiculata*), kudzu beans (*Dolichos bitloris* L.), and broad beans (*Vicia faba* L.), soybean (*Glycine max*), Pea (*Pisum sativum*) and black gram (*V. mungo*).

Causal Organism: *Colletotrichum lindemuthianum* (Sacc. & Magn.)

Symptoms

The disease affects all aerial parts of the bean plant. Leaves, stems and pods of bean plants are susceptible to infection.

Symptoms on Seedlings

Symptoms on the seedlings are manifested by rust coloured specks in the hypocotyl, which enlarge longitudinally resulting in sunken lesions. Diseased areas may girdle the stem and kill the seedling. The fungus may destroy one or both of the cotyledons. Severely infected cotyledons senesce prematurely, and growth of the plants is stunted. Under moist conditions, the lesions are filled with a pinkish to orange coloured exudates, which consists of masses of spores. Spores produced on cotyledon and stem lesions may spread to the leaves. Infected hypocotyls may rot and collapse, resulting in a poor plant stand.

Symptoms on Leaves

Leaf symptoms are often not obvious and may be ignored when inspecting bean fields. Early signs of infection usually appear on the underside of the leaf surface along the veins as linear brick red to purplish red discoloration. As the disease progresses, the discoloration also appears on the upper leaf surface. However, lesions are more common on leaf petioles, the lower surface of leaves, and leaf veins. In petioles, the symptoms are expressed as elongate, somewhat angular, brick-red to purple spots that soon turn dark-brown to black. Simultaneously, brown lesions of various sizes, with black, brown, or purplish red margins, develop around small veins. Throughout the disease progression, vein necrosis appears first, followed by wilting and bleaching of the leaflet tip. Bleaching usually occurs first at the tip of the leaflet before spreading over the margin and finally over the centre of the blade.

Symptoms on Stems

Stem infection is exhibited by dark brown eyespots which develop longitudinally along the stems. Numerous lesions may girdle the stem to the point where it cannot support the top of the plant. The stem may snap away when the eyespots enlarge in young seedlings, but in older stems, the eyespots enlarge to an approximate length of 5-7 mm. The lesions often become sunken with a cankerous center.

Symptoms on Pods

Pod infection is preceded by stem infection. On the pods, symptoms begin with small tan to rust-coloured spots. The most prominent symptoms on pods are small brown flecks on these rust coloured spots. The spots rapidly enlarge and usually reach a diameter of 5-8 mm and become slightly sunken surrounded by a slightly raised black ring with reddish brown coloured border and greyish black centre. In the centre of the spots often a tan to pink coloured exudes, consisting of spores, is often visible as a viscous droplet in humid conditions. The exude dries into greyish brown to black granules. Severely affected pods may shrivel and dry and the seeds they carry are usually infected. Infected seeds have brown to black blemishes and sunken lesions.

Symptoms on Seeds

The fungus can invade the pods and infect the seed coat of developing seed. On the seeds, large or small yellowish-brown sunken spots may form, which turn brown to black. In highly infected seed, the lesions may extend into the cotyledons.

Disease Cycle

Although not all infested and infected seeds are capable of transmitting the disease, seed infection is a major primary inoculum source of bean anthracnose transmission to the next crop generation and provides conditions which enable the fungus to survive unfavourable weather conditions. Seeds also play an important role in the long distance distribution of the pathogen. The fungus remains alive as long as the seed remains viable. The fungus also survives between two crop seasons in infected crop debris and can be disseminated in air and water. The fungus can survive on contaminated seed and on crop debris for at least two years.

Initial infection can take place anytime during the growing season during cool, wet weather. Secondary infections can occur from spores forming on infected plants and spreading in wind and splashing rain, or being transported on equipment. The time from infection to visible symptoms ranges from 4-9 days, depending on the temperature, bean variety, and age of the tissues.

Usually, lesions begin producing fruiting bodies and conidia within 2-4 days of infection. Production of conidia and new plant infections are favoured by temperatures of 13-26° C and relative humidity greater than 92 % and free moisture. The fungus produces cell wall degrading enzymes and low molecular weight phytotoxins that may, by killing cells in advance of the invading hyphae, contribute to the necrotrophic growth of the pathogen. Ultimately conidiophores rupture through the host cuticle and form acervuli (fruiting bodies) on the plant surface. Disease development is favoured by frequent, moderate rainfall, accompanied by winds or splashing rain. The fungal spores are easily carried to healthy plants by wind-blown rain, by people and machinery moving through contaminated fields when the plants are wet.

Epidemiology

The Anthracnose disease is favoured by cool to moderate temperature and prolonged periods of high humidity. A prolonged wet period is necessary for the fungus to establish its infection as well as for development, spread, and germination of the spores and infection of the plant. Temperatures between 13-26°C with an optimum of 17°C, relative humidity above 92% and free moisture favour the germination of spores and initial infection. Symptoms appear between 18-25 °C but are delayed or prevented by temperatures outside the range of 7-33 °C. Cool to moderate temperatures or free water on the foliage and young pods promote typical anthracnose symptoms and lesions develop in all the above ground plant parts. These lesions serve as secondary sources of inoculum that initiate secondary infection in the field when washed down with water to other plant parts. The spores can spread from infected to healthy plants by rain splash, wind-blown rain and through the movement of insects, animals and man, machineries, especially when the foliage is wet. Disease occurrence and further disease development in the field are highly favoured by frequent rainy weather, particularly those accompanied by driving winds and cool temperature which result in subsequent pod and seed infections that can bring on epidemics.

Management

Bean anthracnose is usually introduced to a production field by infected seeds. Management strategies used to minimize seed-borne infection in the seed production field include Cultural, host resistance, chemical and biological control methods.

Cultural Control

The use of disease-free seed is the most important control measure. Either certified seed, approved seed produced in dry areas or seed known to have a long disease-free history should be used. Rotation with non-host crops such as cereals and corn, for 2-3 year and removal or burial in soil of infested bean debris after harvest should be practised to reduce winter survival.

Physical Methods

Soil solarization through covering the soil with transparent plastic sheeting for one month before sowing resulted in the reduction of both severity and incidence of anthracnose. Hot-water seed treatment by soaking at 64 to 72°F for 15 hours followed by another soaking at 117°F for 25 minutes has been reported to kill the fungus in infested seeds without reducing germination.

Biological Control

Seed dressing or application of spore suspension of *Trichoderma viride* as seed dip and soil drench is effective against seed borne infection. Smearing infected seeds with cultures of *T. harzianum, T. viridae, T. hamatum* and *Gliocladium virens* for 15min and drying them overnight before sowing significantly inhibited infection of *C. lindemuthianum* and increased seed germination. Neem (*Azadirachta indica*) seed extract have been shown to effectively inhibit both germination of conidia and mycelial growth of *C. lindemuthianum*.

Host Plant Resistance

Resistance is the most effective and efficient method of anthracnose management. Nine resistance genes have been reported so far in different parts of the world. Cultivar mixtures containing at least 60% of a resistant cultivar have been reported to offer a good control of anthracnose.

Chemical Control

Seed dressing with Benlate (500WP) @ 2g/kg seed followed by foliar spray of Difenoconazole (250 EC) @ of 87.5 g *a.i.*/ha is reported to effectively reduce anthracnose severity and incidence and increase yield per plot. Seed treatment with Mancozeb @ 3g/kg seeds or Carbendazim @ 2g/kg seeds followed by foliar spray with carbendazim @ 0.5 kg/ha have been suggested to reduce anthracnose severity and incidence.

Integrated Disease Management

Due to high pathogenic variability of the pathogen, and new races of *C. lindemuthianum* are reported frequently. Thus, integrated disease management is considered to be the most effective tactic to minimize the yield losses due to anthracnose. Integration of soil solarization, Mancozeb seed treatment at a rate of 3 g/kg seeds and Carbendazim foliar spray at a rate of 0.5 kg/ha were found to be effective in reducing bean anthracnose epidemics. In India, botanicals and biopesticides (10% extracts of *Adenocalymma alliaceae, Azadirachta indica* and *Lawsonia inermis*, 0.4% talc formulation of *T. viride* and *Pseudomonas fluorescens* along with fungicides (Carbendazim (0.2%) and Mancozeb (0.4%) were evaluated in a greenhouse and field experiments giving promising results.

Bean Common Mosaic

Bean Common Mosaic is one of the most common and most destructive diseases that infect common beans (*Phaseolus vulgaris* L.) as well as a range of other cultivated and wild legumes. Yield losses due to this disease can be as high as 100%. Believed to originate in South or East Asia, Bean Common Mosaic has now spread worldwide and can be found wherever legumes are grown.

Causal Organism

Bean Common Mosaic Virus (BCMV)

Bean Common Mosaic Necrotic Virus (BCMNV)

Both BCMV and BCMNV are members of the potyvirus family of plant viruses. BCMNV strains were previously referred to as necrotic strains of BCMV, but it was found that the necrotic strains were actually a distinct virus species. Thus, these strains were formally separated in 1992 and given the name Bean common mosaic necrosis virus.

Host range: It has a wide host range. Six susceptible host families are recognised for BCMV *viz.* Amaranthaceae, Chenopodiaceae, Leguminosae-Caesalpinioideae, Leguminosae-Papilionoideae, Solanaceae, and Tetragoniaceae. However, BCMV is well known for naturally infecting wild and crop legumes and Passion fruit (*Passiflora* spp). The most notable cultivated legume hosts are Phaseolus species (predominately *P. vulgaris*), Faba bean (*Vicia faba*), peanut/groundnut(*Arachis hypogaea*), and Cowpea (*Vigna unguiculata*)

Symptoms

Two main types of symptoms are associated with bean common mosaic disease: common mosaic symptoms and common mosaic necrosis symptoms. The occurrence of either type of symptoms depends on the particular virus present and whether or not the bean variety possesses the dominant I resistance gene. If the variety has the dominant I gene, it is resistant to strains of the Bean common mosaic virus (BCMV), but hypersensitive to strains of the Bean common mosaic necrosis virus (BCMNV). Since both BCMV and BCMNV cause similar symptoms on bean varieties without the I gene, they are distinguished by their reaction on I gene-containing varieties or other tests, such as ELISA or PCR-DNA detection methods.

Symptoms of Bean Common Mosaic

BCMV causes mosaic patterns of light green-yellow leaf tissue, dark green tissue, or both light and dark mosaics together on the trifoliolate leaves. Leaf discoloration is usually accompanied by puckering, blistering, distortion, and a downward curling and rolling as well as vein banding. The intensity and severity of the symptoms depends on the strain of BCMV, the bean variety, and the age of the plant when infected. Plants infected at a young age may be stunted and distorted.

Necrosis symptoms only develop when the virus infects varieties that possess the dominant I gene. The initial symptoms become visible as small, red-brown spots that appear on primary or trifoliolate leaves shortly after the virus has been introduced through an aphid vector. The veins around these spots become brown-black, and this vein necrosis then spreads into the phloem tissue of the plant, causing wilting followed by necrosis of young leaves and the meristem. The entire plant eventually dies. Cross sections of stems and pods reveal a red-brown streaking in the vascular tissue. These symptoms are often referred to as black root rot. Common mosaic necrosis symptoms can be confused with those of Fusarium wilt caused by *Fusarium oxysporum* f. sp. *phaseoli.* However, necrosis in the vascular tissue of the pods, which is diagnostic of bean common mosaic necrosis disease, is absent in plants having Fusarium wilt.

Disease Cycle

Bean common mosaic virus (BCMV) overwinters in infected weed hosts and in infected seeds. The virus can remain alive in bean seeds for many years. In fields, it is most commonly spread by aphids. The virus is also disseminated among fields by pollens, movement of contaminated seed, and by movement of contaminated equipment between fields. The virus can be transmitted mechanically by plant-to-plant abrasion and equipment.

Epidemiology

BCMV can be a challenge when aphids are present and susceptible varieties are grown near infected plants in the same field or nearby fields. Aphids transmit the virus in weed hosts and in seeds in a non-persistent way, which means that just after feeding on a BCM infected plant, the aphid can at once move to a healthy plant and transmit the virus as it feeds again.

The virus is also spread through seed. Seeds of *Phaseolus vulgaris* can retain infectious BCMV for at least 30 years. This stability of virus in the embryo is likely a major contributory factor for seed being the most important path for long-distance worldwide dissemination of BCMV and BCMNV through trade. The percentage of seed infection varies between 0.67 to 98%. However, rates of seed transmission range from 10% to 30%. It can then be spread from plants grown out of infected seeds by aphids. Thus, an outbreak can be triggered by the use of contaminated seed stock and amplified by aphid-mediated virus transmission to generate an epidemic. Alternatively, the viruses may infect a healthy crop population when viruliferous aphids immigrate from infected wild plants. Disease development occurs in temperature between 20 to 25°C.

Management

- Since virus is seed borne, therefore seeds should be collected from plants that do not show symptoms of the disease. Preferably, seeds should be obtained from a commercial company or from local growers whose plants have been monitored for the disease.
- Intercroping beans with other plants like maize helps to reduce spread of the disease.
- All plant debris should be burnt after harvest.
- The use of insecticides for the control of the vector (aphids) is generally not recommended. Since the aphids transmit the virus in non persistent way, by the time the insecticide kills the aphid it already transmits the virus.
- Plant-derived products, such as neem, derris, pyrethrum and chilli (with the addition of soap as adjuvent) or semiochemicals can be used to control aphids in weed hosts before planting of the main crop.

Future Approaches and Conclusion

Plant diseases and their management have always been challenges to mankind. Changes in the global climate, introduction of new varieties and hybrids as well as changes in farming practices, intensifications of resources, monocultures are considered major factors for emergence and outbreak of new pathogens/ diseases as well as epidemics of prevailing diseases. Further, the ever explosion of global population and reducing arable lands, and depletion of natural resources make it even more challenging to increase the potential for increasing agricultural productivity and food security. Disease management through the use of chemicals and other management practices often has been difficult and costly. To meet the challenge, plant disease management strategies must be planned based on the socio-economic and ecological point of view. To better understand the crop health management, profiling, documentation, and quantification of the importance of diseases and their insect vectors are essential. Indiscriminate and non-judicious uses of synthetic pesticides by farmers in vegetable production develop pesticide resistance and harmoligosis in pathogens and insect vectors and also adversely affect beneficial insect and microbes. With this background, Integrated disease management (IDM), which combines biological, cultural, physical and chemical control strategies in a holistic way rather than using a single component strategy proved to be more effective and sustainable. Disease resistance breeding and harnessing potential of bio-agents (BCAs) and microbes are other economically viable option for disease management of diseases in vegetable crops. Knowledge of the biology and variability of the pathogens is also a prerequisite for breeding program aimed at obtaining durable resistance. While going for a management strategy, both innovative and conventional approaches should be used to explore interaction the various interactions occurring among hosts, pathogens and vectors in an ecological and epidemiological context as well as with due consideration of the economic threshold of management. A sustainable plant disease management strategy should also be built upon epidemic and evolutionary patterns of plant disease under changing micro and macro climate and agricultural production philosophies. Disease forecasts and knowledge of the pathogen population genetic structure will further increase the efficiency of the management strategy and reduce negative impacts on the environment.

References

Agrios, G. N. (2005). Plant Pathology Fifth Edition. Burlington, MA: Elsevier Academic Press, pp 629.

Ararsa, L., Fikre, L. and Getachew, A. (2018). Evaluation of Integrated Management of Common Bacterial Blight of Common Bean in Central Rift Valley of Ethiopia. *Am. J. Phytomed. Clin. Ther.* 6 No. 1:3. doi:10.21767/2321-2748.100339.

Arnold, D.L. (2011). *Pseudomonas syringae* pv. *phaseolicola*: from 'has bean' to supermodel, *Mol. Plant. Pathol.* : 617–627.

Barnes, E.H. (1979) *Atlas and Manual of Plant Pathology* Plenum Press, New York, pp 60-61.

Belete, T. and Bastas, K.K. (2017). Common Bacterial Blight (*Xanthomonas axonopodis* pv. *phaseoli*) of Beans with Special Focus on Ethiopian Condition. *J. Plant Pathol. Microbiol.* 8: 403. doi: 10.4172/2157-7471.1000403.

Dutta, P., Das, B. C. and Islam, M. (2008). Eco-friendly strategies for management of *Sclerotinia* rot of french bean *J. Biol. Control,* **22**(2): 405-410.

Elsheshtawi, M., Elkhaky, M.T., Sayed, S.R., Bahkali, A.H., Arif A., Mohammed, A. A., Gambhir, D., Mansour, A.S. and Elgorban, A.M. (2016). Integrated control of white rot disease on beans caused by *Sclerotinia sclerotiorum* using Contans® and reduced fungicides application, *Saudi Journal of Biological Sciences* 24:405–409.

Fetene, S. and Ayalew, A. (2016). Integrated management of common bacterial blight (*Xanthomonas campestris* pv. *phaseoli*) and its effect on seed yield of common bean (*Phaseolus vulgaris* L.). *Int. J. of Life Sci*. **4:** 336-348.

French,R. and Muchovej, R. M. (2016). Texas A&M AgriLife Extension Service; The Texas A&M System.

Gillard, C. L., Conner, R. L., Howard, R.J., Pauls, K. P. and Shaw, L. (2009). The performance of dry bean cultivars with and without common bacterial blight resistance in field studies across Canada. *Can. J. Plant Sci*. **89:** 405-416.

Heffer Link, V. and Johnson, K. B. (2007). White Mold. *The Plant Health Instructor*. DOI: 10.1094/PHI-I-2007-0809-01.

Harveson, R. M., and Schwartz, H. F. (2007). Bacterial diseases of dry edible beans in the central high plains. Online. *Plant Health Progress* doi:10.1094/PHP-2007-0125-01-DG.

Kaiser, W.J. and Vakili, N.G. (1978). Insect transmission of pathogenic xanthomonds to bean and cowpea in Puerto Rico. *Phytopathology* **68:** 1057-1063.

Kakoty, R. K. (2002). Further approach to the integrated management of Sclerotinia sclerotium (Lib.) De Bary-The white rot pathogen. PhD Thesis, Assam Agricultural University, Jorhat, India.

Karavina, C., Mandumbu, R., Parwada, C and Tibugari, H. (2011). A review of the occurrence, biology and management of common bacterial blight, *Journal of Agricultural Technology* **7**(6): 1459-1474.

Kennedy, B.W. and Alcorn, S.M. (1980). Estimated crop losses to prokaryote plant pathogens, *Plant Disease* **64:** 674-676.

Lalfakawma, C. (2015).Carpogenic germination of sclerotia of *Sclerotinia sclerotiorum* (Lib.)de bary and development of an IDM module for management of Sclerotinia rot of French bean. PhD Thesis, Assam Agricultural University, Jorhat, India.

Macnab, A.A., Sherf, A.F. and Springer, J.K. (1983). *Identifying Diseases of Vegetables*. The Pennsylvania State University. pp 3.

Mohammed, A. (2013) An Overview of Distribution, Biology and the Management of Common Bean Anthracnos, *J. Plant. Pathol. Microb*. 4: 193 doi:10.4172/2157-7471.1000193.

Morales, F. J. and Bos, L. (1988). Descriptions of plant viruses: Bean common mosaic virus. PV337. Wellesbourne UK: Association of Applied Biologists. http://www.dpvweb.net/dpv/ showdpv.php?dpvno=337.

Mutlu, N., Vidaver, A. K., Coyne, D. P., Steadman, J. R., Lambrecht, P. A., and Reiser, J. (2008).Differential pathogenicity of *Xanthomonas campestris* pv. *phaseoli* and *X. fuscans* subsp. *fuscans* strains on bean genotypes with common blight resistance. *Plant Dis*. **92:** 546-554.

Natti, J.J. (1971). Epidemiology and Control of Bean White Mold *Phytopathology* 61: 669-674.

Noble, T. J., Young, A. J., Douglas, C. A., Brett, W. and Mundree, S. (2019) Diagnosis and management of halo blight in Australian mungbeans: a review *Crop and Pasture Science*, **70:** 195–203.

Ocamb, C.M. D.H. Gent (2020). Bean, Snap (Phaleolus vulgaris)-White Mold (Sclerotinia rot)In Pschiedt, J.W and Ocamb, C.M (Eds.) *Pacific Northwest Plant Disease Management Handbook*, Oregon State University https://pnwhandbooks.org/node/2258.

Rakesh, Rathi, A.S., Kumar, P, Kumar, A and Kumari, P. (2016) Sclerotinia rot of rapeseed mustard: A comprehensive review, *Journal of Applied and Natural Science* **8** (4): 2325-2336.

Rothmann, L.A. and McLaren, N.W. (2018). *Sclerotinia sclerotiorum* disease prediction: A review and potential applications in South Africa. *S. Afr. J. Sci.* 114(3/4):1-9 http://dx.doi.org/10.17159/ sajs.20180/20170155.

Sahni, S., Kumar, B. and Singh, S. (2018). Field Screening of Rajmash Germplasm for Resistant against *Sclerotinia sclerotiorum* Causes White Mold, *Current Journal of Applied Science and Technology* **31:** (3): 1-5.

Schwartz, H.F. 1991. Anthracnose *In* R. Hall (ed.) *Compendium of Bean Diseases*. APS Press, St Paul, MN. pp. 16-17.

Schwartz, H.F. (2013) Bacterial Diseases of Beans. Colorado State University Extension. 2.913.

Schwartz, H. F. and. Singh, S. P. (2013). Breeding Common Bean for Resistance to White Mold: A Review, *Crop Sci.* **53:**1832–1844.doi: 10.2135/cropsci2013.02.0081.

Sharma, P., Verma, P.R., Meena, P.D., Kumar, V. and Singh, D. (2015). Research progress analysis of Sclerotinia rot (*Sclerotinia sclerotiorum*) of Oiseed Brassicas through bibliography. *J. Oilseed Brassica,* **6:** 45-125.

Singh, S. P. and Schwartz, H. F. (2010). Breeding common bean for resistance to diseases: a review. *Crop Science*, **50:** 2199–2223.

Smoli D. ska, U. and Kowalska, B. (2018). Biological control of the soil-borne fungal pathogen *Sclerotinia sclerotiorum*- a review, *Journal of Plant Pathology* **100:** 1-12.

Strausbaugh, C. A. and R. L. Forster, R. L. (2003). Management of White Mold of Beans, PNW 568- A Pacific Northwest Extension Publication,University of Idaho.

Taylor, J.D. (1979). Epidemiology and strategy for the control of halo-blight of beans, *Annals of Applied Biology*: 167.

Taylor, J.D., Teverson, D.M., Allen, D.J. and Pastor-Corrales, M.A. (1996). Identification and origin of races of *Pseudomonas syringae* pv. *Phaseolicola* from Africa and other bean growing areas. *Plant Pathol.* **45:** 469–478.

Tripathi, A. N. (2018) Emerging diseases and their management in vegetable crops *Plant Physiol. Pathol.* 6: 67 DOI: 10.4172/2329-955X-C2-018.

Tumsa, K. (2007). Integrated management of common bacterial blight of common bean through host resistance and chemical applications in the central rift valley, Ethiopia. M.Sc. Thesis. Haramaya, University, Haramaya, Ethiopia.

Uppal, B.N., Patel, M.K. and Nikana, B.G. (1946). Bacterial blight of *Phaseolus vulgaris* var. white kidney. In: *Proceedings of National Institute of Science New Delhi, India.* **12:** 351-359.

Wohleb, C.H., and L.J. du Toit. (2010). Common Bacterial Blight and Halo Blight: Two Bacterial Diseases of Phytosanitary Significance for Bean Crops in Washington State. Washington State University Fact Sheet No. FS038E.

Worrall, E. A., Wamonje, F. O., Gerardine, M. Harvey, J. J. W., Carr, J. P. and Mitter, N. (2015). Bean Common Mosaic Virus and Bean Common Mosaic Necrosis Virus: Relationships, Biology, and Prospects for Control *Advances in Virus Research*, **93:** 2-35.

Zhang, S., Palmateer, A. J., Pernezny, K. and Jones, J. B.(2018). Common Bacterial Blight of Snap Bean in Florida PP-62 Plant Pathology Department, UF/IFAS Extension http://edis.ifas.ufl.edu.

13

Diseases of Cowpea (*Vigna unguiculata* L., Syn. V. Sinensis Savi) and Their Integrated Management

S. Baruah[1] **and** ***Pranab Dutta***[2]

[1]*KVK-Tinsukia, Assam Agricultural University (AAU), Tinsukia, Assam*

[2]*School of Crop Protection, College of Post Graduate Studies in Agricultural Sciences Central Agricultural University, Umiam, Meghalaya*

Introduction

Cowpea (*Vigna unguiculata* subsp *unguiculata*) is widely grown in Africa, Asia and Brazil. The crop is grown on about 7 million ha in warm to hot regions of the world (Rachie, 1985). It acts as a major source of protein (20-30%) for Indian vegetarians. Besides, it contains high carbohydrate. Cowpea is generally rich in lysine and limiting in sulphur amino acids and important for human food and animal feed. The crop is also valuable source of dietary fibre as well as vitamins and minerals including folate, thiamin and riboflavin. Besides this, it can fix about 240 kg $ha^{-1}$of atmospheric nitrogen and make available about 60-70 kg ha^{-1}nitrogen for succeeding crops grown in rotation with it (Aikins and Afuakwa, 2008).

The crop is prone to several diseases induced by species of pathogens belonging to various pathogenic groups viz., fungi, bacteria, viruses, nematodes, and parasitic flowering plants, and it constitute one of the most important constraints to the profitable Cowpea production in all agroecological zones. Different diseases like anthracnose, fusarium wilt, bacterial blight, Cowpea mosaic virus (and several other less prominent viruses), cercospora leaf spot, rust and powdery mildew etc are reported to be important disease of the crop. In this chapter detail of these diseases and their integrated management approaches are mentioned.

Seedling Mortality

Importance

Seedling mortality is caused by fungal pathogen which are ubiquitous in nature. The pathogen prefers cool, wet and overcast weather. Seedlimg mortality causes 75% mortality of Cowpea seedlings within 21 days after sowing.

Causal organism : *Pythium aphanidermatum* and *Corticium solani* (*Rhizoctonia solani*) are the main causal agent for the disease.

Epidemiology

In India, Sunder *et al*. (1996) studied the effects of inoculum characteristics on pathogenicity of *R. solani.* They showed that an increase in inoculum dose from 50 to 200 mg/pot resulted in an increase in seedling mortality and that the pathogenicity of the isolate declined as the age of the inoculum increased. Their results also revealed that inoculum grown at 25 °C or 30 °C and pH 6.5-7.5 produced higher seedling mortality compared to inoculum cultivated at 20 °C or 35 °C. These pathogens are abundant in the soils in the forest region.

Symptoms

Both pathogens causes pre and post emergence seedling mortality. In the later stage, symptoms can be observed on the hypocotyls. The reddish brown lesions caused by *Corticium solani* are usually limited to the collar region of the hypocotyl at which point the diseased seedling topples. *P.aphanidermatum* however, moves rapidly to the hypocotyls giving it a grey- green, wet appearance and the seedlings undergo a watery collapse.

Management

- Pathogen free certified seeds should be used
- Proper drain should be maintained.
- Seed treatment with *Trichoderma viride* @10g/kg or *Pseudomonas fluorescens* @ 10g/kg of seed or Carbendazim or Thiram 2g/kg of seed.
- Soil drenching with Carbendazim 1g/lit or *P. fluorescens* / *T. viride* 2.5 kg/ha with 50 kg FYM should be done
- Seedling mortality can be effectively controlled with Chloroneb (= demosan) applied as a seed dressing (2g/kg of seed).

***Cercospora* Leaf Spot**

Importance: The disease has been reported from all Cowpea growing regions of the world. It causes yield loss of about 20 %.

Causal organism : *Cercospora canescens* or *C.cruenta.* The conidia are uniform in colour, pale to medium brown, multiseptate, medium to large size, conidial scar present on the rounded apex, thickened hilum.

Epidemiology

The disease is mainly occurred in moderate temperature and high humid condition. Abundant fruiting bodies are found on the lower surface of the leaf. Conidia are formed at 28°C, while at 24°C and 32°C less conidia are formed. The presence of light increases the number of conidia (Mulder and Holliday, 1975). The pathogens survive the no-crop period on infected crop residue and in infected seed (Williams 1975; Patel 1985).

Symptoms

The symptoms are prominent on the leaves alone. Symptoms found on various plant parts are as follows:

Leaves: Subcircular to broadly irregular spots having pale tan to grey centre surrounded by dark brown or reddish margin. The spots coalesce to form round lesions which are brown and necrotic with dark, and slightly depressed edges.

Pods: Damaged pods, infected pods dries up prematurely.

Stem: At later stage lesions may also developed on the stem and cotyledons.

Management

- Diseased plant debris should be burnt down.
- Pthaogen free seeds should be used for healthy plants
- Seed treatment with carbendazim or thirum @2g/kg of seeds should be done.
- Mancozeb (0.2%) or carbendazim (0.2%) should be sprayed at fortnight interval
- Weekly spraying of benomyl, beginning at three weeks after planting, gave the best control of the diseases and the highest grain yield (Amadi,1995).

Anthracnose

Importance : Disease causes economical losses to crops affecting yield, seed quality and marketability of the crop. The disease causes huge losses in temperate and subtropical zones.

Causal Organism

Anthracnose of Cowpea caused by *Colletotrichum lindemuthianum* belong to the group Deuteromycotina, class Coelomycetes and family Melanconiaceae. The fungus produces septate hyphae and conidiophores, conidiophores are light to brown colour. The fungus forms dark coloured acervulli with dark, septate setae. Non septate terminal conidia are borne on conidiophores which are held firmly around the base of the setae in dry weather.

Epidemiology

A conidium takes 6-9 hours to germinate under favourable environmental conditions. The pathogen penetrates the cuticle and epidermis mechanically (Leach, 1923). Following penetration of host cells, when temperatures are favorable, infectious hyphae enlarge and grow between the cell wall and protoplast for 2-4 days without apparent damage to host cells.

C. lindemuthianum is seed borne and seed transmitted. The pathogen was able to survive for at least 5 years on pods and seeds that were air-dried and kept in storage at 4°C or on dry, infected plant materials left in the field in sealed polyethylene envelopes (Tu, 1983).

Symptoms

The fungus infects all stages of the plant, flowering, podding, pre-emergence, seedling and vegetative growing stages and all plant parts including the pods, leaves and seeds. Initial symptoms may appear on cotyledonary leaves as small, dark brown to black lesions. The infected tissues manifest minute rust-coloured specks. The specks gradually enlarge longitudinally and form sunken lesions or eye-spots.

Leaves: Lesions first develop on leaf petioles, the lower surface of leaves and leaf veins as small, angular, brick-red to purple spots which become dark brown to black. Later, the lesions may also appear on veinlets on the upper surface of leaves.

Seedlings: Lesions enlarge on the hypocotyl of the young seedling, causing rotting of young seedlings.

Stem: Eye-shaped lesion develop on stem. In high relative humidity condition rotting of stem occurs.

Pod: Infections appear as rusty brown spots with small, brown specks, sunken cankers delimited by a slightly raised black ring and surrounded by a reddish-brown border. Young pods shrivel and dry up.

Seed: Discolouration, dark brown to black cankers, brown to light chocolate spots on the seed coats.

Management

- The best method is useof plant resistant varieties.
- Certified disease-free seed should be used.

- Good field sanitation should be maintained such as removing crop debris from field after harvest to reduce levels of inoculums.
- Anthracnose incidence and severity is lower in the intercrop relative to the sole crop while reduction in both inter- and intrarow spacing resulted in an increase in the incidence and severity of anthracnose (Adebitan and Ikotun 1996). Adebitan (1996) has also evaluated the effects of weed infestation and application of phosphorus fertilizer on anthracnose. The severity of anthracnose was lowest in plots given 80 kg/ha P_2O_5 and highest in plots that did not receive phosphorus fertilizer. That is why intercropping and application of 80 kg/ha P_2O_5 is recommended.
- Seed treatment with mancozeb @0.2% of seeds or seed treatment with carbendazim or thirum @0.2% of seeds should be done prior to seed sowing.
- Fungicidal field sprays with mancozeb @ 0.2% should be done in the field during active growth stage.
- Water or alcohol extract of *Piper betle*, *Ocimum sanctum*, and *Citrus limon* are effective in checking the incidence and spread of anthracnose in the field.
- Crop rotation with non-host crops for 2-3 years is also found effective.
- Spraying of mancozeb (0.2%) or carbendazim (0.2%) at fortnight interval is also recommended.

Fusarium wilt

Importance

Fusarium wilt of cowpea caused by *F. oxysporum* f.sp. *tracheiphilum* is a destructive disease in the Southern and Eastern USA (Toler *et al.*, 1963). In India, the yield loss caused by the fungus was 26.8-64.5% in 1954 and 74.6% in 1955 (Singh and Sinha, 1955). And in 1996, the wilt incidence in India was recorded as 30% by Ushamalini (1997).

Causal Organism

Fusarium oxysporum f.sp. *tracheiphilum* (E.F.Sm.) Snyder & H.N. Hansen. Morphology of *F. oxysporumf* sp. *tracheiphilum* on oat meal agar, has a cotton wool appearance, with a white to purple aerial mycelium. The microconidia are ellipsoidal, hyaline, and aseptate. The macroconidia are slightly curved, mostly septate. The chlamydospores are ellipsoidal to globose, terminal or intercalary and aseptate. The fungus also produces small, abundant sporodochia on PDA in 8-12 days at 26-30°C.

Epidemiology

The symptoms of vascular wilt are severe at higher temperatures of 27°C (Swanson and Van Gundy, 1985). Swanson(1984) and Harris and Ferris (1991) further reported that the wilt symptom is very severe when the fungus is in association with the nematode, *Meloidogyne javanica* . *F. oxysporum* f. sp. *tracheiphilum* is seed borne and seed transmitted. Seeds from wilted plants stored in the refrigerator for 4 years remained viable (Armstrong and Armstrong, 1950). The fungus was found to be present in the seed coat, cotyledon, and embryo.

Symptoms

Fusarium oxysporum f. sp. *tracheiphilum* attacks young and old plants. Symptoms of the disease are visible on leaves, stem and roots. The vascular bundles show brownish-black discoloration. All plant parts are affected and showing different symptoms such as on:

Leaves: Yellowing, withering, chlorosis, leaf drooping,

Shoots: Dry and naked

Stem: Blackened and swollen symptoms develops on the stema, wilting occurs after 1 week of infection.

Pods: Presence of pinkish-white fungal growth on the infected pods.

Seeds: Discoloration, shrivelling, ashy white and shrunken

Roots: Rotting occurs, infection reduced root system, absence of lateral roots, develops lesions on infected roots

Whole plant: Systemic infection, colonization of xylem vessels resulting in chlorosis, die back, dwarfing & ultimately wilting of the infected plant).

Management

- Use resistant varieties.
- Seed treatment with mancozeb (Ethylene Bisdithiocarbamate) @0.2% of seeds.
- Wilted plants should be rouged.

Scab

Importance

One of the serious diseases of Cowpea causing loss of foliage, pods and seeds. In severe case there is complete loss of yield. Exact yield loss caused by this fungus has not been quantified. Recently, it was reported that *Sphaceloma*occurred on nine out of 14 major weed species found in cowpea fields.

Causal Organism

The disease is caused by *Sphaceloma* sp. or *Elsinoe phaseoli,* Jenkins ; *Elsinoe vignicola*. The mycelium is hyaline, scanty and submerged. Conidia are hyaline to pale coloured. The conidia are produced in pycnidia. The ascospores borne on the asci are hyaline, pale colored oblong to elliptical, and 3 septate.

Epidemiology

Inoculum of the pathogen for primary infection under field conditions originated either from the infected seed or from infected Cowpea debris, with primary lesions appearing on the hypocotyl or epicotyl (but not on the unifoliate primary leaves) about 25 days after sowing (Mungo *et al.* 1998). Secondary spread of the fungus was by rain splash and wind-driven moisture. Cowpea scab development is favoured by moderate temperatures (23–28 °C), three or more consecutive days of wet weather, and consequent high relative humidity.

Symptoms

The fungus attacks the stem, leaves, and pods of *Vigna unguiculata*. Symptoms on the different parts such as on:

Leaves: Spots develops on both leaf surfaces, cupped. Appearance of small grayish scab lesion along the veins. Distortion of infected leaves occurs.

Stems: Oval to elongated silver grey lesions surrounded by red or brown elliptical ring. Lesions coalesce, distortion.

Pod: Sunkened spots with grey centres surrounded by brown borders appear on the infected pods, pods shows symptom of malformation, dark coloured pycnidia formed in the brown spots.

Management

- Use resistant varieties.
- Seed treatment with mancozeb (Ethylene Bisdithiocarbamate) 80g/kg of seeds.

Brown Blotch of Cowpea

Importance: The disease causes, yield losses of 30% - 57% in the fields (Lehman *et al.,* 1976). Brown blotch of maturing plants causes serious losses, particularly during the rainy period when shaded lower branches and leaves die due to severe infection.

Causal Organism

It is caused by *Colletotrichum truncatum* (Schwein.) The hyphae are hyaline, branched and septate. *C. truncatum* has crowded, black acervuli which are borne on well-developed stromata. It produces numerous, black mixed setae in culture, some are long and others short. The conidia are borne singly on conidiophores, bluntly tapered, curved, unicellular, and hyaline. Conidia produce one or two short germ tubes which produce dark, sticky appressoria when in contact with the surface of the host plant. The disease is seed borne and seed transmitted.

Symptoms

C. truncatum affects all plant stages and parts-flowering, podding, seedling stages also the pods, inflorescence, leaves, stems and whole plant. The stems, pods and leaves may be infected without showing symptoms. In the advanced stages of anthracnose in the late reproductive stages, infected tissues are covered with black fruiting bodies (conidiomata) which produce setae (minute black spines).

Symptoms Found on the Different Plant Parts Such as on the

- **Leaves**: Lesions; abnormal colours and forms; fungal growth, necrosis of laminar veins, leaf rolling and petiole cankers. Premature defoliation, girdling of the leaf and petiole Development of shepherd's crook.
- **Seedlings**: Discoloration; canker, premature death, and severe reductions in seedling emergence was recorded.
- **Pods**: Lesions. irregularly-shaped, brown areas, no pod formation or reduction in pod size and number.
- **Seeds**: Small, irregular, grey areas with black specks, brown staining, symptomless. The fungus is confined at first to the seed coat. The infected seeds may die during germination or, if they germinate, produce infected seedlings.
- **Whole plant**: Damping off; dwarfing, early senescens.

Management

- Seed treatment or soil drenching with conidial suspension (1×10^8condia/ml) of *Trichoderma viride* effectively reduced brown blotch infection.

- Seed Treatments with mancozeb @ 0.2% followed by 2 foliar sprays of mancozeb @ 0.1%.
- Wide spacing of Cowpea resulted in lower incidence and severity of brown blotch compared to the closer planted crop, both monocrop and intercrop.
- In addition, foliar application of spore suspension of *T. viride* once or twice weekly, beginning three days after inoculation of seedlings with the pathogen, reduced brown blotch in the field.

Brown rust

Importance

Infection at the early stages of plant growth, results to crop failure. Yield losses reaching 28-54, 8-33 and 13-29% on different cultivars were reported by Gonzalez and Garcia (1996).The disease received the highest research attention among the fungal diseases of Cowpea during the 1995-2000 period. There can be crop failure if infection takes place in the early stages of plant growth.

Causal organism

Uromyces appendiculatus (Pers.) is the causal organism of the diseases. The mycelium of the fungus is hyaline, branched and septate existing internally in host tissue producing various sori. It is autoecious having white pycnidia in small groups. Aecia are very rare but the aeciospores are globose to ellipsoid. Uredinia, sometimes absent but if present are pale brown, solitary or sometimes aggregated, minute,and cinnamon brown. Urediniospores are globose to sub globose. Telia are blackish-brown to black. The teliospores are subglobose to ovoid or ellipsoid and rounded at the apex.

Epidemiology

Overcast conditions and a temperature of 20-25°C favour the pathogen.

Symptoms

All plant parts above the ground are susceptible to infection. Infection appears at all stages of the plant growth.

- **On the Leaves**: Yellow to yellowish-brown small raised blister like spots, as infection progresses, the blisters erupt and powder like uredo spores are exposed. The spores reinfect the plant and defoliation resulting to death of the plant.

- **Stem**: Brown spots, spores and hyphae are borne internally and externally, spores visible to the eye.
- **Pods**: Brown spots appear on the infected pods.
- **Seed**: Malformation, decrease in weight of seed , sometimes symptom less

Management

- Seed treatment with mancozeb @ 0.2% followed by field spray before flowering reduced infection but did not eradicate the pathogen.
- Use resistant varieties for planting.
- The infected crop should be dusted with Sulphur 80 WP @ 37.5-50 kg/ha.
- Spray mancozeb (0.2%) or oxycarboxin (0.1%) or carboxin (0.1 %) at fortnightly interval.

Leaf Smut or Black Leaf Spot

Importance

The disease is of economic importance in Brazil where yield losses of 30-40% were reported.

Causal Organism

Entyloma vignae or *Protomycopsis phaseoli.* Chlamydospores are black and spherical. Germination by formation of basidia.

Symptoms

The symptoms are very prominent on the leaves.

- **Leaves**: Dark ash grey to black, circular, purplish spots surrounded by yellow hallow border. As infection progresses, spots coalesce, become dark purple leaves become ragged and fall off.
- **Whole plant**: Severe infection causes defoliation.

Management

- Infected plants should be rouged out as and when symptom appears in the plants
- Seeds should not be harvested from infected field.

Septoria Leaf Spot

Importance : Disease is distributed in India, East Africa, Savannah zones of tropical Africa, Nigeria. The fungus is worldwide in distribution attacking many crops.

Causal Organism

Septoria vignae Hennor Septoria vignicola Vasant Rao I scausaloerganismopf the disease. Fungus produces conidia in pycnidia. Conidia are septate. It is seed borne and seed transmitted.

Symptoms

- Symptoms on leaves are very pronounced
- Spots coalesce to form dark red circular lesions on both leaf surfaces. Lesions also appear on the pods and stems. Pycnidia are formed in the lesions.

Management

- Seed treatment with mancozeb @0.2%
- Spray mancozeb (0.2%) or carbendazim (0.2%) at fortnight interval.

Podrot

Importance

The fungus is worldwide in distribution attacking many crops. The exact data on yield loss are seldom mentioned. Cowpea losses were estimated at 7-20% by Oladiran (1980).

Causal Organism

Choanephora cucurbitarum (Berk. &Ravenel). *C. cucurbitarum* is the causal agent of the disease and it produces elongate mycelium without septa. The sporangiola, and typical sporangia are sometimes produced but on separate sporangiophores. The zygospores are formed by fusion of two morphologically similar gametes. Zygospores germinate to produce a sporangium containing sporangiospores (French, 2000).

Symptoms

- The pathogen infects plant tissues that arte previously damaged either by insects or by physical means.
- The fungus causes a number of diseases on a range of crops affecting all plant stages: flowering, fruiting, post-harvest, pre-emergence, seedling and vegetative growing and all plant parts; pods, growing points, inflorescence, leaves, seeds, stems and whole plant.

- The fungus causes dieback, stem and leaf blight on Cowpea.
- Visibly field symptoms on the host tissues have a white hairy appearance resulting from the tall sporangiophores that produce a cluster of brown sporangiola at their tips. On Cowpea, symptoms begin as water-soaked lesions at the leaf margins and tips. These lesions became dry and turned olive-green to light brown. Numerous spiny, long sporangiophores developed during dry weather results to total necrosis of the entire plant.

Symptoms on Various Plant Parts

Pods: Visible mould and whitish spine like mycelium, rotting occurs on pods.

Growing points: Infection causes dead heart.

Inflorescence: Lesions develops on infected inflourescens.

Leaves: Lesionsdevelopes with fungal growth like sooty mould, rotting occurs and emits foul odour.

Seeds: Rotting of seeds occurs.

Stems: External discoloration; canker; dieback; sooty mould.

Whole plant: Dieback symptom appear on the infected whole plants.

Management

- Seed treatment with mancozeb @ 0.2%

Charcoal Rot or Root Rot

Importance : Among the fungal diseases the charcoal rot causes significant loss in yield. The fungus is soil-brone and poses great problem in managing the disease.

Causal Organism : The disease is caused by *Macrophomina phaseolina*

Symptom

The seedlings remain stunted and some of them develop rot in the hypocotyls region and extend downwards. Roots also exhibit rotting and seedlings get dried. In mature Cowpea plants, grayish black sunken lesions appear on the lower stem and roots accumulate small black sclerotia. Sometimes stem girdle which may break longitudinally.

Management

- Soil application of neem cake @ 150 kg/ha and Farmyard manure @ 10t/ha reduced the charcoal rot incidence significantly and increased the yield (Ushamalini, 1997).
- Crop rotation with non-host crops for 2-3 years.
- Seed treatment with carbendazim or thiram @ 2g/kg of seeds.
- Seed treatment with *Trichoderma* or other biocontrol agent @ 10g/ kg of seeds and soil application of biocontrol agents along with organic manure.

Powdery Mildew

Importance : Powdery mildew occurs in all areas where cowpea is grown; *Sphaerotheca fuliginea* only reported in India.

Causal Organism : *Erisyphe polygani*and /or *Sphaerotheca fuliginea* are the main causal organism of the disease.

Symptoms

White powdery fungal growth on upper surfaces of leaves; chlorotic or brown patches on leaves; leaves dropping from plant.

Management

- Use resistant varieties.
- Use adequate plant spacing to avoid overcrowding and promote good air circulation around plants
- Dusting with sulphur dust @ 25-30 kg/ha or we can go for spraying of fungicide like Carboxin 50 WP @ 1 gm per litre of water.

Bcaterial Blight

Importance

Bacterial blight is an important disease of Cowpea in tropical Africa, America and India. Seedling mortality of 60 per cent was recorded from seed-borne infection. In some parts of Africa an annual losses of 40 per cent was recorded (Cisse, *et al.,* 2000).

Causal Organism

Xanthomonas vignicola is sthe causal pathogen of the disaese.

Epidemiology

The disease spreads rapidly during heavy rainfall. The pathogen is seed borne, and survives on diseased crop residues.

Symptoms

The disease first appeared on cotyledons, primary and the new trifoliate leaves. Initially water-soaked spots developed on leaves which enlarge and become necrotic; spots may be surrounded by a zone of yellow discoloration; lesions coalesce and give plant a burned appearance; leaves that die remain attached to plant. Circular, sunken, red-brown lesion may be present on pods; pod lesions may ooze during humid conditions. Later on the disease spreads to the stems and covers other parts of the plant resulting the death of the plant.

Management

- Plant only certified seed or use plant resistant varieties
- Spraying of an appropriate protective copper based fungicide before appearance of symptoms is recommended.
- In case of severe infection, spraying with bactericides like Streptomycin 100 @ 0.4 gm per litre of water can be done.

Cowpea Severe Mosaic Disease

Importance: Disease is widespread in tropical and sub-tropical America. In Brazil, yield losses of 60-80 per cent was recorded.

Causal organism: Cowpea severe mosaic virus. It is sap and seed borne and transmitted by several beetles including *Ceratoma* spp., *Phaseolus lathyroides*.

Symptoms

Several viruses cause disease in cowpea. These viruses produce a mosaic pattern (intermixing of green and yellow patches) on leaves. They may be found singularly or in combination with others. They cause irregular light and dark green mosaic patterns on the leaves. Some viruses cause thickened, malformed and distorted leaves. The mosaic patterns are best observed on the younger foliage. Plants may be stunted and fail to produce normal pods. If the disease attacks plants at the early growth stage, no pods should be expected.

Management

- Plant resistant varieties.
- Use disease- free healthy seeds

- Crop rotation with non-legumes for 4-5 seasons
- Remove the infected plants at the first instance, weeds and alternate hosts.
- Management of vectors through insecticide spray to prevent secondary infection
- Spray any one of the systemic insecticide like Chlorpyriphos @ 0.1% to control the vector.
- Spray Metasystox @ 1 litre/hectare dissolved in 1000 litres of water to check the whitefly.

Cowpea Aphid Borne Mosaic Virus (CpMV)

Importance: Among the many species of plant viruses infecting Cowpea, Cowpea mosaic virus (CpMV) is considered as a significant disease as it causes yield loss of 40-60 per cent (Sastry, 2013). A yield loss from CpMV ranges between 87-100 per cent was reported from different parts of the World including India (Shoyinka *et al.,* 1997; Raheja and Leleji, 1974). CpMV was first reported in Italy by Lovisoloand Conti (1966) and later from Nigeria in 1976 (Ladipo, 1976). CpMV is widespread in many countries in different continents including Asia, Africa, Europe, North and South America, and Australia (Pio-Ribeiro *et al.,* 2000). It is a seed borne disease.

Causal Organism

Cowpea aphid borne mosaic potyvirus. The virus is transmitted by sap, seed and by Cowpea mosaic aphid (*Aphis craccivora*).

Symptoms

Typical CpMV symptoms observed on infected plants are mosaic, vein clearing, green vein banding, blistering, stunting, mottling and distortion of the leaves (Bock and Conti 1974) (Plate 9). The pods also get infected, get curved, twisted and reduced in size. The seeds of those infected pods get shrivelled and are in even lesser in number.

Management

- As the disease is seed borne, use of healthy disease free seed is very much important.
- Seed treatment can be done with bioformulations of *Metarhizium anisopliae* @ 5ml/lit of water.
- The disease affected plants should be uprooted and destroyed as soon as the disease appeared.

- Spraying of liquid formulation of *Metarhizium anisopliae* amended with Glycerol (10.0%) + Sunflower oil (0.5%) at 15 days interval for twice is very much effective for preventing secondary spread of the disease by controlling its vector (*A. craccivora*) (Boruah *et al.,* 2015).
- Under chemical control, the insect vector can be controlled by spraying Dimethoate 30 EC @ 1.5 ml per litre of water.

List of Pathogens on Cowpea

1.	*Colletotrichum lindemuthianum*	Anthracnose
2.	*Cercospora canescens* or *C.Cruenta*	Cercospora leaf spot
3.	*Pythium aphanidermatum* and *CorticiumSolani* (*Rhizoctonia solani*)	Seedling mortality
4.	*Fusarium oxysporum* f.sp. *tracheiphilum*	Fusarium wilt
5.	*Sphaceloma* sp. Or *Elsinoe phaseoli*	Scab
6.	*Colletotrichum truncatum*	Brown blotch
7.	*Uromyces appendiculatus*	Rust, bean rust
8.	*Entyloma vignae*	Leaf smut
9.	*Septoria vignae* Henn.	Septoria leaf spot
10.	*Choanephora cucurbitarum* (Berk. & Ravenel) Thaxt	Blight of cowpea or pod rot of cowpea
11.	*Macrophomina phaseolina*	Charcoal rot/ root rot
12.	*Erisyphe polygani* and *Sphaerotheca fuliginea*	Powdery mildew
13.	*Xanthomonas vignicola*	Bacterial blight
14.	*Cowpea mosaic virus* (CMV)	Cowpea severe mosaic
15.	*Cowpea aphid borne mosaic virus*	Cowpea aphid borne mosaic (CpMV)
16.	*Sclerotium rolfsii*	*Sclerotium* stem rot
17.	*Synchytrium dolichi*	False rust or yellow blister
18.	*Dactuliophora tarrii*	*Dactuliophora* leaf spot
19.	*Ascochyta phaseolorurn*	*Ascochyta* blight

Conclusions

Diseases are one of the major constrain in production of Cowpea. To prevent the Cowpea plant from these diseases we need to go for integrated disease management (IDM) approach where pesticides must be integrated with other disease control strategies. Before going for IDM, survey and surveillance should be done to identify the 'hot spots' of Cowpea diseases that will help in the management approach. There is an urgent need to create awareness among the farmers regarding etiology, epidemiology and management approaches of major diseases of Cowpea. There is scope of research in India to develop new technologies and new management skills in order to prevent the major diseases of Cowpea.

References

Adebitan, S.A. (1996). Effect of phosphorus and weed interference on anthracnose of cowpea in Nigeria. *Forage Research* **22**(1): 11–19.

Adebitan, S.A. and T. Ikotun. (1996). Effect of plant spacing and cropping pattern on anthracnose (*Colletotrichum lindemuthianum*) of cowpea. *Fitopatologia Brasileira* **21**(1): 5–12.

Aikins S.H.M. and Afuakwa. J.J. (2012). Effect of four different tillage practices on soil physical properties under cowpea. *Agriculture and biology journal of North America.* 3(1): 17-24

Amadioha, A.C. (1999). Evaluation of some plant leaf extracts against Colletotrichum lindemuthianum in cowpea. *Archives of Phytopathology and Plant Protection* **32**(2): 141–149.

Amadi, J.E. (1995). Chemical control of Cercospora leaf spot disease of cowpea (*Vigna unguiculata* [L.] Walp.). *Agrosearch* **1**(2): 101–107.

Armstrong GM, Armstrong JK. (1950). Biological races of the Fusarium causing wilt of cowpeas and soybeans. *Phytopathology,* **40**:181-193.

Bock, K.R. and Conti, M. (1974). Cowpea aphid-borne mosaic virus. Description of Plant Viruses, No. 134, Kew, Surrey, England.

Boruah. S, Dutta. P, Puzari, K. C. and Hazarika, G. N. (2015). Liquid bioformulation of *Metarhizium anisopliae*is effective for the management of Cowpea mosaic disease. *International Journal of Applied Biology and Pharmaceutical Technology*. **6**(1): 178-185.

Cissé, N. Thiaw, S., Baldé, M., Ndiaye, M., Ndiaye, A. & Wade, M., (2000). *Breading cowpea for the sahelian zone and resistance to striga and disease.* Bean/Cowpea CRSP,Midcourse 2000 Research Meeting, April 9-14, Senegal. p: 9-10.

French ER. (2000). Choanephora blight. In Compendium of Potato Diseases. Second edition. St. Paul, Minnesota, USA: APS Press, in press.

Gonzalez M, Garcia E. (1996). Evaluation of losses due to the rust on bean (*Phaseolus vulgaris* L.) in four sowing times in Cuba. *Agronomia Mesoamericana,* **7**:95-98.

Harris AR, Ferris H. (1991). Interactions between *Fusarium oxysporumf.sp. tracheiphilum* and Meloidogyne spp. in Vigna unguiculata. 1. Effects of different inoculum densities on Fusarium wilt. *Plant Pathology,* **40**(3):445-456.

Ladipo, J.L. (1976). A vein-banding strain of Cowpea aphid-borne mosaic virus in Nigeria. *J. Sci*. **10**(1–2) : 77–88.

Leach JG. (1923). The parasitism of *Colletotrichum lindemuthianum*. Minnesota Agricultural Experiment Station Technical Bulletin, 14.

Lehman PS, Machado CC, Tarrago MT. (1976). Frequency and severity of soybean diseases in the States of Rio Grande do Sul and Santa Catarina. Fitopatologia Brasileira, **1**(3): 183-193.

Lovisolo, O. and Conti, M. (1966). Identification of an aphid-transmitted Cowpea mosaic virus. *Nether. J. Pl. Pathol*. **72** : 265–269.

Mungo, C.N., A.M. Emechebe, and D.A. Florini. (1998). Isolation of Sphaceloma sp. from cowpea plant parts using eight media. *Crop Protection* **17**(4): 341–343.

Mulder JL, Holliday P. (1975). Cercosporacanescens. CMI Descriptions of Pathogenic Fungi and Bacteria, No. 462. Wallingford, UK: CAB International.

Oladiran AO. (1980). Choanephora pod rot of cowpea in southern Nigeria. *Tropical Pest Management,* **26**(4): 396-402.

Patel, P.N. (1985). Fungal, bacterial and viral diseases of cowpea in the USA. Pages 205–213 in Cowpea research, production and utilization, edited by S.R. Singh and K.O. Rachie. John Wiley and Sons, Chichester, UK.

Pio-Ribeiro, G.; Pappu, S.S.; Pappu, H.R.; Andrade, G.P. and Reddy, D.V.R. (2000). Occurrence of Cowpea aphid-borne mosaic *virus* in peanut in Brazil. *Pl. Dis.* **84** : 760-766.

Raheja, A.K. and Leleji, O.I. (1974). An aphid-borne mosaic disease of irrigated Cowpeas in northern Nigeria. *Pl. Dis.* **58** : 1080-1084.

Rachie, K. D. 1985. Introduction In S. R. Singh and K. D. Rachie (eds) Cowpea research production and utilization. John Wiley and Sons. Chichester, England.

Sastry, K.S. (2013). Plant Virus and Viroid Diseases in the Tropics. *Sping. Sci.* **1**: 1-361.

Shoyinka, S.A.; Thottappilly, G.; Adebayo, G.G. and Anno-Nyako, F.O. (1997). Survey on Cowpea virus incidence and distribution in Nigeria. *Intl. J. Pes. Mana.* **43**(2) : 127-132.

Sunder, S., S.K. Sharma, A.S. Taya, and O.P. Sheoran. (1996). Effect of age and quality of inoculum, temperature and pH of substrate on the pathogenic behaviour of Rhizoctoriasolani. *Plant Disease Research* **11**(1): 107–110.

Singh RS, Sinha RP. (1955). Studies on the wilt disease of cowpea in Uttar Pradesh. *J. Indian Bot. Soc.,* **34**: 375-381.

Swanson TA. (1984). Root-knot nematode and Fusarium wilt diseases of cowpea and soybean. Ph D. thesis University of California, Riverside.

Swanson TA, Gundy SDVan. (1985). Influences of temperature and plant age on differentiation of races of *Fusarium oxysporumf.sp. tracheiphilum* on cowpea. *Plant Disease,* **69**(9): 779-781.

Stavely JR. (1984). Genetics of resistance to *Uromyces phaseoli* in a *Phaseolus vulgaris* line resistant to most races of the pathogen. *Phytopathology,* **74**(3): 339-344.

Tu JC. (1983). Epidemiology of anthracnose caused by *Colletotrichum lindemuthianum* on white bean (*Phaseolus vulgaris*) in southern Ontario: survival of the pathogen. *Plant Disease,* **67**(4): 402-404.

Toller, R. W., S. S. Thompson, and J. M. Barber. (1963). Cowpea (southern pea) diseases in Georgia. *Plant Dis. Rep.* **47**: 746-747.

Ushamalini, C., K. Rajappan, and K. Gangadharan. (1997). Inhibition of Macrophominapha seolina and Fusarium oxysporumf.sp. tracheiphilum by antagonists under in vitro conditions. *Plant Disease Research* **12**(2): 167–170.

Williams, R.J. (1975). Diseases of cowpea (*Vigna unguiculata* [L.] Walp.) in Nigeria. *PANS* **21**(3): 253– 269.

14

Diseases of Soyabean (*Glycine max* L.) and Their Integrated Management

Ankita Das **and** ***Pranab Dutta***

Department of Plant Pathology, Assam Agricultural University, Jorhat, Assam

Soyabean (*Glycine max* L. Merril) belongs to the legume family Fabaceae is one of the world's most important legume crop and is the second most important cash crop in in India.

Economic Importance

Soyabean contributes to 25% of the global edible oil and about two-thirds of the world's protein concentrate for livestock feeding (Agarwal, *et al.*, 2013). In many countries such as China, Japan, Korea, Manchuria, Philippines, and Indonesia soyabean is considered to be a source of meat, milk, cheese, bread, and oil and therefore it is called "Cow of the field" or "Gold from soil" (Edwards *et al.*, 2019). Soyabean is a chief source of protein and oil as it contains about 40% protein and 20% oil. Many processed products are made from soyabean such as soy milk, soy flour, soy protein, tofu etc. Due to its amino acids composition, soyabean protein is called a complete protein and is consumed at homes for food purposes. The protein fiber that remains after the oil has been extracted is a good source of animal feed for poultry, pork, cattle, other farm animals and pets. Soyabean is an important foreign exchange earner due to export of soybean defatted oil cake (Agarwal, *et al.*, 2013). Bio diesel is produced from soyabean which is non-toxic, renewable and environmentally friendly and reduces particulate emissions.

Diseases of Soyabean

Pathogens attack soybean and infect all plant parts including root, shoots and stems, leaves, pods and seeds. The important ones are mentioned below:

Bacterial Diseases

Bacterial Blight

This is one of the most important bacterial disease in soyabean which occurs all over the world. Yield loss caused by bacterial blight is estimated to be 4% to 40% depending on the severity of the conditions (Faske, *et al.*, 2014).

Symptom

The symptom begins as small, angular, translucent, water-soaked yellow to light-brown spots. As the disease progresses the center of the spot becomes reddish brown and is surrounded by a water-soaked margin bordered by a yellowish-green halo. These spots then fuse to form large, dead patches on the leaves. These dead tissues may fall out or shredded away after strong winds and beating rains giving the leaves a ragged appearance. When disease is severe, early defoliation may occur.

Casual Organism

Bacterial blight is caused by *Pseudomonas syringae* pv. *glycinea.*

Disease Cycle

The pathogen overwinters in the crop residues that remains in the field and when these bacterial cells are carried away by rain or wind to the healthy leaves of soyabean plant infection occurs. The bacteria enter the plants through stomata or wounds on leaves. But in order to enter through the natural openings the leaf surface must be wet. The pathogen then starts secreting toxin which then inhibits chlorophyll production giving a yellowish green halo appearance in the infected leaves.Also, the pathogen can spread from infected leaves to uninfected leaves when the leaves rub against one another during cultivation (especially when there is dew), rain, or wind.

Epidemiology

Cool, windy, wet weather conditions with a temperatures of range 75 to 79°F favours the disease development while hot, dry weather suppress it.

Management

The severity disease can be prevented by the following methods:

- Crop rotation with non-host crop such as corn, sorghum, alfalfa, clover, or cereal grains will help to reduce the disease.
- High quality, disease-free seed should be used for planting.
- Spraying a mixture of streptocycline and copper oxychloride on young plants can help to reduce the disease (Faske, *et al.*, 2014).

Bacterial Pustule

Bacterial pustule of soybean is another major bacterial disease causing a yield loss of 18% to 20% (Horvath, 1926).

Symptoms

Initial infection starts as small, light green spots with raised centers on leaves on either surface of the leaf but are more common on the underside of the leaves. With the advancement of the disease small light colour pustules are formed in the center of these spots and these spots merge together to form irregular areas which appear as yellowish lesions and later the pustules become brown in colour.The infected areas may give the leaves a ragged appearance due to loss of tissues. Small raised spots may also develop on pods.

Casual Organism

Bacterial pustule is caused by the bacterium *Xanthomonas axonopodis* pv. *glycines*.

Disease Cycle

The pathogen persists on the winter crop residue in the field and and on seeds. The infection starts when the bacterium is carried away bywindblown rain, rain splash onto healthy leaves and the pathogen then enters through natural openings and wounds causing further infection.

Epidemiology

Unlike bacterial blight, this disease is favoured by warm temperature with frequent rains and the optimal temperatures for disease development ranges from 86-92°F.

Management

The effective strategies to manage the disease are as follows:

- Use of disease free planting material should be used.
- Crop rotation with non host crops also helps to reduce the disease.
- Treating the seeds in hot water also helps to eradicate the disease.
- Seed treatment with Thiram @ 3g/kg of seed and foliar spray of Copper oxychloride @ 0.3 % + Streptocycline @300 ppm can help to reduce the disease (Jagtap, *et al*., 2012).

Fungal Diseases

Anthracnose Stem Blight

Anthracnose disease of soyabean occurs worldwide causing a yield loss of 16% to 26% (Khare,1984).

Symptoms

The symptoms appearwhen plants approach maturity. The symptoms initiate as irregular brown spots which develop in a haphazard pattern on stems and pods which are later on covered with small black dots, which are the fruiting bodies (acervuli) of the fungus. Brown cankers can appear on petioles and cause defoliation. The fungus completely fill the cavity in the infected pods thereby inhibiting or reducing seed production ultimately reducing the yield.Infected leaves may develop brown veins and curl up.

Casual Organism

This disease is caused by *Colletotrichum truncatum.*

Disease Cycle

The pathogen overwinter on crop residue or on infected seeds and these infected seeds may consequently lead to pre-and post-emergence damping off or the fungus may colonise the seedlings thereby producing spores which may be carried away by wind or rain in turn affecting the healthy plants.

Epidemiology

Signs of infection generally appear when weather conditions become warm and moist.

Management

The effective strategies to manage the disease are as follows:

- Use of pathogen free planting material is effective for controlling the disease.
- Crop rotation with non legume crop and burying the infected crop residues help to prevent the pathogen build up.
- Foliar spray with SAAF and Mancozeb @ 2.5 g/l of water helps to reduce the disease (Khare, *et al.*, 2003).

Septoria Brown Spot

Septoria brown spot results in a yield loss of about 8% to 15%.

Symptoms: The symptom develop as brown irregular spots on both the surface of the leaf. Adjacent lesions merge together and form larger, irregularly shaped blotches and they generally turn blackish brown in colour and often the leaf spot develop a yellowish halo around them. Infected leaves turn yellow in colour and fall off prematurely.

Casual Organism

The disease is caused by the fungus *Septoria glycines*.

Disease Cycle

The pathogen overwinters on infected plant debris and occasionally on seed. The pathogen produces conidia on the infected parts of the plant which germinate on leaf surfaces and enter the plants through stomata causing infection. These conidia are dispersed via wind, rain on the leaves, petioles, stems and pods causing secondary infection.

Epidemiology

Warm and wet weather conditions are favourablefor disease development. Optimum temperature for disease development ranges from 79° to 83°F.

Management

Disease severity can be reduced by the following methods:

- Crop rotation with non host crop so as to allow time for soybean straw with fruiting bodies to degrade.
- Tillage practices should be followed to reduce crop residues infected with the pathogen in the field.
- Use of healthy planting material will control the disease infestation.
- Seed treatment with fluopyram also helps to reduce the disease.
- Biological fungicides, based on *Bacillus subtilis* also shows effective results in managing the disease (Lin and Kelly, 2018).

Frogeye Leaf Spot

Frogeye leaf spot causes a significant yield loss of 10-60% (Mantecón, 2008).

Symptoms

Symptoms mostly occur at the flowering stage.The symptoms initiate as small, yellow spots on the leaves which enlarge to a diameter of about ¼ inch. The center of these spots become grey to brown and have a reddish-purple margin. Yellow halo are not produced around the lesions in this case. As the disease progresses these spots may coalesce causing premature leaf drop. Stem develops dark red to brown lesions with flattened centers. Pods develop circular to oblong, reddish, sunken lesions and the pathogen grows through the pod walls and infect the seeds. Infected seeds become discoloured with circular to oblong, reddish, sunken lesions and cracked and flaking seed coats.

Casual Organism

Cercospora sojina is the causal agent of this disease.

Disease Cycle

Epidemiology

Warm and wet weather conditions with heavy dews and temperature of about 77-86pF and relative humidity more than 90% are favourable for disease development.

Management

The effective strategies to manage the disease are:

- Use of disease free, healthy planting material
- Planting resistant varieties also helps to reduce the disease. Three major dominant genes (Rcs1, Rcs2, and Rcs3) confer resistance to frogeye leaf spot.
- Residue decomposition should be practiced to reduce the inoculum level
- Application of strobilurin fungicide also helps to reduce the disease.
- Biocontrol agent such as *Bacillus amyloliquefaciens* is also a preventive method to control the disease (Rupe and Luttrell, 2008).

Cercospora Leaf Blight and Purple Seed Stain

This disease does not directly affect the grain yield but if seed discolouration occurs more than 50% delayed germination may take place.

Symptoms

The initial symptom of this disease is the development of light purple discolouration on the upper surface of the leaves that are exposed to light and later the spots develops on both the surface of the leaves. These spots expand and develop reddish-purple or bronze in colour and the infected leaves appear leathery and sunburned. Discoloured spots may conjoin forming large necrotic areas resulting in defoliation of the infected leaves. The reddish-purple lesions are also visible on the leaf stalks, stems and pods. In case of seeds, the seed coat develop pink to purple discoloration resulting in lower germination and seedling vigour.

Casual Organism

The fungus *Cercospora kikuchii is* the causal agent of *cercospora* Leaf blight and purple seed stain.

Disease Cycle

The pathogen overwinters in the crop debris in the field and the infected seeds wherein they produce spores. These spores are carried away by wind or splashing rain to the healthy plants thereby initiating infection. The pathogen produces a toxin called cercosporin that confers the reddish-purple discoloration to the diseased tissue.

Epidemiology

Infection and disease development is favoured by high temperature (82 to 86°F), rainfall and high humidity.

Management

The severity disease can be prevented by the following methods:

- Use pathogen-free seed and varieties with low susceptibility to this disease should be used.
- Crop rotation with non host crop like corn, cotton, rice or sorghum helps to reduce the disease.
- Using tillage practices that hasten decomposition of crop residue also helps to control the disease.
- The disease can be managed by spraying strobilurins, triazoles and mixtures of both.
- Infected seeds should be treated with a seed treatment fungicide when used for planting material purpose.

Downy Mildew

Downy mildew is distributed worldwide but it rarely causes yield loss.

Symptoms

The initial symptom begins as small, light green spots on the upper surface of the leaves which enlarge into pale to bright yellow spots. These spots later coalesces to form large irregular brown areas with a yellow halo. On the underside of the leaves, grey fungal-like tufts appear when the conditions for the disease is favourable. In the infected pods, fungal outgrowth appears and infects the seeds. Infected seeds have a dull white appearance and is covered with a whitish coating of fungal hyphae and spores.

Casual Organism

This disease is caused by *Peronospora manshurica*.

Disease Cycle

The pathogen overwinters in leaf debris and the on seeds as thick-walled spores i.e. oospores which is the initial source of inoculum. These oospores are carried by wind to the healthy plants where they geminate and cause infection. Once the infection initiates, sporangia are produced on newly infected leavesgiving the fungal tuft like structure on the lower surface of the leaves.

Epidemiology

Disease development is favoured by high humidity, extended period of dew and cool temperature 68° to 72°F.

Management

The best management practices for controlling the disease are as follows:

- Use of disease free, clean seeds helps in managing the disease.
- Growing soyabean for consecutive years in the same field should be avoided.
- Crop residues containing the pathogen should be burned.
- Fungicides such as mancozeb, maneb or zineb provides effective control against this disease (Rupe and Sconyers, 2008).

Soyabean Rust

Soyabean rust is a serious disease which causes yield loss of about of 30% to 80%.

Symptoms

Minute red tobrick red spots appears on the leaves. These spots coalesce killing off large areas in the leaves. As the disease progresses reddish-brown pustules containing the fungal spores develop on the undersurface of the leaves. Leaves on heavily infected plants turn yellow and die, and plants defoliate prematurely.

Casual Organism

The fungus responsible for downy mildew of soyabean is Phakopsora pachyrhizi.

Disease Cycle

P. pachyrhizi is an obligate parasiteso it requires a living host for survival.The pathogen survives on alternate host and overwinters there. The pathogen is carried away by wind to a healthy plant. The pathogen then enters through the natural openings or wounds and colonise the plant causing infection.

Epidemiology

Warm and humid climate favours disease development with a temperature range of 61° to 82°F, prolonged leaf wetness and relative humidity of 75% to 80%.

Management

The effective strategies to manage the disease are:

- Planting early maturing varieties helps to avoid the disease until the crop has been harvested.
- Wider row spacing with lower plant population may result in faster canopy drying
- Adjusting the level of potassium and phosphorus in soil helps to increase the disease resistance in plants.
- Chloronitriles, Strobilurins, and Triazoles are three groups of fungicides that are most effective against rust. Strobilurins is protectant fungicides so itmust be applied before infection occurs. Triazoles on the other hand, damages the pathogen in the plant and prevent pustules and spores formation (Simonetti, *et al.*, 2012).

Target Spot

Target spot of soyabean is an important foliar disease causing yield loss of about 18% to 32%.

Symptoms

Reddish brown round to irregular shaped spots with diameter of about 1cm or more appear in the lower to middle canopy. These spots are surrounded by yellow halo and at maturity they develop concentric rings and cause immense defoliation. Elongated brown lesions appear on the infected stems and petioles. On the infected pods, tiny circular purple to brown spots appear with brown margin.

Casual Organism

Target spot is caused by *Corynespora cassiicola.*

Epidemiology

The favourable conditions for disease development are wet weather conditions with prolonged leaf wetness, extended rainfall and high humidity (> 80%).

Management

The best management practices for controlling the disease are as follows:

- Crop rotation with non host crop is an effective measure to control the disease.
- Use of healthy quality seeds to reduce the disease.
- Wider spacing between two plants should be done so as to hasten canopy drying.
- The most effective fungicides against this disease is mixture of Fluxapyroxad + Pyraclostrobin (FLUX_PYRA), Epoxiconazole + FLUX_PYRA, prothioconazole + Trifloxystrobin (Subedi, *et al.*, 2015).

Pod and Stem Blight

It is a common soyabean disease which reduces the seed quality and effects the yield.

Symptoms

The most characteristic symptoms is the appearance of small, black, flask-shaped fungal fruiting bodies called the pcynidia which occurs in a rows in stems, petioles and pods. The infected seeds within the infected pods appear cracked and shriveled, and are covered with white mold and are smaller in size which have lower germination rate or produce weak and diseased seedlings.

Casual Organism

The pathogen responsible for this disease is *Diaporthe phaseolorum* var. *sojina*

Disease Cycle

The pathogen overwinters in the infected seeds or crop residues. The pycnidia present on the infected parts produces conidia which are source of inoculum for infection. This conidia are transported to the healthy plants via rain splashes or wind.

Epidemiology

Warm, wet weather during pod development favours disease development with seed moisture content of 19-35% and temperature of about 77 F.

Management

The best management practices for controlling the disease are as follows:

- Using high quality disease free seeds helps to combat the disease
- Tillage practices for decomposing the crop residues should be followed.
- Crop rotation with non host crop should be followed.
- Use of fungicide like Trivapro helps to control the disease and improve seed quality.

Viral Diseases

Soyabean Mosaic virus

An important soyabean disease which causes a yield loss of about 8% to 35% and upto 94% in some areas.

Symptoms

Leaves develop light and dark green mosaic pattern which may develop blistered or raised with time and the leaf margin curling downwards thereby causing premature defoliation. Infected seeds are mottled brown or black and are smaller in size.

Casual Organism

This disease is caused by Soyabean Mosaic Virus (SMV).

Disease Cycle

The pathogen is spread from plant to plant via soyabean aphids (*Aphis gylcines*). Once an aphid feeds on an infected soybean plant, it acquires the virus in a very short time and this infected aphid then feed on healthy plants spreading the disease.

Epidemiology

The disease is most pronounced during cool weather and disappear during hot weather.

Management

The effective way to combat this disease are:

- Use of clean, virus free seeds at the time of planting.
- Early planting helps to control the disease as late planting may coincide with higher aphid population.

- Dominant genes *Rsv1*, *Rsv3*, *Rsv4* and *Rsv5* confer resistance to SMV.
- RNA silencing plays an important role in seed transmission and virusinduced seed coat mottling.

Bean Pod Mottle Virus

Another important viral disease of soyabean is bean pod mottle virus causing a yield loss of about 10-60% (Yeh, *et al.*, 1991).

Symptoms

Younger leaves develop yellow to green mottling and as the disease progresses the leaves may become distorted. Pod formation may be reduced and infected seeds are mottled and discoloured. The characteristic symptom of this disease isgreen stem syndrome where in the stems and leaves retain the green colour even after the pods and rest of the field has matured.

A co-interaction may occur between Soyabean Mosaic Virus Bean pod mottle virus that may lead to a yield loss of 66% to 86% and causes severe dwarfing, foliar distortion, leaf necrosis and mottling.

Casual Organism

Bean pod mottle virus (BPMV) is the casual organism of the disease.

Disease Cycle

BPMV is transmitted by bean leaf beetle in a non persistent manner. The beetle feeds on the infected parts and then feeds on the healthy plants spreading the infection. The virus enters the host cell and replicates, then spreads to the surrounding cells, making its way through the plant systemically, as the plant develops symptoms.

Epidemiology

Cool temperature favours the disease development.

Management

The best management practices for controlling the disease are as follows:

- Use of disease free quality seeds for planting.
- Delayed planting should be done to avoid the beetle population.
- Remove and destroy all the infected plants.
- Use of foliar insecticide such as lambda-cyhalothrin helps to provide long term control against the beetles.

Tobacco Ringspot Virus

Tobacco ringspot virus causes a yield reduction of 25-100%.

Symptoms

The characteristic symptom of this disease is bud necrosis, proliferated growth of buds and leaves, dark and thick leaves, leaf vein discolouration shepherd's crooking, stunting, wilting, delayed maturity, underdeveloped pods and low seed formation rate. Infected stem show brown discolouration and may remain green for a long time as compared to the healthy plants.

Casual Organism

This disease is caused by Tobacco ringspot virus (TRSV).

Disease Cycle

Seed transmission is the most important mean of long-range dissemination and carry-over from season to season. Dagger nematode *Xiphinema americanum* and nymphs of *Thrips tabaci* also transmit the disease.

Epidemiology

Hot and dry weather is most favourable for disease development.

Management

The effective way to combat this disease are:

- Removal of the indigenous weeds of soyabean like Palmer amaranth (*Amaranthus palmeri*) and lambsquarter (*Chenopodium album*) that may serve as the source of inoculum.
- Disease-free, high-quality seed should be used for planting.
- Planting soyabean next to corn fields also helps to prevent the spread of the disease.
- Avoid planting of soyabean where there is infestation of dagger nematodes.

Soybean Vein Necrosis virus

It is a recently reported disease and the yield loss caused by this particular virus is under investigation.

Symptoms

The symptoms initiate as clearing of the vein along the main vein along with vein yellowing and as the disease progresses the leaves become distorted along with the development of lesions and the severely affected leaves dies off.

Casual Organism

This disease is caused by soyabean vein necrosis virus (SVNV).

Disease Cycle

The pathogen is transmitted by soyabeanthrips (*Neohydato thripsvariabilis*) in a persistent propagative manner i.e. the virus is acquired by the thrips in larval stage and carry it till they are adult. Adult thrips then feed on healthy plants and cause infection. Ivyleaf morning glory (*Ipomoea hederacea*), a weed of soyabean plant is an alternate host for the virus. Seed is another source of infection.

Epidemiology

Cool temperature and mild winter followed by warm spring is favourable for disease development.

Management

The best management practices for controlling the disease are as follows:

- Use of clean, disease free seeds to control the disease.
- Ivyleaf morning glory should be destroyed to control the disease.
- The population of the thrips should be under control.

Future Approaches

Future management strategies include improved diagnostic and investigation procedures to forecast the disease timely, evaluating the genes from wild host to develop resistance against certain diseases, gene editing or exploiting allelic diversity, development of novel resistance gene as well as markers and mechanisms that contribute to the expression of quantitative disease resistance or partial resistance.

Conclusion

Soyabean is affected by many diseases worldwide and their proper management strategies should be followed to combat the disease. Integrated strategies should be included such as seed treatment, use of resistance variety, crop rotation, proper use of fungicides. The ideal practice in combating the diseases of soybean is to combine the best of all the strategies. This will ensure the sustainability of the management practice with considerably lower input cost.

References

Agarwal, D.K., Billore, S.D., Sharma, A.N., Dupare, B.U. and Srivastava, S.K. (2013).Soybean: Introduction, Improvement, and Utilization in India-Problems and Prospects. *Agric Res.*, **2:** 293–300.

Edwards Molina, J. P., Paul, P. A., Amorim, L., da Silva, L. H. C. P., Siqueri, F. V., Borges, E. P., Campos, H. D., Nunes Júnior, J., Meyer, M. C., Martins, M. C., Balardin, R. S., Carlin, V. J., Grigolli, J. F. J., de R Belufi, L. M. and Godoy, C. V. (2019). Meta-analysis of fungicide efficacy on soybean target spot and cost–benefit assessment. *Plant pathol.* **68**(1): 94-106.

Faske, T.,Kirkpatrick, T., Zhou, J. and Tzanetakis, I. (2014). Soybean diseases. *Arkansas Soybean Production Handbook*, **11**: 1-18.

Horvath, A.A. (1926). Changes in the blood composition of rabbits fed on raw soy beans. *J. Biol. Chem.*, **68**:343–355.

Jagtap, G.P., Dhopte, S.B. and Dey,U. (2012). Bio-efficacy of different antibacterial antibiotic, plant extracts and bioagents against bacterial blight of soybean caused by *Pseudomonas syringae*pv. *glycinea*. *Sci J Microbiol.*,**1**(1): 1-9.

Khare U K. (1984). Studies on *Xanthomonas campestris*pv. *glycines* (Nakano) Dye, 1980, incitant of bacterial pustule of soybean (*Glycine max*). M. Sc. (Ag) thesis JNKVV, Jabalpur, India.

Khare, U. K., Khare, M. N. and Ansari, M.M. (2003). Bacterial Pustule Disease of Soybean-Present Scenario and Future Strategies.*Soybean Res.* **1**: 43-57.

Lin, B. and Kelly, H. (2018). *Frogeye Leaf Spot of Soybean*. The Plant Health Instructor. DOI: 10.1094/PHI-I-2018-1018-01.

Mantecón, J. D. (2008). Efficacy of chemical and biological strategies for controlling the soybean brown spot (*Septoria glycines*). *Ciencia e investigaciónagraria*, **35**(2): 211-214.

Rupe, J. and Luttrell, R. G. (2008). Effect of pests and diseases on soybean quality. In., *Soybeans*,AOCS Press, pp. 93-116.

Rupe, J. and Sconyers, L. (2008). Soybean Rust. *The Plant Health Instructor*. DOI: 10.1094/PHI-I-2008-0401-01.

Simonetti, E., Carmona, M. A., Scandiani, M. M., Garcia, A. F., Luque, A. G., Correa, O. S. and Balestrasse, K. B. (2012). Evaluation of indigenous bacterial strains for biocontrol of the frogeye leaf spot of soya bean caused by *Cercosporasojina*. *Lett. Appl. Microbial.* **55**(2): 170-173.

Subedi, S., Gharti, D. B., Neupane, S. and Ghimire, T. (2015). Management of Anthracnose in Soybean using Fungicide. *J. Nepal Agri. Res. Council*, **1:** 29-32.

Yeh, C. C., Hartman, G. L. and Talekar, N. S. (1991). Plant protection technology for vegetable soybean.*Vegetable Soybean Research Needs for Production and Quality Improvement,* pp. 86–87. ISBN 9789290580478.

15

Diseases of Pea (*Pisum sativum* L.) and Their Integrated Management

***Amar Bahadur*[1], *Pranab Dutta*[2] and *Durga Prasad Awasthi*[1]**

[1]*Department of Plant Pathology, College of Agriculture, Tripura Lembucherra Agartala-799210 (Tripura)*

[2]*School of Crop Protection, College of Post Graduate Studies in Agricultural Sciences Central Agricultural University, Umiam, Meghalaya*

Pea (*Pisum sativum* L.) is considered as the important pulse and popular leguminous vegetable in India. Pea belongs to the genus *Pisum*, a member of family papilionaceae. *Rhizobium* bacterium found in root nodules that fixing atmospheric nitrogen and helps in soil fertility. Pea (*Pisum sativum* L.) originated in the near East and Mediterranean regions. Pea diseases are affect the crop both quantitatively as well as qualitatively. Pea crop diseases caused by the fungal (rust, powdery mildew, downy mildew, aschochyta blight, septoria blight, fusarium root rot, aphanomyces root rot, alternaria blight, wilt, anthracnose, cercospora leaf spot, damping off, seedling rot), bacterial (bacterial blight and brown spot), nematode (root-knot nematode, cyst nematode and lesion nematode) and viral diseases (pea enation mosaic, pea mosaic, pea seed borne mosaic, pea early browning virus, pea streak, pea stunt and cucumber mosaic virus). India is the largest producer, consumer and importer of pulses in the world.

Powdery Mildew

Economic Importance

This disease was firstly reported in New Zealand in 1855 (Berkeley 1855). Powdery mildew is most prevalent in the late season and greater yield loss than early planted, damaging late sowings and late maturing varieties. Pea powdery mildew is an air-borne disease of worldwide distribution and very important disease of pea crop. Occurs in sporadically when warm humid conditions favour its growth late in the season. The disease is more serious than the downy mildew of peas. Their effect on reduction of seed size and reduce yields if disease develops during early to mid-pod stage. In susceptible varieties cause severe infections and significantly reduce yield. The disease can cause 25-50% yield losses, reducing total yield biomass, number of pods per plant, number of seeds per pod, plant height and number of nodes. Powdery mildew appears in epidemic form almost every year when the plants are in the pod stage towards the end of January and in February. Yield reduction due to disease is very high in a short period of time. Heavy reduction in pod formation occurred due to severe infection of powdery mildew as reported in Bombay (Uppal *et al.*, 1953). The losses in yield in a 100% infected crop were estimated by Munjal *et al.* (1963) to be 21-31% in pod number and 26-47% in pod weight.

Symptoms

Powdery mildew is most prevalent late in the season when warm days and cool nights result in dew formation. First symptoms appear of powdery mildew in the lower canopy leaves in the form of white floury patches then spreads to other green parts such as tendrils, pods, stems etc. The patches on the leaves originate in the form of minute specks from powdery mass. In the advanced stages of the disease large areas of the host get coveted with white floury patches. The disease usually begins as small distinct white tufts. Once it appears, the disease can spread very fast and the white fungal growth can quickly cover entire leaves and green tissues of plants (Singh, 2000). Powdery mildew causes white powdery fungal growth over all above-ground parts of the plant. As the disease progresses, black specs (cleistothecia) as reproductive structures often develop within the white fungal growth.

Infected plants impart dirty appearance. In extreme severe infections the infected leaves are shed leaving stem devoid of the leaves. In India, the disease generally appears during November-December.

Causal Organism

Powdery mildew is caused by the pathogen *Erysiphe pisi (Syd.).* The fungus is an obligate parasite having septate, hyaline, profusely branched superficial mycelium sending finger shaped haustoria to the host cells. Physiologic races of the fungus attack beans, urd, lucerne, coriander, turnip, cabbage and many other plants.

Disease Cycle

The primary infections get through ascospores released in soil by the disintegration of the wall of cleistothecium. Under favourable conditions, the ascospores come in contact with the lower most leaves of the host, germinate by germ tube and cause primary 'infection'. Later the symptoms in the form of white powdery mass contain conidia and conidiophores.

The secondary infection of the host takes place by conidia which are produced in large numbers and disseminated by wind. Upon reaching the host the conidia germinate and cause infection.

During the growing season, more than one secondary infection may occur because the establishment of infection and conidia production is a quick process. This results in severe large scale infection on the crops.

The cleistothecia-the perennating structures are produced on fallen infected debris in soil. These remain buried in the soil waiting for the favourable conditions. These contain asci and ascospores which serves as primary inoculums of the disease infection.

Powdery mildew infection and development is favored when dry, warm weather are accompanied by nights that are cool enough for dew to develop. The disease can develop very rapidly when environmental conditions are favorable. The fungus over-winters on infected pea trash and produces spores which are blown by wind into new crops.

Epidemiology

The disease powdery mildew, caused by the pathogen *Erysiphe pisi*, oversummers on infected pea trash and produces spores which are blown by wind into new crops. Under favourable conditions, the disease may completely colonise a plant in five to six days. Once a few plants become infected, the disease rapidly spreads to adjacent areas. Warm (15-25°C), humid (over 70 per cent RH) conditions for four to five days late in the growing season, during flowering and pod filling, favour disease development. However, heavy rainfall is not favourable for the disease as it will actually wash spores off plants. Night time dews are sufficient for the disease to develop.

Integrated Management

Fungicides function as protectants and eradicants. A protectant fungicide prevents new infections from occurring whereas an eradicant can kill an existing infection. Apply protectant fungicides to highly susceptible plants before the disease appears. Use eradicants at the earliest signs of the disease. Fungicide applications are highly effective against powdery mildew disease. Field sanitation, rotation of crop and destruction of diseased plant debris effectively control the disease. Sulphur dusting @ 25-30 Kg/hectare gives effectively control the disease. Wettable Sulphur such as Sulfex and Thiovit at 3 kg/ha, 0.03% Calixin followed by Karathane (0.2%) and Elosal (0.5%) and Morocide (0.1%) three time at 10 days interval have been found effective against disease. Do not apply sulfur when air temperature is near or over 90°F and do not apply it within 2 weeks of an oil spray. Plant-based oils such as neem oil and jojoba oil, work best as eradicants and also have some protectant activity. Early maturing and resistant varieties (Arka Ajit) should be cultivated, avoid late planting. Control volunteer field peas which can harbour disease.

Biological fungicides a bacterium, *Bacillus subtilis* commercially available when sprayed on the plant, destroy fungal pathogens and helps prevent the powdery mildew from infecting the plant. Neemazal a natural product of neem, induce resistance to pea powdery mildew (Singh and Prithiviraj, 1997).

Downey Mildew

Economic Importance

Downy mildew is one of the most common fungal diseases of pea and occurs in cool wet seasons. The disease is most severe early in the growing season when the crop canopy wet for long periods following rain or dews. This disease is most common and affect at any stage of growth during the moist and cool weather. Severely infected seedlings often die. A significant reduction in grain yield when the disease is severe. Systemic infection can lead to the appearance of the disease late in the season if conditions are conducive, but yield losses due to downy mildew arise from the stunting of plants early in their growth, or death of seedlings in more extreme instances.

Symptoms

The disease is most common after emergence and affect plants at any growth stage during periods of moist, cool weather. Downy mildew causes leaf damage and stunting of plants. Early infection causes systemic infection in plants that are a sickly yellowish-green and severely stunted and distorted. Symptoms appear as greyish white, downy growth appears on the lower leaf surface, and brown scattered patches appears on the upper surface of the leaf. Infected leaves can turn yellow and die if weather is cool and damp. Brown blotches appear on pods. The infected leaflets and stipules become reduced in size with their margins curled downwards. This growth is composed of branches sporangiophores, arising from the mycelium of the fungus which has en-tered within the leaf tissues. Sporangia are borne at the tips, which spared and cause secondary infections. Secondary infection results in the appearance of isolated greenish yellow to brown blotches on the upper surface of leaves and underside of the leaf, masses of mouse-grey coloured spores are produced.

Causal Organism

Downy mildew of pea is caused by *Peronospora pisi* Syd. but the *Peronospora viciae*, is responsible in Victoria, South Australia and Tasmania.

Pathogen mycelium consists of aseptate branched hyaline hyphae, confined to the intercellular spaces of the host tissues. The hyphae produce branched, finger-shaped haustoria which penetrate into the adjoining cells of the intercellular spaces.

The sporangiophores arise directly from the internal hyphae and emerge in clusters from the stomata of the under surface of the leaflets.

Sporangiophores are dichotomously branched 2 to 10 times branched at acute angles. The sporangia are nearly oval but narrowed a little below, greenish-yellow to pale-violet in mass and 22 to 27μ by 15 to 19μ in diameter. They are germinating by a germ tube in presence of moisture. The sporangia are blown about freely by the wind and spread the disease rapidly from plant to plant.

They germinate directly by germ tube behaving like conidia. In the old withered infected leaves oospores are found in the host tissues singly within a thin-walled oogonium, which soon disappears. The oospores are roundish, light-brown having a thick epispore marked by a large, raised reticulation. The oospores germinate by a germ tube. They can stand unfavourable conditions, like desiccation. They measure 26 to 43μ, in diameter.

Disease Cycle

The fungus that causes downy mildew survives in the soil and on pea trash and can be seed-borne. Infected seed can act as a primary source for systemic and local infections. Systemic infection of plants can lead to the disease developing late in the season if conditions are favourable. The disease can develop quickly when conditions are cold (5-15^{0}C) and humid over 90 per cent for 4-5 days, when seedlings are in the early vegetative stage. When individual seedlings become infected and act as foci of infection from which the disease spreads. Heavy dews will promote release of spores, while rain is the major means of spore dispersal, and secondary infection.

Infection starts in the leaflets and stipules from the inoculums produced by the germination of oospores of previous year. The secondary cycle started when the pods become infected from the sporangia. Host pene-tration is through stomata by the germ tube produced by sporangial germination. The sporangia are light and are developed profusely. They spread the disease very rapidly. Whereas the oospores are responsible for carrying the infection from year to year.

Epidemiology

Downy mildew is one of the most common fungal diseases of peas and is favoured by wet, cool seasons. Night temperatures below 10°C and morning dew promotes the disease. Dry, warm weather is unfavourable for the disease. Cool humid condition is very favourable for the development of the disease, while the disease incidence is slowed down with warm condition. The sporangia which behave as conidia spread the disease, but cannot stand desiccation.

Integrated Management

Pathogen survives from season to season in the form of oospores in the plant debris, the destruction of previous year's plant litter and following crop rotation of two to three years are very effective control measures and minimise the risk of disease. Spraying and dusting pea plants with fungicides are effective in limiting the spread of the disease. Growing a resistant variety is the most effective means of controlling downy mildew. There are two strains of the downy mildew fungus. The parafield strain that is considered as a non-virulent strain infects all conventional type tall field peas such as parafield. Whereas the new kaspa strain is more virulent and can infect conventional type, older field pea varieties as well as newer semi-leafless varieties such as kaspa. There are no commercial varieties with resistance to both strains of the fungus. Seed treatments reduce the number of seedlings with primary infection, thereby reducing the amount of air-borne spores that cause secondary infection in the surrounding crop.

Rust

Economic Importance

Rust of pea caused by *Uromyces fabae* (Pers) de Bary is considered the most important under warm and humid conditions and reported from different part of Europe, Africa; Canada; Australia. Yield losses in pea due to rust was also reported by Upadhyay and Gupta (1998). Losses depend on different growth stages of plant. Early infections completely destroy the crop. Occurrence of disease at pod formation stage reduced the grain yield. The disease is serious in northern India. It is serious on the late sown crop. Pea rust caused by *Uromyces fabae* (Pers.) de-Bary is a major disease of peas and is responsible for significant yield losses especially in the sub-tropical regions characterized by warm humid weather conditions (Kushwaha *et al.,* 2006). These conditions usually coincide with the reproductive phase of pea and favour rust outbreak (Kushwaha *et al.* 2007). One of the best possible ways to stabilize the productivity of pea crop is to grow rust resistant varieties. Resistance to rust in pea is reported to be governed by single dominant gene (Tyagi and Srivastava 1999), a oligogene

(Vijayalakshmi *et al.*, 2005) showing partial dominance along with some minor genes and 2-3 additive genes (Singh *et al.*, 2012).

Symptoms

The symptoms of this rust appear first in the month of January. All the green parts of the plant including pods are affected. Under cool and wet climate in northern India entire plant may be killed. Pustules develop which are powdery and orange brown in appearance. The earliest symptoms are the yellow spots having aecia in round or elongated clusters. Pea rust is characterized by the appearance of two types of symptoms in India.

Early symptoms develop on abaxial side of older leaves and form round to oval aecidia. Initially aecidia form creamy white to light yellow to bright orange colored pustules on the leaf and stem. Under favorable environment these pustules further developed and spread to other parts of the plants. The aecial and pycnial cups are produced on the same host plant. An aecidia is a cluster of several small cups like structure on the plant. Aeciospores released from the aecial cups are deposited as yellow powder. Small aecidial pustules are mostly confined to the leaf. However it can be seen on stem also. In tendril genotypes of pea it can also be seen on the stipules and tendril. Uredial pustules developed on both the surface of leaf but mostly confined to the stem. The uredosori are small, oval or round and brown in colour are present on the surfaces of leaves. The ruptured epidermis on infected portion of host exposes black to brown powdery mass. As the disease progresses, the dark brown coloured telia are formed on stem and leaves. Telial symptoms appear after aecial/uredial infection late in the same season. Teliospores are formed in the aecial or uredial pustules. Telia are mostly formed on the stem and tendril.

The first symptoms appear with the development of aecia. The yellow aecia appear first on the undersurface of the leaves, stems and petioles. The formation of aecial stage is preceded by a slight yellowing which gradually turns brown. The uredopustules are powdery light brown in appearance. All the four stages develop on every green part of the host including the pods. The teleutopustules occur in the same sources as the uredia and develop from the same mycelium (Singh, 1973).

Causal Organism

Rust of pea caused by *Uromyces fabae* de Barry is worldwide occurrence and attacks number of host species belonging to different genera of the family leguminoceae. It is a biotrophic, macrocyclic, autocious rust. Biotrophic plant pathogenic fungi distinguish specialized infection structures called haustorium. Pathogenic variability has been reported in field collection of *Uromyces fabae*.

The urediospores of *Uromyces fabae* was the only infective spore in temperate climate. Aeciospores acts as as repeating spore in case of *Uromyces fabae* and play important role in the out-break of disease under warm humid conditions.

Two species of *Uromyces* have been reported to cause rust of pea. One of them *U. pisi* (Persoon) de Bary, has been reported from several European countries (Deutelmoser, 1926; Mayer, 1947; Palter and Stetbiner, 1957). It is a heteroecious species having its aecial stage in *Euphorbia cyparissias* and rarely occurs in India.

In India, another species *U. fabae* (Pers.) de Bary has been found to cause pea rust (Butler, 1918; Prasada *et al.,* 1948; Kapooria *et al.,* 1966). *Uromyces fabae (Uromyces viciae- fabae)* the rust of pea was first reported by Persoon in 1801. Later de Bary (1862) changed the genus and renamed it as *Uromyces fabae* (Pers.) de Bary. The pathogen *U. fabae* is described as autoecious rust with aeciospores, urediospores and teliospores found on the same host plant (Arthur and Cummins, 1962; Gaumann, 1998). *Uromyces fabae* is an autoecious and heterothallic fungus forming all the four type of spores *viz.,* pycniospores/ spermatiospores, aeciospores, urediospores and teliospores on pea only.

Butler (1918) reported the occurrence of rust pathogen *Uromyces fabae* on pea and other leguminous crops from India. Prasada and Verma (1948) found several species of *Vicia, Lathyrus, Pisum* and *Lentil* susceptible to *Uromyces fabae* in India and abroad. In India species of *Vicia, Lathyrus* and *Pisum* are described as host plant for *Uromyces fabae* (Pers. de Bary). Conner and Bernier (1982) reported a total of 52 species oiVicia faba and 22 species of Lathyrus infected by *Uromyces viciae-fabae.* They also found this pathogen on pea, lentil and faba bean. Prasada and Verma (1948) also reported the occurrence of *Uromyces fabae* on lentil crop from Delhi. Roy (1949) in his list of fungi of Bengal recorded the prevalence of *Uromyces fabae* on the leaves and stems of *Pisum sativum.* Mitter and Tandon (1930); Pavgi and Upadhyay (1966) and Kapooria and Sinha (1966) reported the distribution of this pathogen in the regions of Uttar Pradesh, respectively. Bilgrami (1979) reported the occurrence of this pathogen on various host species of Pea, Lentil and Lathyrus. Baruah (1980) reported that rust infection on the pea plants is caused by both *Uromyces fabae* and *U. pisi. U. pisi* is of rare occurrence in India. Occurrence of *Uromyces fabae* have also been reported from Canada, Europe, Ethiopia and Australia in mild to severe forms on pea, lentil and faba bean.

Disease Cycle

Pea rust *(U. fabae)* is of worldwide occurrence and attacks number of host species belonging to different genera of the family *Leguminosae* in the Indo-

Gangetic plains (Butler, 1918). *Uromyces fabae* is a macrocyclic, autoecious rust fungus all spores are produced by single host (Mendgen, 1997). After overwintering on residual plant material, diploid teliospores germinate in the spring with a metabasidium. After meiosis, the latter produces four haploid basidiospores with two different mating types. These spores after landing on a leaf of a host germinate and produce infection structures. Pycnia are produced which contain pycniospores. Pycniospore are exchanged between pycnia of different mating types and after spermatization, dikaryotization occurs in aecial primordial. An aecium differentiates and dikaryotic aeciospores are produced. These aeciospores germinate and form infection structures from which uredia develops, which produce urediospores. Urediospore is the major asexual spore form of rust fungi produced in huge amount through repeated infection of host plants. Urediospores are dispersed aerially and can travel thousands of kilometers (Brown and Hovmoller, 2002). Rust disease of pea caused by *Uromyces fabae* is very severe under warm and humid conditions in Tarai region. Prasada and Verma (1948) reported that relatively low temperatures, 17-22^{0}C result in formation of secondary aecia while at 25^{0}C development of uredia takes place.

Epidemiology

The aeciospores were produced abundantly at the temperature range 10-15^{0}C and 20-25^{0}C at all the growth stages of crop. During pod formation the number of aecidia/pustules/leaf was highest. At higher temperature (20-25^{0}C) promoted more numbers of aecial pustules than the lower temperature. Teliospores were produced, when plants entered into the senescence stage. The aeciospores and urediospores of *Uromyces fabae* can't survive at a temperature more than 30°C for one week. Teliospores survives during the intervening season and germinates to produce basidiospore that produced pycnia, and subsequently causes infection in pea. Wide host range pathogen Conner and Bernier (1982) have suggested, the role of collateral host in the occurrence of disease on pea. They reported *Vicia* and *Lathyrus* species as a collateral host to *Uromyces fabae* and pathogen survive on these hosts during the absence of the main crop. Pea is grown through-out the year in the different part of the India its role in the multiplication and further spread of the *Uromyces fabae.* The optimum condition for aeciospore germinat at 25^{0}C with 100% relative humidity. The percent germination of the aeciospores decreased gradually with the decreasing temperature below 25^{0}C. None of the aeciospores germinated at relative humidity of 88.5% at a temperature below 25^{0}C. Longer leaf wet duration increased the rust severity. The mean temperature during the initiation of the disease ranged between 15-20^{0}C. Infection on leaves depended on moisture.

Integrated Management

A studied the effect of alteration in date of sowing on rust severity and grain yield in field pea. Delayed sowing is not only increased disease severity but also lower grain yield in plants having narrow spacing as compared to wider row spacing. Minimum disease severity recorded in pea + mustard inter cropped plants followed by the pea + wheat, pea + linseed and pea + rajma. Upadhyay *et al.,* (2018). Effects of different dates of sowing, inter-row spacing and intercropping on disease severity and grain yield of field pea (Tripathi and Rathi 2003).

The use of host plant resistance is the best means of rust control (Bayaa and Erskine, 1998). Screening of field pea germplasms under field conditions for resistance to rust has been reported in India (Singh *et al.,* 1995). Many factors such as biotic (previous pathogen attack) and various chemical and environmental stimuli (abiotic) may act on plants to induce systemic acquired resistance (SAR) to subsequent pathogen attack (Kauss *et al.,* 1992; Kessmann *et al.,* 1994; Dann and Deverall, 1995; Barilli *et al.,* 2010). SAR has been reported to be effective against a broad spectrum of pathogens including viruses, fungi, bacteria, nematodes and parasitic weeds (Beckers and Conrath, 2007).

Hiremath and Pavgi (1971) obtained complete inhibition of aeciospores germination of *U. fabae* with aureofungin to 20 g/ml and recommended early application of higher aureofungin concentrations to control rust disease. Sugha *et al.,* in 1994 reported sensitivities of aeciospores and urediospores to benzimidazole and triazole fungicides. They concluded that benomyl, carbendazim, thiobendazole and thiophanate methyl have very good potential for suppressing the early establishment of pea due to aeciospores, whereas benomyl, flutriafol and mycobutanil should be effective in suppressing the late infection due to urediospores. Management of the pea rust disease through a suitable control measures and loss can be minimized.

Ascochyta Blight

Economic Importance

Ascochyta blight is a serious disease of pea worldwide and is one of the most important diseases affecting field peas. The disease occurs in almost all pea-growing regions of the world and can cause significant crop losses when conditions are favourable for an epidemic. The causal fungi, *Didymella pisi. Mycosphaerella pinodes* and *Phoma pinodella* that occur as a complex in the field and cause a single disease, where the symptoms caused by each pathogen are undistinguishable. They can all be found on a single diseased plant. Severity of the disease varies from crop to crop and between seasons. In wet seasons when conditions are conducive, yield losses up to 60 percent. . It was first described in Europe by Libert who named the pathogen *Ascochyta pisi* in 1830 (Libert, 1829-1830). Ascochyta (Mycosphaerella) blight is the most widespread and economically damaging foliar disease in field peas. Infections can reduction in, productivity and even seed yield, if infection occurs early in the growing season. Along with weather conditions, timing of initial infection influences the effect of mycosphaerella blight on crop yield. Mycosphaerella blight symptoms in field peas should occur from the 10th node stage during the vegetative pea stages to the beginning bloom stage.

Symptoms

All parts of the pea plant can become infected by any of the three fungi causing Ascochyta blight but the pathogens are very difficult to distinguish based on symptoms in the field. Symptoms include the development of purplish black to brown spots or lesions on stems, leaves, tendrils, and pods. Black spore-producing structures may form in these lesions. Pod lesions may become sunken. The fungi can overwinter in seed, infected crop residue, and in the soil. Although symptoms may vary depending upon the fungus causing the infection, generally Asocochyta blight appears as a blackened stem, yellow foliage with brown blotches and bud drop. Both pods and seeds may be afflicted, and severe infections kill off seedlings.

The pathogen infects all above–ground parts of field pea plants as well as the crown below ground level. Symptoms consist of necrotic spots that coalesce into large lesions on stems, leaves and pods and a root rot may occur in severe cases. Early infection can cause seedling death. Lesions are purplish-black in colour and cause streaking of the lower stem. Conspicuous spotting of the leaves and pods also occurs. Spots on the pods may coalesce to form large, sunken, purplish-blackish areas. Infected seeds may be discoloured and appear purplish-brown .

Early symptoms (purple-brown irregular flecks) are first observed under the plant canopy on lower leaves, stems, and tendrils, where conditions are more humid. These flecks enlarge and coalesce; resulting in the lower leaves becoming completely blighted and falling off. Severe infections on the stem may lead to girdling near the soil line, which is known as foot rot. Typical foot rot lesions are purplish-black in color and may extend above and below the soil line. Girdling lesions weaken the stem and can lead to lodging and yield loss (Anonymous. 2008). In addition to girdling foot lesions, black to purplish streaks may develop on stems. Pod spots are gray to purplish, lack concentric rings, and are sunken.

Lesions caused by *Didymella pisi* differ from *Mycosphaerella pinodes*. *Didymella pisi* lesions are typically tan or brown in color with a distinct dark brown margin and visible pycnidia within the lesion. Pycnidia will also develop in lesions caused by *Mycosphaerella pinodes*, but will be less obvious as they blend into the already dark lesion.

Causal Organism

Asocochyta blight is composed of fungi, *Didymella pisi. Mycosphaerella pinodes* and *Phoma pinodella* which survive through the winter months in plant debris, wind and rain transmit spores onto healthy plants. The causal fungi, *Didymella pisi. Mycosphaerella pinodes* and *Didymella pinodella* that occur as a complex in the field that together can cause leaf, stem and pod spot, stem lesions and foot rot symptoms (Liu, *et al.,* 2013), Where the symptoms caused by each pathogen are undistinguishable.

Disease Cycle

Ascospores are the main source of primary infection, whereas the secondary infection is caused by production of conidia. Discharge of both types of spores needs rainfall or dew therefore epidemics are more severe in wetter conditions. Spores produced on infected foliage are transferred onto adjacent healthy plants by wind and rain splash. The fungi that cause ascochyta blight may either be seed borne, soil borne or survive in pea trash.

The disease usually becomes established when sexual ascospores of the fungus (*Didymella pinodella*), produced in perithecia on old pea stubble, are carried into the new crop by rain and wind causing early infection. Asexual conidia are produced by other pathogens in pycnidia (fruiting bodies) and can infect pea plants at any stage of plant growth. Pycnidia and Perithecia develop on infected plants throughout the growing season and after harvest on pea stubble and infected volunteer plants.

Field peas are the single host crop of mycosphaerella blight, caused by a pathogen that can be stubble, air, soil and seed-borne. *Mycosphaerella pinodes* overwinters on pea stubble and residue, the primary source of inoculum, and can survive on stubble or in the soil. Air-borne spores are released and spread by rain splash to plants nearby, or by wind to plants up to several kilometres away. Plant shoots can also be directly infected through exposure to resting spores in soil or from fungus on seeds that infects emerging seedlings. Foot lesions develop from infected seed, though seed is considered a minor inoculums source and risk of mycosphaerella blight infection transferring from seed to seedling is low.

Epidemiology

Disease normally will not develop at temperatures below 4°C and above 35°C Ascochyta blight disease development is favorer able temperatures between 20 to 21°C and high relative humidity. Cool, wet conditions encourage the initiation of infection and disease development. Field pea plants are infected throughout the growing season, with the production and release of new spores during wet periods. Mycosphaerella blight progresses upwards from the bottom of the plant, where symptoms appear on lower leaves, branches and the stem. Frequent precipitation and humid conditions in the lower canopy often cause greater disease severity.

Integrated Management

Management strategies include crop rotation, even though it has minimal impact reducing Mycosphaerella pinodes and using pathogen-free seed. Seed infection can negatively affect emergence and vigor (Kraft, and Pfleger 2001), but seed-to-seedling transmission in the field is low (Anonymous. 2008).

Ascochyta blight is best controlled by destroying infected pea trash and self-sown plants. The severity of disease may also be reduced by crop rotation, by the use of disease-free seed, resistant varieties, fungicidal seed dressing and foliar fungicides. Seed should be tested before sowing. Avoid early sowing at high seeding rates as this increase of pea seedlings to the ascochyta pathogens as increased high humidity and increase the risk of developing disease. Application of foliar fungicides before or during the early stages of mycosphaerella blight development can minimize.

Destroying pea stubble and reduce disease risk by minimising the number of spores to infect new crops. Avoid planting of crop near old field pea stubble. Previous pea crop residues can harbour the ascochyta blight pathogens. Fungi survive in soil and on old pea trash. Peas should not be sown on land planted to peas the previous year. The crop rotation should be fallow four to five years.

Septoria blight

Economic Importance

Septoria blight is a minor disease occurring sporadically and little effect on the yield of most of the pea varieties. The disease particularly occurs sporadically and disease is often seen on old foliage, pods and stems late in the growing season. Septoria blotch of field peas is usually a minor disease occurring sporadically but can cause major losses in some time. The disease is often seen on senescing plants late in the growing season. The disease occurs in most pea growing regions but seldom causes significant yield losses. In susceptible pea varieties, grain yield losses of up to 40 per cent have been reported.

Symptoms

The disease is mainly found on the lower, senescing parts of the plant and pods. Diseased areas on leaves are of indistinct size and shape and yellow then straw coloured to light brown often with a lighter centre and pale yellow halo on the margin. The disease is characterised by yellow blotches on plant tissue, which become necrotic and covered in numerous brown spots. Lesions vary in size, are roughly circular and have no distinct margin. First they appear yellow later becoming straw-coloured. Several such blotches may join to cover the entire leaf. As the blotches dry out many pinpoint-sized black pycnidia may be seen scattered widely on infected plant parts, including pods, diseased tissues may dry off prematurely.

Causal Organism

Septoria blight is caused by the fungus *Septoria pisi*.

Disease Cycle

The fungus survives from one season to the next on infected pea waste and seed. Spores of the fungus are carried by wind from infected waste into the new crop. Infection starts on the lower foliage where the humidity high and following rain or heavy dews. The disease is favoured by warm temperatures between 21 to 27°C with high humidity. Secondary spread occurs during the growing season. Rain splash helps in spreading the disease within a crop. Seed transmission can also occur .

Epidemiology

Disease development favours prolonged high humidity at least 24 hours and moderately warm temperatures of 21 - 27°C. Splattering of water assists in spreading the disease within a field. Seed-borne transmission can occur it is less important.

Integrated Management

Septoria can be managed by using an integrated approach that includes crop rotation, stubble management and fungicides. The septoria blotch fungus survives in soil and on old pea trash. Septoria blight disease can be controlled by destroying infected stubble and by crop rotation. Destroying pea stubble by forage, burning will help in reducing the amount of inoculums to infect new crops. It is safe to re-crop an area pea after pea whereas debris has decomposed. Crop rotation should be follow at least 2-3 years between successive field pea crops will minimise the risk. Almost all field pea varieties are moderately susceptible to septoria, however, the disease appears to be most severe on the short, semi-leafless varieties.

Aphanomyces Root Rot

Economic Importance

Aphanomyces root rot is a disease that affects in the legume family. *A. euteiches* is a root-infecting pathogen; primary symptoms occur on roots and subterranean stem tissues. Infected root tissue appears gray and water-soaked, becoming soft and honey-brown or blackish-brown in appearance. Ultimately roots are reduced in volume and function. It is common for symptoms in advance from roots into the stems, which characterize by chlorosis of the cotyledons and necrosis of epicotyls or hypocotyls. Primary symptoms of roots and stems will finally lead to secondary symptoms of chlorosis, necrosis, and wilting of the foliage. Pre-emergence damping-off is not commonly associated with Aphanomyces root rot. Roots infected by *A. euteiches* commonly have decreased nodulation.

Symptoms

Symptoms are common among hosts of *A. euteiches.* In pea lesions progress upwards on epicotyls and extend above the soil line, epicotyls eventually turn black and the tissues will collapse, resulting in a "pinched" appearance originating above the cotyledons. Sexual reproductive structures, oogonia and oospores can be readily seen in the cortical tissues of the root and hypocotyl with the compound microscope. Asexual reproductive structures, sporangia, primary spores, and zoospores of *A. euteiches* typically develop on, or near, the surface of host tissues. Primary spores are spherical and found loosely clustered as similar to a bunch of grapes, on the tips of a sporangium. The arrangement of primary spores on the sporangium is a characteristic feature of the genus Aphanomyces. Primary spores give rise to biflagellate (whip-like structures) zoospores which can be observed swimming in the water. Zoospores are attracted to roots in response to chemical signals released from roots.

Causal Organism

Aphanomyces euteiches is an Oomycete in the kingdom Chromista (Straminipila). *Aphanomyces* is in the order Saprolegniales, and is the only genus to contain species pathogenic to plants. *Aphanomyces cochlioides*, a pathogen of sugar beet, and *A. raphani*, a pathogen of radish, are also economically significant pathogens within the genus. *A. euteiches* has coenocytic (no septa) hyphae cellulose in its cell wall (chitin), and produces motile spores (zoospores). It is a soil-borne pathogen that spends a part of lifecycle in the soil and a part of in its host. The sexual spore of *A. euteiches* is the oospores. Oospores' is a spherical structure with double cell wall that remains dormant in the soil for several years in the absence of a host. A haploid female gametic nucleus formed in the oogonium and fertilized by a second haploid nucleus male gametic nucleus that has migrated from antheridium. Once fertilization occurs, a single oospore develops within the oogonium. *Aphanomyces euteiches* is characterized as homothallic meaning both the oogonium and antheridium arise from the same hypha and are compatible (self-fertile).

Disease Cycle

In the soil, oospores are stimulated to germinate by chemical signals exuded by host roots. Oospores form a germ tube that can either proliferate as hyphae or function as sporangia. Sporangia, which are indistinguishable from vegetative hyphae.

Spherical primary spores aggregate at the apex of the sporangium and will release zoospores that emerge through a pore in the cell wall of the primary spore. Zoospores are motile, kidney-bean shaped spores possessing one long "tinsel" and one short "whiplash" flagellum both attached laterally. Zoospores migrate through water-filled soil pores towards host roots. Once a zoospore reached the rhizoplane (root surface) loses both flagella, encysts, and germinates by formation of a germ tube, germ tube directly penetrate epidermal host cells and colonize throughout root and subterranean stem tissues. Within host tissues, hyphae differentiate into antheridia and oogonia and oospores are formed. The host range of A. euteiches is several legume species. *A. euteiches*, isolates express a diversity of pathogenic. The host specificity is based on the ability of the pathogen to progress into hypocotyls or epicotyls and initiate symptom development. The *forma specialis* (f. sp.) based on host range has been applied to isolates pathogenic to pea (*A. euteiches* f. sp. *pisi*) and isolates pathogenic to bean (*A. euteiches* f. sp. *phaseoli*). An isolates of *A. euteiches* are capable of infecting roots of several legume species. However, progression of the pathogen into the hypocotyl or epicotyl and the degree of pathogenicity and the host species from *A. euteiches* isolated. Within a host species, cultivars may react differently to isolates of *A. euteiches*, suggesting the existence of races.

Epidemiology

A. euteiches occur at any time during the growing season although it occurs during the seedling emergence. The symptoms of Aphanomyces root rot are sometimes difficult to distinguish from those caused by other root-infecting organisms viz; *Pythium, Rhizoctonia*, and *Fusarium*. Aphanomyces root rot is a monocyclic disease, as only one infection cycle per season. Oospores decaying host residue or in the soil germinate when chemical signals from host roots. Germination can occur directly as a germ tube that ramifies hyphae, or indirectly through the production of sporangia, which release primary spores and zoospores. Hyphae are resulting from an oospore or an encysted zoospore, infection can occur at any stage during the host's development and within a large range of temperatures. Infection and disease development of seedlings are best temperatures range 22-28° C. Zoospores require water for movement to host roots. Although water-saturated soils favor infection, symptoms development greatest, if warm and dry soil conditions occur post-infection. Symptoms of disease are observed within 10 days following infection. Oospores remain in host tissue or released into the soil as roots decompose. Once formed, oospores remain dormant and persist in the soil for more than 10 years, serving as primary inoculums of host crop.

Integrated Management

The effective management of Aphanomyces root rot through the use of resistant cultivars, improved soil drainage and avoidance of highly infested fields. Crop rotation has been used to slow the rate of inoculum build-up in pea production areas. Soil fumigants and biological control have been explored for use in commercial production. Pendimethalin and trifluralin are active ingredients of the dinitroaniline herbicides that have suppressive effect on Aphanomyces root rot of pea. Cover crops and green manure crops have been studied as lower the severity of Aphanomyces root rot of pea. Using of oats as a green manure has been successful, cruciferous plants have shown the greatest potential for controlling of Aphanomyces root rot. Isothiocyanates, thiocyanate ions, nitriles and epithionitriles are released from *Brassica* leaf or seed tissues by hydrolytic enzymes. In pea fields' incidence and severity of Aphanomyces root rot showed negatively correlated with the concentration of calcium in the soil. the application of calcium carbonate or calcium sulfate (gypsum) to soil can significantly reduced disease severity.

Bacterial Blight

Economic Importance

The disease is seed borne and more prevalent after frost and cause epidemics resulting in significant yield loss.

Symptoms

Bacterial blight disease first appears as small, dark green, water-soaked spots on leaves and stipules, and near the leaf base. The spots enlarge and merge often limited by the veins. The leaf spots turn yellowish and later brown and papery, spots on pods are sunken and olive brown, spots can develop on the stem near ground level. These begin as water-soaked areas that later turn olive-green to dark brown. Stem lesions may coalesce, causing the stem to shrivel and die. Stem lesions may spread upwards to the stipules and leaflets. Spots can merge causing the stem to shrivel and die.

Causal Organism

Disease of bacterial blight of pea caused by the bacteria *P. syringae pv. syringae,* is a serious disease of peas and considered as the main cause of the disease in pea crops.

Disease Cycle

Bacterial blight commonly becomes established within a field by sowing infected seed or from infected pea trash that is nearby. Infection may occur at any stage of plant growth. The pathogens can remain on the surface of plants without causing symptoms. Symptoms can develop damage by rain, heavy dew, and frost of plant tissues. Damage to field peas provides entry of bacteria into the plant tissue . Because the disease depends on wet conditions, bacterial blight is most severe in wet seasons. Pathogen has a wide host range including clover, common beans, faba beans, lentils, chickpeas and vetch which act as alternate hosts.

Epidemiology

During wet weather, bacteria spread from infected to healthy plants by rain splash and in wind-borne water droplets. A combination of excessive rainfall and strong winds provides the most favourable conditions for spread of the disease within crops. Early infections may lead to epidemics but later infection can also cause yield losses. Physical damage allows bacteria to enter plant tissue. The severity of bacterial blight can increase if plant tissue is damaged.

Integrated Management

Bacterial blight can be avoided by using an integrated approach to management that planting disease-free seed, crop rotation, variety selection and avoiding early sowing. Stubble is a significant source of inoculums. Destroy by burying, burning infected stubble. The survival time of inoculums is significantly reduced by burying pea trash 10 cm below the soil surface. Peas should not be grown on the same land more than once in three years. Early sown crops are more vulnerable to bacterial blight infection than late sown crops. If disease occurs the rotation should be extended to once in four years. The use of clean seed can minimise the possibility of disease. Do not use seed from bacterial blight during field inspections. Bacteria remain viable on seed for at least 2 years. Equipments used in an infected crop should be cleaned and washed thoroughly with disinfectant. Farm workers should move from crop to crop taking precautions against the spread of bacteria.

Future Approaches

Pea (*Pisum sativum* L.) crops are important source of protein and considered as the popular leguminous vegetable in India. Root nodules bacteria (*Rhizobium sp.)* that fixing atmospheric nitrogen and increase soil fertility. Pea is suffering several diseases that caused by fungus, bacteria, virus and nematodes. Diseases can affect on nodulation of crop and decrease fertility of soil. Pea is good sources of nutrient in human dies it may increase their quality through manage their disease.

Conclusions

Pea crops are suffering number of diseases during their crop growth that can reduce yield. These diseases can be minimized through preventative management. The pathogens generally seeds soil and air born in nature, spared and initiate the infection and caused diseases on crop plans. Disease management strategies through integrated management approaches can minimize the inoculums and manage the disease. The environment and health issue is one most factors for human in life, as world population gradually increasing food demand also increasing, in this aspect the crop diseases management required for increase crops yield, Pea crop disease in this aspect diseases awareness and their management strategies are necessary.

References

Anonymous (2008). Ascochyta blights of field peas. Online. Ministry of Agric., Gov. of Saskatchewan, Regina, SK, Canada.

Arthur, J.C. and Cummins, G.B. (1962) Manual of rusts in United States and Canada. Hafner Publishing Co. 438pp.

Butler E.J. (1918). Fungi and diseases in plants. Thatcher, Spink Co. Calcutta, 547 pp.

Berkeley, M. (1855). Vibrio forming cysts on the roots of cucumbers. *Gardener's Chronicle and Agricultural Gazette*, 14: 220.

Bayaa B and Erskine W. (1998). Diseases of lentils. In: Allen DJ and Lenné JM (eds.) The Pathology of Food and Pasture Legumes, CAB International and ICRISAT, Wallingford, UK. pp. 423-471.

Barilli, E. Rubiales, D. and Castillego M.A. (2010). Comparative proteomic analysis of BTH and BABA-induced resistance in pea (*Pisum sativum*) toward infection with pea rust (*Uromyces pisi*). *J. Proteomics.* **75**: 5189-5205.

Beckers GJM and Conrath U. (2007). Priming for stress resistance: from the lab to the field. *Curr. Opin. Plant Biol.* **10**: 425-431.

Baruah, H.K. (1980). Text book of Plant Pathology. Oxford and IBH, New Delhi.

Bilgrami, K.S., Jamaluddin, S. and Rizvi, M.A.(1979). The fungi of India part I (list and references). Today and Tomorrow's Printers and Publishers, New Delhi.

Brown, J.K. and Hovmoller, M.S. (2002). Aerial dispersal of pathogens on the global and continental scales and its impact on plant disease. *Science.* **297:** 537-541.

Conner, R.L. and Bernier C.C.(1982). Host range of *Uromyces viciae-fabae*. *Phytopathoogy.* **72:** 687-689.

Deutelmoser, E.(1926). Plant protection using natural defence system of plants. *Adv. Plant Patho.* **11:** 211-228.

Dann, E.K. and Deverall, B.J. (1995). Effectiveness of systemic resistance in bean against foliar and soilborne pathogens as induced by biological and chemicals means. *Plant Pathol.* **44:** 458-466.

Gaumann, E. A.(1998). Comparative morphology of fungi. Translated by Caroll Willian Dodge, Biotech Books, Delhi. pp: 563.

Hiremath, R.V. and Pavgi, M.S.(1971). In-vitro assay of aureofungin against some rust fungi, *Hind. Anti. Bull.* **13:** 83-86.

Kushwaha, C., Chand, R., Srivastava, C.P.(2006) Role of aeciospores in outbreak of pea (*Pisum sativum* L) rust (*Uromyces fabae*) . *Euro J Plant Pathol.* **115:** 323–330.

Kushwaha, C., Srivastava, C.P., Chand, R., Singh, B.D. (2007) .Identification and evaluation of a critical time for assessment of slow rusting in pea against *Uromyces fabae*. *Field Crop Res.* **103:** 1–4.

Kapooria, R.G. and Sinha, S.(1966). Studies on the host range of Uromyces fabae (Pers.) de Bary. *Indian Phytopathol.* **95:** 229-230.

Kauss, H., Theisinger-Hinkel, E., Mindermann, R. and Conrath, U. (1992). Dichloroisonicotinic and salicylic acid, inducers of systemic acquired resistance, enhance fungal elicitor responses in parsley cells. *Plant J.* **2:** 655-660.

Kessmann, H., Staub, T., Hofmann, C., Maetzke, T., Herzog, J., Ward, E., Uknes, S. and Ryals, J. (1994). Induction of systemic acquired disease resistance in plants by chemicals. *Annu. Rev. Phytopathol.* **32:** 439-459.

Kraft, J. M., and Pfleger, F. L. (2001). Compendium of Pea Diseases and Pests. 2nd Edn. *American Phytopathological Society*, St. Paul, MN.

Liu, J., Cao, T., Feng, J., Chang, K.F., H, S.F., and Strelkov, S.E. (2013). Characterization of the fungi associated with Ascochyta blight of field pea in Alberta, *Canada. Crop Protection* **54:** 55-64.

Libert, M. A. (1829-1830). Mémoire concernant les plantes cryptogames qui peuent être réunies sous le nom d'Ascoxytacei. Mem. *Soc. Sci. Agr. Lille* **1831**:174-176.

Mitter, J.H. and Tondon, R.N. (1930). Fungi flora of Allahabad, India. *J. of Indian Botanical Society,* **9:** 190-196.

Mendegen, K. (1997). The Uredinales. The Mycota Vol. V Plant Relationships Part B (Esser K & Lemke PA, eds), pp. 79-94. Springer-Verlag, *Berlin*. **57:** 267-276.

Mayer, E. (1947). Mycological notes- XII. Bull. Soc. Neuchatel, Sci. Nat. IXX: 33-60., Rev. Appl. Mycol., 158.

Munjal, R. L., Chenulu, V. V. , and Hora, I. S. (1963). Assessment of losses due to powdery muldew *Erysiphe polygoni* on pea. *Indian Phytopath.,* **16:** 260-267.

Pavgi, M.S. and Upadhyay, H.P. (1966). Parasitic fungi from North India VI. *Mycopathologia et mycologia applicata.* **30:** 527-260.

Palter, J. and Stetbiner, M. (1957). The principal pests and diseases affecting legume crops in spring. *Rev. Appl. Mycol.,* **530**.

Prasada, R. and Verma, U.N. (1948). Studies on lentil rust, *Uromyces fabae. Indian Phytopathol.* **1:** 142-146.

Persoon, D.C.H.(1801). *Synopsis Methodica Fungorum.* **1:** 224.

Singh, U. (2000). Pea-powdery mildew-an ideal pathosystem. *Indian Phytopathology.* **53**(1), 1-9.

Singh, U.P. and Prothaviraj, B. (1997). Neemazal-aproduct of neem (*Azadiracta indica*) induces resistance in pea against Erysiphe pisis . *Physiol. Mol. Plant Pathol.* **51:** 181-194.

Singh, A.K., Rai, R., Srivastava, C.P., Singh, B.D., Kushwaha, C., Chand, R. A. (2012). Quantitative analysis of rust (*Uromyces fabae*) resistance in pea (*Pisum sativum*) using RILs. *Ind. J. Agric Sci.* **82:**190–192.

Singh,V.K; Sangar, R.B.S. and Singh, R.N. (1995). Effect of varieties and sowing dates on disease incidence and productivity of field pea (*Pisum sativum*). *Indian J.Agron.* **4:** 451-453.

Sugha, S.K., Chauhan, R.S. and Singh, B.M. (1994). Sensitivity of aeciospores and urediospores of pea rust pathogen to selected systemic fungicides. *Trop. Agric.* **71:** 27-30.

Singh, R.S. (1973). Plant Diseases. Oxford and IBH, New Delhi. 512p.

Tyagi, M.K., Srivastava, C.P. (1999). Inheritance of powdery mildew and rust resistance in pea. *Annals Biol.* **15:** 13–16.

Tripathi, H.S. and Rathi, Y.P.S.(2003). Studies on epidemiology and management of rust of field pea. Final Technical Report, CAS in Plant Pathology, G.B.P.U.A & T., Pantnagar, Uttarakhand. pp 125.

Uppal, B. N., Patel, M. K. and Kamal, M. N. (1953). Pea powdery mildew in Bombay. Bombay, Dep. *Agr. Ball.,* p. 103.

Upadhyay, V., Kushwaha, K.P.S., and Pandey, P.(2018). Evaluation of potential fungicides for the management of pea rust under field condition. International journal of chemical studies, **6** (2): 3085-3090.

Vijayalakshmi, S., Yadav, K., Kushwaha, C., Sarode, S.B., Srivastava, C.P., Chand, R., Singh, B.D. (2005). Identification of RAPD markers linked to the rust (*Uromyces fabae*) resistance gene in pea (*Pisum sativum.*) *Euphytica.* **144:** 265–274.

16

Diseases of Onion (*Allium cepa* L.) & Garlic (*Allium sativum* L.) and Their Integrated Management

Shweta Singh, Chandramani Raj, Chandan Kapoor* and *Sangay Chuzem Lepcha

ICAR-National Organic Farming Research Institute, Tadong, Gangtok-737102 Sikkim

Onion (*Allium cepa*) and garlic (*Allium sativum*) are monocots originating from Asia. These are collectively called as Alliums and formerly classified with the Liliaceae (lilies and relatives). But, now they have been classified in their own family the Alliaceae. The plants of Alliaceae family are mostly used for seasoning and cooking. Some common crops in the family are bulb onion (*Allium cepa*), spring or salad onion (*A. cepa*), bunching onion (*A. fistulosum*), garlic (*A. sativum*), leek (*A. porrum*).

Onions and garlic are prone to several different diseases, especially in wet growing seasons. These often get started on their leaves, and if severe, can reduce bulb growth and yield. Another way onion diseases can cause damage is when they infect bulbs later in the season, which may lead to losses in storage. Some important diseases of onion and garlic are being discussed in this chapter:

	Disease	Causal organism
A.	**Fungal Diseases**	
1.	Stemphylium leaf blight and stalk rot	*Stemphylium vesicarium* (Wallr) E. Simmons
2.	Onion blast	*Botrytis* spp
3.	Downy mildew	*Peronospora destructor*
4.	Powdery mildew	*Leveillula taurica*
5.	Smut	*Urocystis cepulae*
6.	Smudge	*Colletotrichum circinans*
7.	Purple blotch	*Alternaria porri*
8.	Rust	*Puccinia allii* (synonym: *P. porri*)
9.	Black Mold	*Aspergillus niger*
10.	Neck rot	*Botrytis allii* (=*B. aclada*)
11.	Fusarium basal plate rot	*Fusarium oxysporum* f. sp. *cepae* (H.N. Hans.)

12.	Penicillium Mold/Blue Mold	*Penicillium hirsutum*
13.	White rot	*Sclerotium cepivorum*
B.	**Bacterial Diseases**	
14.	Sour skin	*Pseudomonas cepacia* (Burkholder)
15.	Bacterial soft rot	*Erwinia chrysanthemi*
16.	Slippery skin	*Pseudomonas gladioli* pv. *alliicola*
C.	**Viral Diseases**	
17.	Onion yellow dwarf	*Onion yellow dwarf virus* (OYDV)
18.	Garlic Mosaic	*Onion yellow dwarf virus* (OYDV) and *Leek yellow stripe virus* (LYSV)
19.	Iris Yellow Spot	*Iris Yellow Spot Virus*

Fungal Diseases

Stemphylium Leaf Blight and Stalk Rot

The disease is one of the most serious and devastating disease of onion limiting the quality and quantity of both bulb and seed. It is a common late-season problem in processing onions grown in most of the onion growing areas and commonly occurs in onions that are maturing and starting to lodge. Infection of seed stalks can reduce seed yield and quality. Significant yield loss can occur in onion fields severely affected by SLB-such outbreaks have been reported to reduce yield by as much as 60%.

Symptoms

Disease is distinguished as small, light yellow to brown and water soaked lesion develop on leaves. These small lesions grow into elongated spots that frequently coalesce resulting in blighted leaves. Lesions usually turn light brown to tan at the centre and later give olive brown to black color on the leaf. The fungus overwinters in the plant debris. In growing season, it infects plants after long warm periods when leaves remain wet.

Causal Organism: *Stemphylium vesicarium* (Wallr) E. Simmons

Tel: *Pleospora allii* (Rabenh) Ces. & de Not

Disease Cycle and Epidemiology

The disease cycle of *Stemphylium vesicarium* is characterised by sexual and asexual phases, and the SLB fungal spores are blown into onion fields from nearby plants. Ascospores are released from pseudothecia on dead plant tissues from the previous season and conidia (asexual spores) are released from leaf infections of nearby onions. Infection of onion leaves by conidia can occur by various mechanisms such as through stomatal openings, wounds caused by other diseases, insect pest feeding, or physical damage such as wind, hail and herbicide damage and direct infection of dead or dying leaves. The onion plants subjected to heat stress are more susceptible to SLB.

Stemphylium vesicarium is a weak pathogen, so it often infects onion plants secondarily, after other diseases develop or when the plant is otherwise stressed. These fungi are introduced into onion fields by windblown spores from nearby plants. Within the fields, the fungi persist by surviving on infected plant debris. Optimum temperatures for infection are between 25 to 30°C for *Alternaria* and 18 to 25°C for *Stemphylium*. Both pathogens require wounds caused by other diseases (e.g. botrytis), thrips feeding, or hail, to enter the plant. In severe cases, lesions enlarge and coalesce to blight the entire leaf. The airborne spores are produced on the lesions throughout the growing season and disperse to adjacent leaves and plants. As leaves get older, they become more susceptible.

Epidemiology

Moderate temperatures and larger canopy humidity favors development of disease. Heavy dew favors their development in desert areas, but foggy and rainy weather promotes their development in other regions. The optimal temperature for disease development is in the 20-25 (°C). In most areas, these diseases are usually not a problem after the end of the rainy season. However, in some areas they can cause damage in the absence of rain, when there are conditions of high humidity, night time dew, and moderate temperatures.

Management

- Selection of resistant varieties like VG-18, VG-19, Bhima Shweta, Bhima Dark Red etc.
- Crop rotation for a period of 3- to 4-year rotation can help to reduce the amount of inoculum present and reduce disease incidence.
- Avoidance of excessive nitrogen applications which can increase disease severity.
- Burying crop residues at the end of the onion growing season by deep tillage reduces SLB by facilitating plant decomposition, increasing the action of pathogen antagonists, and preventing aerial release of ascospores after pseudothecia mature Proper disposal of culls and other plant debris. Culls and plant debris can be a source for both pathogens and insects that cause wounding on new onion plants. Debris can be buried or disposed of in the trash.
- Methods to reduce leaf wetness duration to reduce the incidence and severity of SLB include increasing plant spacing in seedbeds to facilitate air movement

and rapid drying of the foliage, aligning rows of plants to follow the direction of the prevailing wind, and irrigating crops during the late morning or early afternoon to allow leaf surfaces to dry rapidly, thereby reducing the potential of infection

- Reduction in plant density to have a good air flow
- Adequate and proper drainage in the field should be ensured before planting.
- Use of fungicides like Tebuconazole 25%EC, Propiconazole 25% EC, Azoxystrobin 25% SC can be used to control both pathogens.
- Use of biocontrol agents like *Trichoderma viride* and *Pseudomonas fluorescens* will also be helpful in controlling the disease.

Future Approaches

The disease spread can be monitored and managed by laying emphasis on development of disease forecasting systems. Development of such mechanisms can help in reducing the number of applications of fungicides required to keep the disease below the economic threshold level. Advancements in technology such as remote sensing to monitor crop characteristics have been Stemphylium disease will be a potential tool to detect diseases more quickly and with higher accuracy and precision using spectral changes in crop canopy.

Onion Blast

The disease is commonly called as leaf blight disease also. This common foliar disease is an important cause of quality loss in green onion, where the condition of the foliage is important for market. It can cause serious loss of foliage in bulb onion, and can adversely affect the maturity and quality of the bulbs thus impacting the yields.

Symptoms

The first symptoms are leaf spots that are small, white, and have pale green halos. Spots are usually elliptical in shape and can increase to 5–10 mm in diameter. Multiple spots coalesce and result in bleached foliage and general collapse and death of leaves.

Leaf dieback can be found 5 to 12 days after the first spotting symptoms. Older leaves are more susceptible than younger foliage. The white spot symptoms may be confused with hail damage or physical abrasion from wind-blown soil. However, these abiotic problems are usually confined to the exposed side of the leaf, are more irregular in shape, and lack the diagnostic halo.

Causal Organism: *Botrytis allii* Munn

Botrytis byssoidea Walker

Botrytis squamosa Walker

Botrytis cepa Hanzawa

Botrytis leaf blight or Botrytis blast disease is caused by *Botrytis squamosa. Botrytis squamosa* has larger conidia (14–24 x 9–18 µm) than the other common *Botrytis* pathogens on onion; *B. cinerea* has conidia measuring mostly 9–14 x 7–10 µm, and *B. allii* conidia are 7–11 x 5–6 µm. *Botyrtis squamosa* is difficult to isolate from young lesions, so examine senescent leaves and older lesions for *B. squamosa* spores. This pathogen produces black sclerotia, measuring 3–10 mm, on leaf and bulb tissues. *Botrytis squamosa* appears to be host specific to *Allium* species. In some cases the fungus produces a perfect stage, *Botryotinia squamosa*. It is not clear what role this teleomorph has in disease development.

Disease Cycle

The fungus survives in the soil on infected plant debris. In wet weather, the spores that come into contact with susceptible plants germinate and enter through wounds. Initial inocula are airborne spores that originate from nearby crops, debris of previously infected crops, and sclerotia in the soil. Inoculum production within the crop usually increases after the oldest leaves have declined. Leaf wetness for a period of atleast 24 hours is necessary for infection to occur. However, at least 6 hours of leaf wetness is required for infection by conidia, and the number of lesions that develop increases as leaf wetness duration is prolonged. The extent of the blight is directly related to the length of the time, the foliage remains wet and long wetness period result in severe outbreaks. Optimum conditions for development are temperatures of 12–24°C. Disease development is favored by high crop density; hence spring onion plantings are more prone to problems than the wider spaced bulb onion.

Epidemiology

The fungus may over-winter in infected plant material or may survive in the soil as small, dark brown sclerotia. During moist periods with moderate temperatures, fungal spores are dispersed from sclerotia, infected leaves and debris to initiate infection. This disease can spread rapidly when environmental conditions are favorable for development.

Mangement

- Planting of healthy onion seeds or sets in well drained soil where air circulation is good.
- Crop rotation for atleast 3 to 4 years. It is advisable not to plant successive crops, especially of spring onion, where Botrytis leaf blight occurs regularly.
- Early season onion crops should not be placed close to overwintered onion.
- Use of tolerant cultivars as this pathogen is most severe on *A. cepa* varieties. If possible plant other types of allium crops for example, chives are reported to be immune.
- Increase row spacing to improve airflow through the crop.
- Avoid excessive nitrogen fertilizer, as this causes thick leaf canopies and increased susceptibility.
- Irrigate early in the day so that leaves dry quickly.
- If available, use forecasting systems to identify periods of disease risk and hence guide timing of fungicides.
- Destruction of infected plant debris after harvest
- Multiple fungicide applications may be required for spring onion, while bulb onion may not need any treatment because of low susceptibility.
- Incorporate crop residues after harvest so that inoculum levels are reduced.

Downy Mildew

Symptoms

Typically the first symptom observed is the brownish-purple velvet-like sporulation of the pathogen on healthy green leaves. As the disease progresses lesions which are slightly paler than the normal leaf color, enlarge and may girdle the leaf. These lesions progress to a pale yellow followed by brown necrosis resulting in collapse of the leaf tissue. Infected seed stalks tend to remain pale yellow and, as with the foliage, are often invaded by other fungi, typically *Stemphylium* or *Alternaria* species. Field infections usually begin in small patches and progress rapidly throughout the field. Bulbs can be infected and may either rot in storage, or if planted, give rise to pale green foliage.

Causal Agent: *Peronospora destructor*

Peronospora destructor (Berk.) Casp. the causal organism of onion downy mildew is an obligate parasite in the family Peronosporaceae under the order Peronosporales of the class Oomycetes. As other members of the class it reproduces sexually by oval shaped oospores measuring 15.8 - 43.2 /u which under favourable conditions germinate by germ tubes. Asexual reproduction is by lemon shaped sporangia measuring 53.4 - 61.7 x 24.2 μ. Sporangia are produced on dichotomously branched and determinate sporangiophores. These sporangia always germinate by germ tubes. *Peronospora destructor* parasitises onion tissues intercellularly deriving nutrients using haustoria which when mature curl and divide and measure 80 - 100 x 2.5 – 5 μ. This absorption of nutrients can result into a green weight reduction of upto 53% per night, and this green weight reduction has a positive correlation with the reduction in yield mentioned earlier.

Epidemiology

The fungus survives in volunteer onion plants, onion sets, plant debris or in the soil. The fungal spores are disseminated onto plants by winds and splashing rain during cool wet weather, which is essential for disease development. Rain, dew or high humidity (>95%) is required for fungal spore germination and infection. The fungus grows internally and continues to produce spores as long as the weather remains cool and wet.

Disease Cycle

Downy Mildew fungus over winters primarily as mycelium in infected onions that remain in onion fields or in nearby cull piles. The pathogen also can over winter in perennial varieties of onion in home gardens. It is suspected that spores of the fungus that persist in the soil may directly infect the roots of young onion plants. These plants become systemically infected and serve as focal points for infection in commercial onion fields. When favorable environmental conditions occur, the over wintering fungal mycelium in systemically infected plants produces spores. After dissemination through the air, these spores infect the leaves of onion plants in commercial fields. Spores are formed at night when high humidity and temperatures of 4–25°C occur, with an optimal temperature of 13°C. The spores mature early in the morning and are disseminated during the day. Spores remain viable for about 4 days. Germination occurs in free water from 1–28°C with an optimal range of 7–16°C. Rain is not needed for infection if heavy dews occur continuously during the night and morning hours. The mycelium of DM in leaves of infected onion plants in commercial bulb production fields produces a new crop of spores called conidia in cycles of approximately 11–15 days. As the upper portions of a leaf are killed, the fungus infects the next lower part of the leaf in each successive cycle of spore formation. Such cycles can be repeated several times until the leaf may be completely killed. These repeated cycles of spore formation can result in severe and continued epidemics of DM if disease favorable environmental conditions persist.

Management

- A regular fungicide spray program based on climatic conditions can reduce crop losses. Three spraying with Mancozeb 0.2 % is effective. Spraying should be started 20 days after transplanting and repeated at 10-12 days interval.
- Avoid planting onion sets that are contaminated with the fungus.
- Eliminate plant debris and cull piles.
- Plant rows in the direction of the prevailing winds and use furrow irrigation rather than sprinkler irrigation.
- A 3-4 year rotation out of onions in areas where the disease is present can help reduce losses.

Powdery Mildew

Symptoms

Circular to oblong chlorotic lesions 5-20 mm (0.2-0.8 in.) in diameter develop on older leaves and rarely on younger leaves prior to bulb initiation. Sporulation gives lesions a gray to white powdery appearance. Chlorosis and eventually necrosis may develop around areas of sporulation. Lesions may coalesce to cover large areas of the leaf surface. This disease appears to be most common on varieties with glossy leaves, which are associated with thin cuticular waxes.

Causal organism: *Leveillula taurica*

Epidemiology

Leveillula taurica overwinters in crop residue and many alternate hosts. Conidia are spread primarily by wind. Environmental conditions that favor infection include relatively hot, dry conditions of summer months that have followed cool, wet spring weather and low relative humidity.

Disease Cycle

The polycyclic disease cycle of *L. taurica* is similar to that of other powdery mildew species. It overwinters (as chasmothecia) in crop residues above the soil surface. Under favorable climatic conditions, the chasmothecia open and release ascospores, which are wind-dispersed. The ascospores enter the host through its stomata, germinate, and colonize the host's tissues with its mycelia. The pathogen then begins to produce its asexual conidia, either singly or on branched conidiophores. The conidia exit through the host's stomata and serve as a secondary inoculum to spread disease after initial infection. In the fall, the pathogen undergoes sexual reproduction and again produces chasmothecia, its dormant, overwintering structure.

Management

- Planting seed or high quality seedlings transplanted into clean fields that were rotated out of onion production for atleast 1-2 years
- Control of volunteer onions will prevent the pathogen from carrying over from one crop to next.
- Following harvest, removal of crop residue, deep tillage and rotation to a non-host crop for at least one year will help eliminate the pathogen.
- Fungicide sprays like Pyraclostrobin will help to control this disease.
- Avoid excessive nitrogen fertilization and moisture stress.

Smut

Smut occurs in many onion growing areas and is occasionally important. It affects bulb and salad onion, leek, shallot, and chives. Garlic appears to be immune and resistance is present in *Allium fistulosum*.

Symptoms

Symptoms can appear as early as seedling emergence, when cotyledons show infections. Cotyledons, true leaves, and leaf sheaths develop oblong to elongated, dark, raised blisters on outer surfaces. These growths can cause downward curling of the leaves. Mature blisters break open to expose black, powdery fungal growth. There is progressive spread of the disease inwards, which can kill seedlings within 3 to 4 weeks. Infected bulb tissue remains firm, though secondary decay organisms can penetrate damaged areas and cause rot. Another disease, smudge (caused by *Colletotrichum circinans*), also produces black fungal growth on onion, leek, and shallot, so differentiating smut from smudge in the field may be difficult.

Causal Organism

Smut is caused by *Urocystis cepulae* (synonymous with *Urocystis colchici* var. *cepulae*). It is a basidiomycete fungus belonging to the Ustilaginales. This pathogen produces distinctive black spore masses that consist of chlamydospores. These are spherical, single-celled, brown to black, 12–15 µm in diameter, with an outer layer of small sterile cells, 46 µm in diameter.

Disease Cycle

The pathogen is soil-borne and can survive for up to 20 years in soil. The pathogen may be introduced on sets or transplants. Seedborne infection is not considered important. Chlamydospores can be spread by winds and water, and optimum temperatures for germination are 13–22° C. Most plant infection occurs at 10–12°C and disease activity is greatly reduced above 25°C.

Epidemiology

The fungus can over-winter as resting spores in the soil for several years. Spread of the fungus occurs through infected onion sets, transplants and when spores are transported by wind, equipment and water. Onion seedlings are susceptible to infection from just after germination until they reach the first true leaf stage. As each new leaf emerges it goes through a growth phase where it is susceptible to infection. After that growth phase, infection does not occur. Optimum temperatures for spore germination and growth are 13-22°C while both are decreased above 25°C.

Management

- Healthy onion sets and transplants that are planted into infested soil may escape infection as these are able to resist infection from soilborne inoculum.
- Chemical seed treatments can protect seedlings through the susceptible stage.
- A crop rotation out of onions for three or more years also reduces disease.
- Planting to be undertaken when soil temperatures are higher.

Smudge

Smudge is a common disease of onions occurring both in the field and in storage or transit. It is confined for the most part to the bulbs and is characterized by dark green to black spots of variable size and shape on the outer scales. The spots may be homogeneous in appearance or may consist of numerous individual stromata scattered miscellaneously or arranged in concentric rings. The disease is most common on the white Varieties of onions and damages materially the appearance and market value of the crop.

Symptoms

The disease occurs late in the season as the crop matures and continues to develop on bulbs in storage. The fruiting bodies of the fungus turn from dark green to black as they mature, and form concentric rings around the neck and on the surface of dry outer bulb scales. If the humidity is high, the disease may spread to the inner scales, causing small, yellow lesions. If the disease continues to develop, the bulb may shrivel and sprout prematurely. Under warm, wet conditions this fungus can cause damping-off and leaf spotting.

Causal Agent: *Colletotrichum circinans*

Colletotrichum circinans is an ascomycete parasite of the Allium family, which includes leeks, onions, garlic, and shallots among its hosts. It has large, dark setae along its acervuli 80-315µm thick and bears cream-colored spores, 14-30µm long and 3-6µm wide.

Epidemiology

The fungus can over-winter in the soil on colonized onion debris for years after an outbreak and can be introduced on infected bulbs. During spring and summer months when conditions become warm and wet it favors conidial production and wind and rain splash spread the conidia by colonizing the outer scales. These conidia infect mature bulb scales and cause disease when free moisture and optimum temperatures [20-26°C] for infection occur. It is possible for *C. circinans* to complete its life cycle in less than a week. Its sexual stage is rarely seen in the wild.

Management

- Rotate crops on a 2-3 year basis as a solution for white cultivars.
- Maintain good drainage.
- Do not harvest during wet weather.
- Remove affected bulbs from group when found.
- Keep stored bulbs dry.
- Use only healthy transplants.
- Use colored varieties of onion.
- Cultivate in soil free of disease for 2-3 years if possible.

Purple Blotch

This disease occurs throughout the world but is most damaging in areas with warm, humid climates. It is an important concern on onion, garlic, and leek. Among the foliar diseases, purple blotch is one of the most destructive diseases, commonly prevailing in almost all onion growing pockets of the world, which causes heavy loss in onions under field conditions. For the first time the purple blotch of onion caused by *Alternaria cepulae* was observed by Ponnappa (1970) in Karnataka.

Symptoms

The initial symptoms of disease are small, water-soaked lesions on leaves or seed stalks that rapidly develop white centers, which gradually enlarge, turn yellow, and develop a pale tan center. As the lesions grow in size, they become zonate and brown to purple in colour, are oblong in shape, and may be several centimeters long. The margin of the lesion is often a shade of red or purple and is surrounded by a yellow zone. Under moist and humid conditions, the surface of the lesions may be covered with brown to dark grey masses of spores dark fungal spores. Large lesions are produced on flower stalks and these can be dark purple. When numerous lesions occur on leaves, dieback of the foliage can take place. Seed stalks, flowers of seed onions and at harvest, bulbs may become infected. The main source of perennation of the pathogen is dormant mycelium which remains alive in diseased crop residue.

Causal Organism: *Alternaria porri*

Purple blotch is caused by two pathogens in the fungi imperfecti group. Purple blotch lesions can be caused by either fungus alone or by both fungi together. *Alternaria porri* forms conidia that are brown to gold-brown, with the main spore body being ellipsoidal with 8–12 transverse septa and usually a few longitudinal septa. Conidia have a long, tapering beak and are borne singly. Overall conidial dimensions are 100–300 x 15–20 μm. The other purple blotch pathogen is *Stemphylium vesicarium*. This fungus produces olive-brown to gold brown, oblong to oval conidia that have up to four longitudinal septa and varying numbers of transverse septa. Conidia measure 25–48 x 12–22 μm and have a length-to-width ratio of 1.5–3.0. Conidia are borne singly on conidiophores that have a distinctly swollen tip. In some regions the pseudothecia and ascospores of the sexual stage, *Pleospora allii*, can be found associated with crop debris.

Disease Cycle and Epidemology

Infection requirements are similar for both pathogens, and disease severity is worse with warm and humid or wet conditions. Spores are airborne and infect the leaves via stomata or directly through the epidermis. Inoculum comes from infested crop residue in the field. The optimum temperature for *A. porri* is 18–25°C for germination and 15–25°C for infection. The pathogen can still be active at lower temperatures (4–13°C), however, enabling disease to develop in autumn and winter. Conidia are produced at night and released early in the day as the humidity decreases. More than 8 hours of leaf wetness at 15–25°C are required for infection. When dew duration lasts for 16 hours or more, *A. porri* conidia infect leaves and result in typical lesions; if there is less than 12 hours of dew, infections result only in small leaf flecks. The older leaves are more susceptible to infection than the young leaves, although damage by thrips may increase the susceptibility of young leaves. As symptoms appear in 1–4 days and new spores are produced after only 5 days, epidemics develop quickly under favorable conditions.

Management

- Plant resistant or tolerant varieties. For example, avoid using sweet Spanish onions because these are reportedly very sensitive to *A. porri*.
- Plow under crop residues after harvest to reduce spread and survival of inoculum.
- Rotate crops with non-hosts.
- Select sites and practice irrigation so that foliage drying is enhanced.
- Apply fungicides, especially as the crop canopy ages and becomes more dense and if leaf-wetness periods favor infection.

Rust

Rust diseases of allium crops occur wherever these crops are grown and are important production factors that cause significant crop damage to onion, leek, garlic, and chives. However, the rust pathogen is comprised of genetically distinct sub-groups; therefore, rust in one part of the world may or may not be the same pathogen as rust elsewhere.

Symptoms

Initial symptoms consist of small, 1–2 mm in diameter,leaf flecks and spots that are irregularly shaped and white or light tan in color. These lesions increase to 3–5 mm spots that develop the typical, bright orange pustules of rust diseases. Pustules erupt through the leaf surface between the veins, develop on both upper and lower leaf surfaces, and release copious amounts of dusty orange spores. There may be chlorotic halos around the pustules. Rust usually first occurs on the older foliage and subsequently spreads to newer leaves. Severely infected leaves can be almost covered with pustules, turn chlorotic, and then finally become tan, dry, and dead. Reduced plant size and yield loss accompany severe rust. In garlic, rust causes severe stunting of plants and bulbs. At harvest the outer garlic bulb sheaths can be split and weak, resulting in shattering of the bulb and overall poor quality. Late in the infection cycle the teliospore phase can also be seen in darker brown pustules on diseased foliage.

Causal Agent: *Puccinia allii* (synonym: *P. porri*)

Rust is caused by the fungus *P. allii*, which is an autoecious macrocyclic (full-cycled) rust. However, isolates with shortened life cycles, which lack pycnia and aecia stages, are common and referred to as having hemiform cycles. The prominent orange pustules are the uredinia stage that produces roughened, one-celled urediniospores measuring 23–29 x 20–24 µm. The dark brown, two-celled teliospore is produced later and measures 28–45 x 20–26 µm. In several places, teliospores are very common on garlic and onion. Pycnia and aecidia stages occur on some hosts such as chives and *A. fistulosum*. Because of variation in rust isolate host ranges and morphologies, over the years the allium rust pathogen has been classified into various genera and species, including the following: *P. allii*, *P. porri*, *P. mixta*, *P. blasdalei*, *Uromyces ambiguous*, or *U. duris*. Currently, it appears justified to group all garlic, chives, and leek isolates under *P. allii*, and to consider *P. allii* as a species complex.

Disease Cycle and Epidemiology

The overwintered crops are a significant source of rust inoculum for spring planted alliums. The rust pathogen is normally active from July onwards and is only inhibited by cold periods in winter. Weedy alliums can also be hosts and provide inoculum for commercial crops. Urediniospores are the primary inoculum and are spread long distances via winds. Optimum conditions for infection are 15ºC and 100% relative humidity for 4 hours. The pathogen is active between temperatures of 10–24º C. Rust infection is favored by high nitrogen applications and low potash levels. The fungus does not survive in the soil. Stressed plants are more severely affected by this disease than are healthy plants.

Management

- Crop rotation with non-hosts can be helpful to reduce the disease, but it may not eliminate, rust pressure.
- Control allium weeds and plough under infected crop residues to help reduce inoculums
- Apply fungicides, such as morpholine or triazole products, where rust occurs. Frequent applications are required to maintain good control in long season crops.

Black Mold

This disease is most prevalent in subtropical and tropical production areas where high temperatures favor its development. While black mold can cause some problems in the field, most losses occur in storage. Black mold concerns in temperate areas, such as the UK in the 1980s, have been associated with poor storage conditions and elevated storage temperatures.

Symptoms

Black mold generally develops at the neck of the bulbs on injured or necrotic leaf tissue. However, it can develop on injured or diseased roots, or on bruised or split outer scales along the side of bulbs. Infected bulbs may develop a black discoloration at the neck. Clusters of black spores generally form along veins and on or between the outer papery scales of bulbs. Infected tissue first has a water-soaked appearance and over time will dry and shrivel. No external symptoms may be visible on some infected bulbs. Soft rot bacteria can follow infection by this fungus.

Causal Organism: *Aspergillus niger*

The pathogen is dark brown to black in culture. The fungus produces a readily recognized conidiophore that consists of an erect unbranched stalk, a single swollen spherical vesicle, prophialides and phialides that line the surface of the vesicle, and densely packed chains of spherical black, spiny conidia (2–5 µm in diameter). Conidia are dry and readily dispersed in the air.

Disease Cycle

The pathogen is seed borne in onion. In addition, the fungus occurs widely in field soils and is a saprophyte that colonizes dead plant foliage. *Aspergillus niger* first grows on senescing onion leaves, then progresses to the neck of the bulb and finally into the bulb itself. Bulb infection in the field is usually associated with damage to the fleshy scales, such as when growth splits occur. During harvesting operations, spores on infected foliage are dispersed and provide another source of the pathogen within the crop and to nearby fields.

Epidemiology

Spores of this fungus are very common in the air and soil. Free moisture for six hours or longer on the onion surface is necessary for infection to occur. Optimum temperatures for growth are 28–34°C and minimum temperatures for spore germination are 17°C. Humidity greater than 80% is required for spore germination. The presence of free water for 6–12 hours on the onion surface is required for infection.

Control

- Use seed that does not have significant levels of the pathogen. If clean seed is not available, broad-spectrum treatments such as thiram have shown good efficacy as seed treatments.
- Foliar fungicides can reduce black mold when applied towards the end of the crop.
- Harvest the crop under dry conditions and minimize damage to the neck tissue of the bulbs.
- Maintain temperature and relative humidity at appropriate levels during storage.
- Monitor storage conditions when using high temperatures to dry out the onions after harvest; temperatures should not exceed 30–32°C and the relative humidity should be kept below 80% as long as possible. Long-term storage at low temperatures stops growth of *A. niger*, but the fungus becomes active at temperatures above 15°C.

Neck Rot

Neck rot is mainly a storage disease of allium crops. The disease caused significant losses for the bulb onion industry for many years until improvements in harvesting and storage techniques were developed. Producers who lack modern drying facilities still incur significant losses. Neck rot affects garlic, shallot, multiplier onion, and leek, but it is most important on bulb onion where losses can sometimes be greater than 50%.

Symptoms

Typical symptoms are a water-soaked or light brown decay at the neck of stored bulb onion. Dense gray mycelium is produced around the neck and between the bulb scales. Numerous black sclerotia, measuring approximately 5 mm in diameter, also develop on the shoulders of the bulb. This fungal growth only becomes apparent after several weeks in storage. The pathogen continues to spread from the neck into the bulb and eventually most of the bulb becomes soft and rotted. A brown rot may occur on the fleshy scales of maturing bulb onion in the field, and this is usually associated with splitting or damage of the bulb. The pathogen can affect any part of the bulb and is not limited to the bulb necks. It is possible to find *B. allii* sporulating on senescent foliage prior to crop harvest.

Causal Organism: *Botrytis allii* (=*B.aclada*)

Neck rot is caused by *Botrytis allii*, which is synonymous with *Botrytis aclada*. This fungus produces conidia that measure 7–11 x 5–6 μm; these spores are smaller than those produced by most other *Botrytis* species found on alliums. The conidia of *B. allii* are readily produced in culture. In contrast, the Botrytis leaf blight pathogen *B.squamosa* does not readily sporulate in culture and has significantly larger conidia (14–24 x 9–18 μm). Another *Botrytis* pathogen, *B. porri*, is important on garlic and leek. *Botrytis byssoidea* is another species that causes a neck rot disease on onion. This pathogen is distinct from *B. allii*, and causes a disease known as gray mold neck rot.

Disease Cycle

Infections occur during the growing season in the field but the visual symptoms are not obvious until the bulbs are in storage. Both species survive in the soil as sclerotia which are hard black balls of fungal tissue or can be seed-borne. Sclerotia are viable in the soil for more than two years without a host present. When onions are planted the sclerotia germinate and grow hyphae that infect onion tissue through wounds. If the sclerotia are associated with dead onion tissue they produce spores that can infect living tissue. Infections occur from contaminated seed when the fungus grows from the seed to leaves and colonizes them. In all cases, the onion tissue remains symptomless which is called a latent infection. Infection from spores produced on dead onion tissue (cull onions) occurs during moist cool weather when temperatures are between 10-24°C. The spores are spread by wind and if the spores land on bulbs with wounds or bulbs with succulent necks at harvest, they germinate and enter the bulb and moist necks tissues (Lacy and Lorbeer, 2008). After the bulbs are placed in storage, it takes 1-2 months before rot develops under storage conditions. The disease is most severe when there is poor ventilation, high humidity and temperatures above 40F in storage. Botrytis frequently sporulates on the surface of infected bulbs in storage but new infections of healthy bulbs in storage do not occur because the two species cannot enter a healthy bulb (Delahaut and Stevenson, 2004).

Epidemiology

Under prolonged wet conditions the fungus can sporulate on dead and decaying tissue in the field as well as from sclerotia. Wind readily disseminates these conidia to other plants where they can infect the neck of the plant through wounds or cuts. Disease spread is most rapid during moderate temperatures with high humidity, rainfall or overhead irrigation. The condition of plants at harvest is important since infection can be more severe if necks are still succulent.

Also, storing uncured onions at temperatures and humidity that are too high can promote disease development and spread. Soil-line rot is often more severe when onions are transplanted and during cool, moist weather.

Control

- Use seed that does not have significant levels of the pathogen.
- Use disease-free sets that are grown from such seed. A 1% seed infection is regarded as an economic threshold.
- Treatment for seedborne *B. allii* has been very effective.
- Use heat treatments to control the pathogen in sets.
- Plant resistant types or cultivars of alliums.
- When topping onions at harvest, leave a relatively long neck (10 cm) that can be thoroughly dried before the fungus spreads down the neck and into the bulb. Bulb necks can be dried by using forced air drying or by harvesting and field drying the onions under warm, dry conditions.
- After drying, maintain stored commodities below 75% relative humidity and at 0–1°C. Fungicides applied during the growing season may provide some suppression of neck rot. Rotate with non-hosts so that soil-borne sclerotia of *B. allii* are reduced in number.

Fusarium Basal Plate Rot

Fusarium basal rot (FBR) [*Fusarium oxysporum* Schlechtend.: *Fr*. f. sp. *cepae* (H.N. Hans.) W.C. Snyder & H.N. Hans] is a root and bulb fungal disease of onions grown in temperate and subtropical regions. This disease is important worldwide in bulb onion, chives, garlic, and shallot. It is becoming increasingly important in the India, UK and USA.

Symptoms

Foliar symptoms develop at any stage during the growing season and include a general yellowing, necrosis of the leaves from the leaf tip downward, and wilting. There may be a tan to pink root rot on onion and red purple discoloration of the stems and bulbs of garlic. For all affected *Allium* species there is a rot of the basal plate tissue where roots are attached to the crown. Such a rot is initially water-soaked, light tan to darker brown, with tissues remaining firm. With further development, the basal infection turns into a soft rot, extends up into the fleshy scales, and causes the plant to collapse. Under humid conditions, a fluffy, white mycelium is produced on the affected tissues. Bulbs showing no obvious symptoms may still rot in storage. On leek, affected roots are initially gray and water-soaked in appearance and later become pink, soft, and rotted. In addition to the basal plate discoloration, leek can develop a tan to pink lesion on outer leaf sheaths in contact with soil.

Causal Organism: *Fusarium culmorum, F. oxysporum* f. sp. *cepae, F. proliferatum*

Fusarium basal plate rot is caused by several species of the fungus *Fusarium*. *Fusarium oxysporum* f.sp. *cepae* causes basal plate rot on onion and *F. culmorum* is the causal agent of the same disease on garlic and leek. *Fusarium proliferatum* is reported to affect the bulbs of onion and garlic. *Fusarium oxysporum* f. sp. *cepae* has morphology and colony characteristics that are similar to other *F. oxysporum* fungi. The fungus forms one- or two-celled, oval to kidney shaped microconidia on monophialides, and four- to six-celled, fusiform, curved macroconidia. Macroconidia are usually produced in cushion-shaped structures called sporodochia. Chlamydospores are formed in culture. This pathogen is apparently host specific to *Allium* species. *Fusarium culmorum* produces abundant macroconidia that are stout, thick walled, and have prominent septa. Macroconidia foot cells are indistinctly to slightly notched. Conidiophores are monophialides. Microconidia are absent and chlamydospores are present in culture. Colonies on potato dextrose agar produce abundant, dense, white aerial mycelia, and the undersurfaces of these cultures are distinctly carmine red. *Fusarium proliferatum* produces mostly microconidia in culture. These are one- or two-celled, borne in long chains from polyphialides, and have slightly flattened bases because of their formation in the chains. Chlamydospores are not formed. Semi-selective media like Komada's medium can help isolate all *Fusarium* pathogens if secondary rot organisms are present.

Disease Cycle

Fusarium basal plate rot is a soilborne disease; the pathogens persist in soil by means of chlamydospores. Pathogens may be carried on onion sets and garlic cloves. Seedborne inoculum may also exist: in California, leek transplants in a soilless medium in trays can develop extensive basal plate rot – circumstantial evidence of seedborne pathogen introduction. The optimum temperature for disease development is 25–28°C and few problems occur when soil temperatures are less than 15°C. These *Fusarium* species invade the roots and spread to the basal plate, but can also infect the bulb directly. Damage from insect feeding can increase problems. Likewise, *Fusarium* infection can attract pests and allow for extensive secondary decay from soil microorganisms. Wet weather close to harvest favors disease development, particularly in garlic.

Control

Use resistant cultivars, as this is the most important component in the control strategy. Rotate crops to nonhosts for at least 4 years to help reduce soilborne inoculum and lessen potential disease pressure. It would be desirable to use seed that is tested and found to not have detectable levels of the pathogen; however, seed assays for these pathogens are not yet fully developed. Plant only transplants which appear to be disease free. The storage of bulbs at temperatures below 4°C will reduce further development in storage. In some regions fungicides have been successfully used on seed or sets and as a dip treatment on seedlings.

Penicillium Mold/Blue Mold

This is mainly a storage disease, but field symptoms can occur on maturing bulbs. Blue mold can be an important problem on garlic.

Symptoms

On garlic, Penicillium mold can cause the death of planted cloves prior to emergence. Infected cloves first show a water-soaked, irregularly shaped lesion that later becomes tan to light brown. The fungus can produce a blue-green, clustered mycelial growth in the lesions. In advanced stages the cloves are extensively rotted, and the blue-green sporulation is widespread on the affected tissues and around the base of young garlic plants. Sporulating *Penicillium* can be present between the scales of onion extending from the neck into the shoulder of the bulb where it can cause a rot. If the pathogen infects commodities near harvest, the presence of sporulating blue-green mold is unsightly and can lead to crop rejection.

Causal Organism: *Penicillium hirsutum*

Penicillium mold is caused by *Penicillium hirsutum*, previously known as *P. corymbiferum*. Several other *Penicillium* species are regularly found on allium bulbs. On diseased tissue and in culture, *P. hirsutum* produces conidiophores that are shaped like a brush (penicillus) and bear long chains of dry, windborne conidia. Conidiophores are conspicuously roughened, form from surface hyphae or in fascicles, and have complex branching patterns (terverticillate or quaterverticilliate). In culture the colonies appear gray-green to dull green.

Disease Cycle

For garlic, the fungus is carried on the cloves that are planted in the field. In addition, inoculum can be present on crop residue in the field; infection of bulbs takes place if there has been damage or growth splitting. The pathogen is spread during harvesting, particularly if there is damage to onion bulbs or garlic cloves. High temperatures and dry soil conditions favor disease development. The optimum conditions for pathogen development are temperatures of 21–25°C and high relative humidity.

Epidemiology

Penicillium spp. can be found in soil, on plant and animal debris or on senescing tissues. Infection of bulbs is usually through tissues damaged by bruising, freezing injury or sunscald. The pathogen grows well at 21-25°C and under moist conditions.

Management

- Carefully handle bulbs and cloves during harvest to minimize damage and prompt drying of harvested bulbs is recommended.
- Avoid high humidity levels in storage. During storage or shipment, a combination of low temperatures (less than 5°C) and low relative humidity are required to restrict disease development.
- Fungicide treatment of bulbs can be effective in controlling this disease.

White Rot

This is a major root rot disease of alliums, with onion and garlic being particularly affected. White rot can also be a common problem in garden settings. In some regions white rot is an increasing problem due to intense crop rotations and shortage of uninfested land.

Symptoms

White rot affects roots and crowns, and overall symptoms are usually first noticed only after root infection is well established. Foliage of infected plants turns yellow, wilts, collapses, and eventually dies and becomes brown and dry. In badly affected areas, foliage growth is poor and patches of plants die rapidly. White, persistent mycelium develops on diseased roots and the base of bulbs in contact with soil. Numerous tiny, black, spherical sclerotia, measuring less than 1 mm in diameter, form on mycelium and diseased tissue. In advanced stages of disease, the roots and bulbs become soft and rotted due to activity from secondary decay organisms. Symptoms on leek are usually less severe than on onion or garlic.

Causal Agent: *Sclerotium cepivorum*

The cause of white rot is *Sclerotium cepivorum*, which has no known perfect stage. Conidia are likewise not produced, though small spermatia occur on germinating hyphae but appear to have no role in disease development. *Sclerotium cepivorum* reproduces, survives, and infects by sclerotia that consist of a smooth black rind that is two to five cell layers deep and an inner medulla of closely packed hyphae. The pathogen can be readily isolated on solid media such as potato dextrose agar. The white rot fungus is host specific to allium crops. The other species, *S. rolfsii*, causes a watery rot of alliums and produces brown, spherical sclerotia that are usually significantly larger (1–2 mm in diameter) than those of *S. cepivorum*. *Sclerotium rolfsii* has a very broad host range.

Disease Cycle

White rot is associated with soilborne inoculum, so affected areas and disease incidence increase as alliums are cropped in infested fields. Sclerotia can persist in soil without a plant host for over 20 years. These propagules remain dormant until allium root exudates, such as propyl and alyl cysteine amino acids that are specific to this group of plants, are present in soil. Soil microorganisms metabolize the amino acids into the stimulatory compounds alkyl and alkenyl thiols and sulphides. Sclerotia germinate and the resulting mycelium grows 1–2 cm through soil and invades the host root and basal plate. Secondary spread can occur by mycelial growth from plant to plant if their roots are in close proximity. Temperature is a key factor in disease development as sclerotia show little activity below 9°C or above 24°C; the optimum range is 14–18°C. Crops planted in the autumn may therefore experience reduced disease severity. Sclerotia germinate under moist conditions (not below –100 kPa [= –1bar]), but germination is inhibited in very wet soils. Bulb onion crops established from transplants are more severely affected than direct seeded crops because of more vigorous root systems that develop earlier in the season and hence encounter more soil borne inoculums.

Epidemiology

This disease is most severe in cool soils when soil moisture is favorable for root growth. The fungus can survive as sclerotia in the soil for many years and it can over-winter in infected onion debris and in diseased onion sets. Within rows this disease can spread laterally from root system to root system. The fungus is spread by movement of infested soil, infected onion sets and transplants.

Management

- Selection of fields and plants with no history of disease
- Phytosanitation to prevent introduction of infested soil and contaminated equipment into clean fields
- Use only disease-free transplants
- Soil fumigation with dazomet, metam sodium, methyl bromide, and chloropicrin on heavily infested soils can be useful treatments, though the fungus is not eradicated
- Drenches with Triazole fungicides are useful, for white rot control.
- Soil inoculum can be reduced if sclerotia can be stimulated to germinate in the absence of host plants. Such germination is triggered by diallyl disulphide (DADS), which mimics the stimulatory activity of allium roots.
- Flooding of infested fields is used in some regions to manage white rot. This is most effective when temperatures are above 20°C. Solarization reduced sclerotial populations in areas where climatic conditions are suitable and might form part of an integrated control strategy.

Bacterial Diseases

Sour Skin

Sour skin, first described in 1950, has been reported from onion-growing areas all over the world. Losses often appear in stored onions, but infection usually begins in the field. The disease can be serious in individual fields, with yield losses of 5–50%. Sour skin is primarily a disease of onions, but other *Allium* species are reported to be hosts.

Symptoms

Field symptoms often appear as one or two leaves that have turned a light brown color. A watery rot develops at the base of the leaves and proceeds into the neck, allowing the leaves to be easily pulled from the bulb. As the disease progresses the outer bulb scales are infected. However, the outer most bulb scales and inner bulb scales may not become infected, which distinguishes sour skin from slippery skin where inner bulb scales are infected first. Infected scales develop a slimy pale yellow to light brown decay and may separate from adjacent scales allowing the firm center scales to slide out when the bulb is squeezed. Infected bulbs often have an acrid, vinegar-like odor due to secondary invaders, especially yeasts, colonizing decaying bulbs.

Causal Organism: *Pseudomonas cepacia* (Burkholder) Palleroni & Holmes.

The cause of sour skin is the gram-negative bacterium *Pseudomonas cepacia* (Burkholder) Palleroni & Holmes, a versatile organism found as an inhabitant of soil and water or as a pathogen of plants and animals. Bacterial cells are rods that measure 1.6–3.2 × 0.8–1.0 µm; they occur singly or in pairs; and they are motile by means of tufts of polar flagella. Most strains produce nonfluorescent, yellowish or greenish pigments, but the pigments may be of a variety of colors. *P. cepacia* is capable of using a wide range of nutrients. A large number of organic compounds are used as sole carbon and energy sources for growth, including a large variety of carbohydrates, monocarboxylic and dicarboxylic acids, monoalcohols and polyalcohols, aromatic compounds, amino acids, and amines. Substrates that are of diagnostic value (used by a majority of strains of *P. cepacia* but used only infrequently by other *Pseudomonas* species) include d-arabinose, d-fucose, cellobiose, saccharate, mucate, sebacate, citraconate, and tryptamine. No organic growth factors are required. Cells accumulate poly-b-hydroxybutyrate as a carbon reserve material. *P. cepacia* is obligately aerobic. The optimum growth temperature is 30–35°C. No growth occurs at 4°C, and most strains grow at 41°C. Denitrification is negative while nitrate is reduced to nitrite. It is oxidase positive and arginine dihydrolase negative and can liquefy gelatin.

Disease Cycle

Apparently, onions are relatively resistant to *P. cepacia* prior to bulbing, or the environment does not become favorable for bacterial multiplication until after bulbing. Infection generally occurs through a wound when free water from rain, overhead irrigation, or flooding causes water congestion of the host tissue. The bacterium can gain entrance to the plant when onion tops are cut at harvest or through other wounds in the neck when the foliage falls over at maturity. Infection can also begin when water contaminated with bacterial cells strikes the younger upright leaves and flows down into the neck in the leaf blade axil. Young leaves are much more susceptible than mature leaves, which are usually symptomless. Infection can remain latent in the growing onion, and symptoms sometimes do not develop until the plant begins to bulb. Bacteria spread more rapidly in water-soaked tissue and when temperatures exceed 30°C. Infection advances into the bulb via the infected leaf and corresponding scale. The infection does not move into adjacent scales.

Inoculum of *P. cepacia* has been associated with contaminated irrigation water. Splashing water from rain or overhead irrigation may carry water- or soil-inhabiting bacterial cells onto the neck of the plant.

Epidemiology

Burkholderia cepacia is commonly spread by heavy rains, overhead irrigation and flooding which splash the bacteria onto young or wounded foliage. Infection typically occurs through wounds including those made when onions are cut at harvest. Infection can also occur when water lands on upright leaves and flows into leaf blade axils carrying the bacterium with it. Sour skin is favored by rainstorms and warm weather, and develops rapidly at temperatures above 30°C (86°F).

Managament

- Proper maturing of the crop and quick drying after topping and harvest.
- Since contaminated irrigation water has been implicated in the spread of the pathogen, the use of recycled or irrigation runoff water should be avoided. The method of irrigation has a substantial impact on the incidence of sour skin. Season-long overhead irrigation provides a favorable environment for infection by *P. cepacia*, whereas furrow irrigation results in almost complete absence of the disease. In experimental plots, the final four or five sprinkler irrigations were accompanied by increases in sour skin of 150–300%. Where sour skin is a potential problem, changing from sprinkler to furrow irrigation, atleast from bulbing to the end of the season, is advisable where feasible.
- Do not damage foliage prior to harvest or bulbs during harvest since *B. cepacia* enters the plant primarily through wounds.
- Onion crops should be harvested at maturity and the bulbs dried quickly.
- Storing onions at cool temperatures 0°C with adequate ventilation to prevent condensation on the bulbs will reduce storage losses resulting from this disease.

Bacterial Soft Rot

Bacterial decay, known as soft rot, is one of the most widespread and destructive storage diseases of onion. Soft rot generally starts in the field just before or during harvest. Several bacteria are known to cause soft rot of onion leaves and internal breakdown of bulbs in the field as well as in storage. These include species in the genera *Pectobacterium* and *Dickeya* (syn. *Erwinia*), *Pseudomonas*, and *Enterobacter*. In addition, *Pseudomonas aeruginosa*, *Serratia marcescens*, *Bacillus cereus*, a *Klebsiella* sp., a *Lactobacillus* sp., an *Enterobacter* sp., and an *Escherichia* sp. are included. These have been found either to be associated with rotting onion bulbs or as part of resident endophytic microflora of symptomless onion bulbs, which may become opportunistic pathogens or secondary invaders under certain environmental, cultural, and storage conditions, harvest practices, or specific physical or physiological states of the bulb tissues.

Symptoms

D. chrysanthemi infects leaves primarily at the base of the plant close to the soil. The affected tissues are water-soaked and soft, turn mushy, and may involve several leaves in the whorl. The rot progresses along the leaf bases through the neck and into the bulb. The bulb tissues develop watery soft rot involving most of the internal fleshy scales. When bulbs with advanced soft rot are pulled from the ground, the stem plate tends to remain in the soil, with watery fluid from rotten internal bulb tissues draining out from the base of the bulbs. Infected bulbs may continue to develop rot even after harvest. *Pectobacterium carotovorum* subsp. *carotovorum* causes whitening of the leaves and wilting of affected plants in the field or may cause bulb rot in the field or during storage. The rot usually initiates from the neck tissue or wounds and spreads to the internal bulb scales. The affected fleshy scale tissues are water-soaked and pale yellow to light brown and become soft as the rot progresses. The whole interior of the bulb may break down, and a watery, foul-smelling, viscous liquid may ooze from the neck if the affected bulb is squeezed.

Causal Organism: *Erwinia chrysanthemi* (*D. chrysanthemi*) *Pectobacterium carotovorum* subsp. *Carotovorum* (syn. *Erwinia carotovora* subsp. *Carotovora)*

The pathogen belongs to the soft-rot-causing group of plant-pathogenic bacteria. They are gram negative rods with peritrichous flagella, catalase positive and oxidase negative, fermentative, facultatively anaerobic, and pecto- lytic. These two pathogens produce grayish white to creamy white colonies on most media and can be distinguished based on several physiological, biochemical, serological, and mo- lecular characteristics. Some of the differentiating cultural and physiological tests include lecithinase and indole production, phosphatase production, acid production from lactose and trehalose, growth at 37°C, growth in 5% NaCl, sensitivity to erythromycin (15 μg), and utilization of malonate and sodium tartrate.

Disease Cycle and Epidemiology

The primary sources of inoculum are infested plant residues and contaminated soil and irrigation water. Splashing rain, ir- rigation water, and insects spread the pathogens. The bacteria gain entry into the bulb through the leaves and neck tissues of plants, through mechanical injuries, or through areas damaged by the onion maggot *Delia antiqua*, storms, or other diseases. *Pectobacterium carotovorum* subsp. *carotovorum* can persist in the intestinal tract of onion maggot larvae and adult flies, which can spread the pathogen from onion to onion. High rates of nitrogen fertilizer application and nitrogen fertilization late in the growing season tend to increase the incidence of soft rot in stored bulbs. Physiological maturity of the onion foliage at the time of lifting, the length of the field curing period, and the prevailing moisture during the curing period influence the severity and incidence of soft rot in bulbs. Bulbs with mechanical injuries, bruises, or sunscald are particularly susceptible to soft rot, especially if they are held under warm (20–30°C) and humid conditions. Infection continues to advance if temperatures during storage or transit are above 3°C.

Management

- Nitrogen fertilization and irrigation should be stopped early enough in the season to allow physiological maturity of the bulb.
- Onion tops should be allowed to mature (more than 90% lodged) before harvest.
- Bulbs should be cured well to dry the necks thoroughly and not leave any moist or green tissue. Bruising of bulbs during harvest and handling should be avoided.
- Onions should be stored only after they have been well dried. Storage at 0°C and less than 70% relative humidity with good ventilation prevents the condensation of moisture on the surface of bulbs and helps reduce the rate of development and the spread of soft rots.

Slippery Skin

Slippery skin of onion was first described from the United States in 1899 and has since been reported in Australia, Bulgaria, England, Hungary, India, Indonesia, New Zealand, Spain, the former Soviet Union, and Thailand.

Symptoms

In the early stages of the disease, the affected bulb may not show any external symptoms except softening of the neck tissue. If the bulb is cut longitudinally, one or two inner fleshy scales that are soft with a cooked or water-soaked appearance and a creamy yellow or yellow-brown discoloration are revealed. The rot progresses from the top of the infected scales downward, usually without spreading across to the adjacent scales. The infection may spread from the basal plate to other scales, and eventually the whole internal tissue may rot, often with an acetic acid odor. In advanced stages, the affected tissue may dry out, and the bulb may shrivel. The disease is referred to as slippery skin because the center core of the affected bulb may slip out the top when the base of the bulb is pressed. The bacterium that causes slippery skin was also reported to cause long, sunken, white, tongue-shaped lesions on seed stalks and leaves of onion in Hungary and a stalk rot and inflorescence decay of onion seed crops in India.

Causal Organism: *Pseudomonas gladioli* pv. *alliicola*

The bacterium that causes slippery skin is *Burkholderia gladioli* pv. *alliicola* (syn. *Pseudomonas gladioli* pv. *alliicola*). It is a gram-negative rod that does not form spores; produces a diffusible, nonfluorescent green or yellow pigment on King's medium B; contains intracellular inclusions of poly-β-hydroxybutyrate; and is polarly flagellated, oxidase positive, and arginine dihydrolase negative. It hydrolyzes gelatin but not starch and does not produce levan from sucrose.

Disease Cycle

The bacterium is primarily a wound pathogen and infects leaves and maturing bulbs in the field or bulbs after harvest. Young leaves are only slightly susceptible to the bacterium. Infection probably occurs just before or at harvest. The disease is usually more severe if tops are damaged by high wind or hail and subjected to wet or rainy conditions prior to harvest. Mature bulbs are very susceptible and may rot completely within 10 days at room temperature.

Epidemiology

This bacterium requires moisture for infection and grows in the temperature range of 5-41°C. Severe disease can occur during periods of high rainfall combined with strong winds or hail. Heavy irrigation and persistent dews are also conducive to this disease. This bacterium is soil-borne and can be readily water-splashed to the foliage and necks where it can enter through wounds. As the plant matures it increases in susceptibility with the mature plant being highly susceptible. In warm weather, approximately 30°C, infected bulbs can decay within 10 days. However, in storage decay moves slowly, often requiring 1-3 months for a bulb to decay completely.

Management

- Onion should be harvested at proper maturity as soon as tops lodge. Do not store bulbs until they have been properly dried.
- Minimizing stem and bulb injury and avoiding overhead irrigation when the crop is approaching maturity can reduce losses from this disease.

- Bulbs should be stored at 0-2°C with adequate ventilation to prevent condensation from forming on the bulbs.

Viral Diseases

Onion Yellow Dwarf

Onion yellow dwarf was first reported in 1919 in the United States in West Virginia but was first named and described as a virus disease in 1929 in Iowa. The disease has been reported from most countries where onion is cultivated and probably occurs wherever onion is grown. The virus causing onion yellow dwarf has also been identified in many garlic cultivars. In onion, the disease can reduce yield and seed and bulb quality. In garlic, it is an important component of garlic mosaic.

Symptoms

The first symptoms in young onion plants are yellow streaks at the bases of the first leaves. All leaves emerging after development of the initial infection show symptoms ranging from yellow streaks to complete yellowing of leaves. Leaves are sometimes crinkled and flattened and tend to fall over. Bulbs remain firm but are undersized. Scapes of infected plants show extensive yellowing, twisting, and curling; flower clusters are smaller and have fewer flowers than those of healthy plants. Seeds from infected plants are of poor quality. In garlic, infection causes a severe mosaic in combination with other viruses.

Causal Agent

Onion yellow dwarf is caused by *Onion yellow dwarf virus* (OYDV), a threadlike potyvirus 722–820 nm long and about 16 nm in diameter. Round or slightly elongate, granular inclusions are readily seen by light microscopy in cells of OYDV infected plants. Pinwheels, scrolls, and virus particles associated with vesicles are apparent in electron micrographs. The virus has a narrow host range, including onion, garlic, a few ornamental alliums, and shallot, on which it causes severe mosaic and stunting.

Disease Cycle

The virus survives in bulbs, onion sets, and volunteer onions. It is transmissible during vegetative propagation and in a non- persistent manner by the green peach aphid *Myzus persicae* and several other aphids. It is not transmissible through seeds or pollen. Losses vary according to the time of infection. Infected seedlings may form very small bulbs or fail to form bulbs at all, while plants infected midseason may produce well-formed bulbs that are only slightly smaller than those of non-infected plants.

Epidemiology

The virus is carried by infected seed bulbs, onion sets and volunteer onions. Many aphid species can transmit this virus from infected to healthy plants. Plants that are infected at a young stage may form small bulbs or fail to form bulbs, whereas plants infected during mid-season may produce slightly undersized bulbs.

Management

- Production of virus-free bulbs and sets in areas free of the virus and the production of commercial crops away from infected crops or volunteers. Because the virus is limited to *Allium* spp. and is transmitted by aphids in a non-persistent manner, an allium-free period in a growing region can break the disease cycle.
- The use of true seeds rather than sets for onion production results in virus-free plants since OYDV is not spread by seeds.
- Other control measures include rouging infected plants and indexing for the virus in vegetatively propagated stock.
- Some onion cultivars are more tolerant than are others and can be used to reduce losses. Insecticides are probably not helpful because the aphids quickly transmit the virus in a nonpersistent manner as they move through the crop in search of more preferable hosts. In garlic, virus-free planting stock produced through indexing and meristem tip culture eliminates the virus.

Garlic Mosaic

A mosaic disease of garlic was first reported in 1946 in the United States and has since been reported wherever garlic is grown. Because garlic is propagated vegetatively, many garlic cultivars are infected by one or more viruses. Before genetic- based methods to describe viruses were employed, attempts to identify the virus or viruses that cause mosaic in garlic often led to confusion because characterization of the causal agent or agents was sometimes based on mixtures of viruses. Today, direct sequencing of partial virus genomes or, in some cases, complete genomes has clarified the worldwide situation, but a number of synonyms for some viruses remain in the literature. In general, the label "garlic mosaic" refers to a mosaic disease caused by a potyvirus, usually in combination with one or more other viruses.

Symptoms

Primary symptoms include mild to strong mosaic, chlorotic mottling, striping, and streaking of leaves. Symptoms are usually more pronounced in young leaves. Infected plants are stunted when compared with virus-free plants grown under identical conditions, and bulb size is significantly increased in crops grown from virus-free stock. Garlic is also infected by one or more latent viruses. While there may be no visual symptoms caused by the latent viruses, the effects of combinations of the many viruses found in garlic on bulb size are largely unknown.

Causal Agents

The two primary potyviruses found in garlic are *Onion yellow dwarf virus* (OYDV) and *Leek yellow stripe virus* (LYSV). OYDV forms filamentous particles 16 nm in diameter and 722–800 nm long; while LYSV forms flexuous particles that are 815–820 nm long and aggregate end-to-end. Both are common, but their prevalence differs among garlic-growing areas of the world. Infection by either of these potyviruses causes a severe mosaic of garlic when the plants are co-infected with other viruses. LYSV alone probably causes few, if any, symptoms on garlic. Similarly, OYDV may or may not cause symptoms by itself but is considered one of the primary causes of garlic mosaic since infected plants are usually co-infected by other viruses, and their combination with OYDV may result in severe mosaic symptoms, bulb size reductions, and economic losses. Apparent synonyms in the literature for one or the other of these potyviruses include *Garlic mosaic virus*, *Garlic yellow stripe virus*, and *Garlic yellow streak virus*.

A number of other viruses have been reported from garlic. Generally, these cause no, or very mild, symptoms if not present in combination with one of the potyviruses. These include the carlaviruses *Shallot latent virus* (SLV; syn. *Garlic latent virus* [GLV]), which is common in Europe and Asia, and *Garlic common latent virus* (GCLV), which occurs in most garlic- producing areas of the world. In addition, there are many viruses in a relatively new genus of viruses, *Allexivirus*, that are particularly important in garlic because it is believed that their presence with one (or both) of the potyviruses accounts for severe mosaic symptoms. The garlic-infecting viruses in the genus *Allexivirus* include *Garlic virus X* (GVX), *Shallot virus X* (ShVX), *Garlic virus A* (GVA), *Garlic virus B* (GVB), *Garlic virus C* (GVC), *Garlic virus D* (GVD), and *Garlic mite- borne mosaic virus* (GMbMV). The latter may be a strain of *Garlic virus C*.

Disease Cycle and Epidemiology

Because garlic is propagated exclusively by vegetative means, there is ample opportunity to maintain multiple infections and to move viruses from one geographic region to another. The primary causes of garlic mosaic, the potyviruses, are also transmitted by various non-colonizing aphids in a non-persistent manner. The carlaviruses are also transmitted by aphids in a non-persistent manner but are less efficiently transmitted than are potyviruses. Allexiviruses are transmitted by the eriophyid mite, *Aceria tulipae*, and like all garlic viruses, during vegetative propagation. Spread of allexi viruses probably takes place during bulb storage, where mites move freely among bulbs. Spread of these viruses rarely occurs in the field. In many instances, aphid vectors move potyviruses into a field too late in the growing season to cause any current-season economic losses. However, cloves from infected bulbs, even if asymptomatic, should not be used for seeds in sub- sequent crops if optimum yields in those crops are desired. In general, the host range of allium infecting viruses is limited to *Allium* spp. No allium infecting virus is transmitted through true seeds, so most either do not occur in onion or occur rarely. So called "virus-free" garlic actually may be free only of allium potyviruses and possibly the carlaviruses, since the mite-borne viruses are apparently more difficult to eliminate. Virus-free garlic can be produced by culturing meristem shoot tips that are less than 1 mm long, but it is difficult to maintain its virus-free status in the field because plants are quickly re-infected by insect vectors. Virus-free stock must be multiplied in areas free of commercial garlic to delay the re-infection. This "seed" garlic is then used to plant commercial fields on an annual basis.

Management

- Virus-free stocks produced from meristem tip culture and multiplied under virus-free condition either in isolated areas away from commercial garlic production or in insect-proof houses, can result in substantially higher yields.
- Bulb size and weight are increased, and virus-free bulbs contain larger cloves than do infected bulbs.
- Up to 50% losses have been reported in field trials comparing infected and virus-free garlic. Reduction in losses to garlic mosaic can also be accomplished to some extent by planting large cloves, even when the propagation material is virus-infected. However, it is preferable to plant virus free cloves.

Iris Yellow Spot

Iris yellow spot is an emerging disease of onion in the United States and the world. A tospovirus was first observed on onion seed crops in the western United States in the late 1980s, but the pathogen *Iris yellow spot virus* (IYSV) was not described until 1998. The presence of IYSV by itself or with *Tomato spotted wilt virus* (TSWV) in diseased onions was reported from Georgia, United States, in 2004. In addition to the United States, IYSV has been reported from Australia, Brazil, Chile, India, Israel, Peru, Reunion Island (France), Slovenia, and Spain. It is possible that IYSV is present in other onion-growing regions of the world as well. Reductions in bulb yield and grade or size vary in relation to host resistance, infection timing, and other factors but may range from undetectable to 100%. The disease can be very damaging to onion seed crops, in which direct bulb yield and seed quality losses may approach 100%.

Symptoms

Disease symptoms vary depending on infection timing and whether lesions form on leaves or seed scapes. Typically, symptoms include dry, straw-colored, spindle- or diamond-shaped lesions on leaves and scapes. Lesions often develop on the mar- gins of the youngest, fully developed leaves or on the swollen portion of scapes. Lesion centers often, but not always, have green centers. Concentric rings of alternating green and straw-colored tissue are sometimes apparent. Initially, lesions are small (approximately 5–10 mm long) but enlarge with age, becoming 25–50 mm or larger. As the disease progresses, the number and size of lesions increase, and lesions become necrotic. Necrotic lesions are often colonized by saprophytic organisms or become scalded by sunlight, imparting a red or faint purple color to the lesions. Such lesions are sometimes confused with purple blotch (*Alternaria porri*) or downy mildew (*Peronospora destructor*). Large areas of the foliage may become necrotic, causing a slowing or cessa- tion of plant development. Seedling infection often kills plants or so severely stunts growth that the field must be replanted or abandoned. Large, necrotic lesions often develop on scapes and cause scapes to lodge. Few seeds are produced in umbels from lodged scapes. Symptomatic plants may be scattered or widespread within fields, but disease gradients are observed in many fields. The highest incidence of symptomatic plants is often found along field margins.

Causal Agent: Iris Yellow Spot Virus

IYSV belongs to the genus *Tospovirus*, family *Bunyaviridae*. Typical of many tospoviruses, IYSV forms spherical, enveloped particles 80–120 nm in diameter, with two surface glycopro- teins (G1 and G2). The IYSV genome consists of one negative (L RNA) and two ambisense (S and M RNA) single-stranded RNAs. The genomic RNA is encapsulated in viral nucleocapsid (N) protein. Nucleocapsids are enveloped in membrane-bound vesicles. IYSV is pathogenically and genetically diverse, and strains from different regions of the world vary in host range. The genetic diversity among IYSV strains approaches that of different virus species. The virus is poorly transmitted to onion by mechanical inoculation but can be efficiently transmitted mechanically to *Nicotiana benthamiana*. The host range of IYSV appears narrow. Reported natural hosts include alstroemeria (*Alstroemeria* sp.), amaryllis (*Hippeastrum hybridum*), a *Cycas* sp., garlic, garlic chive (*Allium tuberosum*), iris (*Iris hollandica*), leek (*Allium porrum*), lisian- thus (*Eustoma russellianum* and *Eustoma grandiflorum*), *Pel- argonium* × *hortorum*, onion, a *Rosa* sp., a *Scindapsus* sp. (syn. a *Epipremnum* sp.), shallot (*Allium cepa* var. *ascalonicum*), and wild onion (*Allium altaicum*, *Allium pskemense*, and *Allium vavilovii*). Experimental hosts may include *Chenopodium amaranticolor*, *Chenopodium quinoa*, *Datura stramonium*, *Emilia sonchifolia*, *Gomphrena globosa*, *Nicotiana benthamiana*, *Nicotiana rustica*, *Petunia* × *hybrida*, and *Portulaca oleracea*.

Disease Cycle and Epidemiology

IYSV persists in onion production regions where overlapping bulb crops, seed crops, or both are produced year round. IYSV also may survive in association with volunteer onions, infected onion transplant seedlings, ornamentals, and weed hosts. The virus is not known to be seedborne. Dissemination of IYSV under field conditions is thought to be exclusively by onion thrips (*Thrips tabaci*; order Thysanoptera, family Thripidae). *Frankliniella occidentalis* and *F. schultzei* are not known to be vectors of IYSV but are vectors of TSWV. A high proportion of *T. tabaci* may carry and successfully transmit IYSV in onion fields. Like other tospoviruses, it is thought that IYSV is transmitted by thrips larvae and adults after acquisition, circulation, and replication of the virus by, and in, first and second larval instars.

Management

- Iris yellow spot management is difficult and requires the integration of as many management strategies as possible to reduce disease incidence and severity.
- The use of transplants free of IYSV and viruliferous thrips is essential.
- Elimination of weeds, alternative hosts, and volunteer onions should reduce reservoirs of IYSV and viruliferous thrips.

- Qualitative resistance to iris yellow spot is not known to occur, but commercially acceptable cultivars vary widely in their apparent resistance and tolerance to iris yellow spot, thrips or both.
- Cultivars with some degree of resistance, tolerance, or both should be planted.
- Uniform and dense plant populations reduce the incidence of iris yellow spot.
- Separation and isolation of onion seed and bulb production fields should help break the green bridge for IYSV inoculum and dissemination by thrips between fields.
- Overhead irrigation tends to suppress thrips populations and reduce iris yellow spot incidence and severity.
- Disease severity can be high when plants are stressed physiologically. Maintaining vigorous plant growth reduces iris yellow spot incidence and severity.
- Straw mulch has improved moisture availability to plants and reduced thrips populations.
- Insecticides appear to provide some iris yellow spot suppression when applied early in the season. Similar results have been observed with acibenzolar-*S*-methyl (inducer of systemic acquired resistance).

References

Abad, J. A., Speck, J., Mohan, S. K., & Moyer, J. W. (2003). Diversity of the Iris yellow spot virus N gene in the USA. *Phytopathology*, **93**(1).

Abawi, G. S., & Lorbeer, J. W. (1972). Several aspects of the ecology and pathology of *Fusarium oxysporum* f. sp. cepae. *Phytopathology*, **62**(87): 876.

Armengol, J., Vicent, A., Sales, R., Garcia-Jimenez, J., & Rodrķguez, J. M. (2001). First report of basal rot of leek caused by *Fusarium culmorum* in Spain. *Plant Disease*, **85**(6): 679-679.

BasalloteUreba, M. J., PradosLigero, A. M., & MeleroVara, J. M. (1999). Aetiology of leaf spot of garlic and onion caused by Stemphylium vesicarium in Spain. *Plant Pathology*, **48**(1): 139-145.

Bertolini, P., & Tian, S. P. (1997). Effect of temperature of production of Botrytis allii conidia on their pathogenicity to harvested white onion bulbs. *Plant Pathology*, **46**(3): 432-438.

Bisht, I. S., & Agrawal, R. C. (1993). Susceptibility to purple blotch (*Alternaria porri*) in garlic (*Allium sativum*). *Annals of Applied Biology*, **122**(1), 31-38.

BOITEUX, I., Lima, M. F., Sobrinho, J. M., & Lopes, C. A. (1994). A garlic (Allium sativum) leaf blight caused by Stemphylium vesicarium in Brazil. *Plant Pathology*, **43**(2): 412-414.

Bos, L. (1982, September). Viruses and virus diseases of Allium species. In *IV Conference on recent Advances in Vegetable Virus Research* **127** (pp. 11-30).

Bos, L., Huijberts, N., Huttinga, H., & Maat, D. Z. (1978). Leek yellow stripe virus and its relationships to onion yellow dwarf virus; characterization, ecology and possible control. *Netherlands Journal of Plant Pathology*, **84**(5): 185-204.

Brammall, R. A. (1989). Resistance to benomyl in isolates of Penicillium sp. causing clove decay of garlic. *Canadian Journal of Plant Pathology*, **11**(4): 409-414.

Burkholder, W. H. (1942). Three bacterial plant pathogens: Phytomonas earyophylli sp. n., Phytomonas alliicola sp. n., and Phytomonas manihotis (Arthaud-Berthet et Sondar) Viégas. *Phytopathology*, **32**(2): 141-149.

Clarkson, J. P., Kennedy, R., & Phelps, K. (2000). The effect of temperature and water potential on the production of conidia by sclerotia of Botrytis squamosa. *Plant Pathology*, **49**(1): 119-128.

Cortes, I., Livieratos, I. C., Derks, A., Peters, D., & Kormelink, R. (1998). Molecular and serological characterization of iris yellow spot virus, a new and distinct tospovirus species. *Phytopathology*, **88**(12): 1276-1282.

Cother, E. J., & DOWLING, V. (1986). Bacteria associated with internal breakdown of onion bulbs and their possible role in disease expression. *Plant Pathology*, **35**(3): 329-336.

De Visser, C. L. M. (1996). Field evaluation of a supervised control system forBotrytis leaf blight in spring sown onions in the Netherlands. *European Journal of Plant Pathology*, **102**(8): 795-805.

Delahaut, K., & Stevenson, W. 2004. Onion disorders: Botrytis leaf blight, leaf fleck, and neck rot. A3803 University of Wisconsin Extension publication.

Dugan, F. M., Hellier, B. C., & Lupien, S. L. (2003). First report of Fusarium proliferatum causing rot of garlic bulbs in North America. *Plant Pathology*, **52**(3).

Ellerbrock, L. A., & Lorbeer, J. W. (1977). Sources of primary inoculum of Botrytis squamosa. *Phytopathology*, **67**: 363-372.

Ellerbrock, L. A., & Lorbeer, J. W. (1977). Survival of sclerotia and conidia of *Botrytis squamosa*. *Phytopathology* **67**: 219–225.

Everts, K. L., & Lacy, M. L. (1990). The influence of dew duration, relative humidity, and leaf senescence on conidial formation and infection of onion by *Alternaria porri*. *Phytopathology*, **80**(11): 1203-1207.

Everts, K. L., & Lacy, M. L. (1996). Factors influencing infection of onion leaves by Alternaria porri and subsequent lesion expansion. *Plant Disease*, **80**(3): 276-280.

Everts, K. L., Schwartz, H. F., Epsky, N. D., & Capinera, J. L. (1985). Effects of maggots on occurrence of fusarium basal rot of onions in Colorado. *Plant disease*, **69**(10): 878-882.

Gent, D. H., du Toit, L. J., Fichtner, S. F., Mohan, S. K., Pappu, H. R., & Schwartz, H. F. (2006). Iris yellow spot virus: an emerging threat to onion bulb and seed production. *Plant Disease*, **90**(12): 1468-1480.

Gent, D. H., Martin, R. R., & Ocamb, C. M. (2007). First report of Iris yellow spot virus on onion and leek in western Oregon. *Plant disease*, **91**(4): 468-468.

Gent, D. H., Schwartz, H. F., & Khosla, R. (2004). Distribution and incidence of Iris yellow spot virus in Colorado and its relation to onion plant population and yield. *Plant Disease*, **88**(5): 446-452.

Gera, A., Cohen, J., Salomon, R., & Raccah, B. (1998). Iris yellow spot tospovirus detected in onion (Allium cepa) in Israel. *Plant Disease*, **82**(1): 127-127.

Gladders, P. (1981). Purple blotch of leeks caused by *Alternaria porri*. *Plant Pathology*, *30*(1).

Gladders, P., Carter, M. A., & Owen, W. F. (1994). Resistance to benomyl in Botrytis allii from shallots. *Plant pathology*, **43**(2): 410-411.

Gupta, R. B. L., & Pathak, V. N. (1988). Yield losses in onions due to purple blotch disease caused by Alternaria porri. *Phytophylactica*, **20**(1): 21-23.

Hall, J. M., Mohan, K., Knott, E. A., & Moyer, J. W. (1993). Tospoviruses associated with scape blight of onion (Allium cepa) seed crops in Idaho. *Plant Disease*, **77**(9).

Hayden, N. J., & Maude, R. B. (1992). The role of seedborne Aspergillus niger in transmission of black mould of onion. *Plant Pathology*, **41**(5): 573-581.

Hayden, N. J., Maude, R. B., & Proctor, F. J. (1994). Studies on the biology of black mould (Aspergillus niger) on temperate and tropical onions. 1. A comparison of sources of the disease in temperate and tropical field crops. *Plant Pathology*, **43**(3): 562-569.

Hayden, N. J., Maude, R. B., El Hassan, H. S., & Al Magid, A. A. (1994). Studies on the biology of black mould (Aspergillus niger) on temperate and tropical onions. 2. The effect of treatments on the control of seedborne A. niger. *Plant Pathology*, **43**(3): 570-578.

Kobayashi, K., Rabinowicz, P., Bravo-Almonacid, F., Helguera, M., Conci, V., Lot, H., & Mentaberry, A. (1996). Coat protein gene sequences of garlic and onion isolates of the onion yellow dwarf potyvirus (OYDV). *Archives of Virology*, **141**(12): 2277-2287.

Koike, S. T., & Henderson, D. H. (1998). Purple blotch, caused by Alternaria porri, on leek transplants in California. *Plant Disease*, **82**(6): 710-710.

Kritzman, A., Lampel, M., Raccah, B., & Gera, A. (2001). Distribution and transmission of Iris yellow spot virus. *Plant Disease*, **85**(8): 838-842.

Perez-Moreno, L., Cordova-Rosales, Z., Barboza-Corona, E., Ramķrez-Malagón, R., Ramirez-Lua, J., Ruiz-Castro, S., & Silva-Rosales, L. (2006). First Report of Leek yellow stripe virus in Garlic in the State of Guanajuato, Mexico. *Plant Disease*, **90**(11): 1458-1458.

Lacy, M.L., & Lorbeer, J.W. (2008). Neck rot. In: Compendium of onion diseases. Second edition. Schwartz, H.F. and Mohan, S.K. (eds.) APS Press, St. Paul, MN.

Lorbeer, J. W., & Vincelli, P. C. (1990). Efficacy of dicarboximide fungicides and fungicide combinations for control of Botrytis leaf blight of onion in New York. *Plant Disease*, **74**(3): 235-237.

Lot, H., Chovelon, V., Souche, S., & Delecolle, B. (1998). Effects of onion yellow dwarf and leek yellow stripe viruses on symptomatology and yield loss of three French garlic cultivars. *Plant Disease*, **82**(12): 1381-1385.

Maude, R. B., (1983). The correlation between seed-borne infection by *Botrytis allii* and neck rot development in store. *Seed Science and Technology* **11**: 829-834.

Maude, R. B., & Presly, A. H. (1977). Neck rot (Botrytis allii) of bulb onions: I. Seedborne infection and its relationship to the disease in the onion crop. *Annals of Applied Biology*, **86**(2): 163-180.

Maude, R. B., & Presly, A. H. (1977). Neck rot (Botrytis allii) of bulb onions: II. Seedborne infection in relationship to the disease in store and the effect of seed treatment. *Annals of Applied Biology*, **86**(2): 181-188.

Maude, R. B., Shipway, M. R., Presly, A. H., & O'connor, D. (1984). The effects of direct harvesting and drying systems on the incidence and control of neck rot (Botrytis allii) in onions. *Plant pathology*, **33**(2): 263-268.

Melhus, I. E., Reddy, C. S., Henderson, W. J., & Vestal, E. (1929). A new virus disease epidemic on onions. *Phytopathology*, **19**: 73-77.

Meredith, D. S. (1966). Spore dispersal in Alternaria porri (Ellis) Neerg. on onions in Nebraska. *Annals of Applied Biology*, **57**(1): 67-73.

Netzer, D. & Dishon, I. (1967). Selective media to distinguish between two *Botrytis* species on onion. *Phytopathology,* **57**(7): 795.

Nielsen, K., Justesen, A. F., Jensen, D. F., & Yohalem, D. S. (2001). Universally primed polymerase chain reaction alleles and internal transcribed spacer restriction fragment length polymorphisms distinguish two subgroups in Botrytis aclada distinct from B. byssoidea. *Phytopathology*, **91**(6): 527-533.

Pappu, H. R., Hellier, B. C., & Dugan, F. M. (2005). First report of *Onion yellow dwarf virus*, *Leek yellow stripe virus*, and *Garlic common latent virus* in garlic in Washington State. *Plant Disease,* **89**: 205.

Prados-Ligero, A. M., Melero-Vara, J. M., Corpas-Hervias, C., & Basallote-Ureba, M. J. (2003). Relationships between weather variables, airborne spore concentrations and severity of leaf blight of garlic caused by Stemphylium vesicarium in Spain. *European Journal of Plant Pathology*, **109**(4): 301-310.

Roberts, P. (1973). A soft rot of imported onions caused by *Pseudomonas alliicola* (Burkh.) Starr & Burkh. *Plant Pathology*, **22**(2): 98-98.

Sherf, A. F., & MacNab, A. A. (1986). *Vegetable diseases and their control*. John Wiley & Sons.

Song, S. I., Song, J. T., Kim, C. H., Lee, J. S., & Choi, Y. D. (1998). Molecular characterization of the garlic virus X genome. *Journal of General Virology*, **79**(1): 155-159.

Suheri, H., & Price, T. V. (2000). Infection of onion leaves by Alternaria porri and Stemphylium vesicarium and disease development in controlled environments. *Plant Pathology*, **49**(3): 375-382.

Suheri, H., & Price, T. V. (2001). The epidemiology of purple leaf blotch on leeks in Victoria, Australia. *European Journal of Plant Pathology*, **107**(5): 503-510.

Sumi, S., Matsumi, T., & Tsuneyoshi, T. (1999). Complete nucleotide sequences of garlic viruses A and C, members of the newly ratified genus Allexivirus. *Archives of virology*, **144**(9): 1819-1826.

Sutton, J. C., James, T. D. W., & Rowell, P. M. (1986). BOTCAST: a forecasting system to time the initial fungicide spray for managing Botrytis leaf blight of onions. *Agriculture, ecosystems & environment*, **18**(2): 123-143.

Tamietti, G., & Garibaldi, A. (1977). Observations on basal rot of leek caused by Fusarium culmorum (WG Smith) Sacc. *Rivista di Patologia Vegetale*. **13**: 69-75.

Tesoriero, L. A., LA, T., PC, F., & LV, G. (1982). First record of bacterial rot of onion in Australia caused by *Pseudomonas gladioli* pv. *allicola* and association with internal browning caused by *Pseudomonas aeruginosa.*

Tsuneyoshi, T., & Sumi, S. I. (1996). Differentiation among garlic viruses in mixed infections based on RT-PCR procedures and direct tissue blotting immunoassays. *Phytopathology*, **86**(3): 253-259.

Tsuneyoshi, T., Matsumi, T., Deng, T. C., Sako, I., & Sumi, S. (1998). Differentiation of Allium carlaviruses isolated from different parts of the world based on the viral coat protein sequence. *Archives of virology*, **143**(6): 1093-1107.

Van Dijk, P. (1993). Carlavirus isolates from cultivated *Allium* species represent three viruses. *Netherlands Journal of Plant Pathology*, **99**(5-6): 233-257.

Van Dijk, P., Verbeek, M., & Bos, I. (1991). Mite-borne virus isolates from cultivated Allium species, and their classification into two new rymoviruses in the family Potyviridae. *Netherlands Journal of Plant Pathology*, **97**(6), 381-399.

Verbeek, M., van Dijk, P., & van Well, P. M. (1995). Efficiency of eradication of four viruses from garlic (*Allium sativum*) by meristem-tip culture. *European Journal of Plant Pathology*, *101*(3): 231-239.

Walkey, D. G. A., & Antill, D. N. (1989). Agronomic evaluation of virus-free and virus-infected garlic (*Allium sativum* L.). *Journal of Horticultural science*, **64**(1): 53-60.

Walters, T. W., Ellerbrock, L. A., van der Heide, J. J., Lorbeer, J. W., & LoParco, D. P. (1996). Field and greenhouse procedures to evaluate onions for Botrytis leaf blight resistance. *HortScience*, **31**(3): 436-438.

Wright, P. J., Clark, R. G., & Hale, C. N. (1993). A storage soft rot of New Zealand onions caused by Pseudomonas gladioli pv. alliicola. *New Zealand journal of crop and horticultural science*, **21**(3): 225-227.

Wright, P. J., Hale, C. N., & Fullerton, R. A. (1993). Effect of husbandry practices and water applications during field curing on the incidence of bacterial soft rot of onions in store. *New Zealand journal of crop and horticultural science*, **21**(2), 161-164.

Yohalem, D. S., Nielsen, K., & Nicolaisen, M. (2003). Taxonomic and nomenclatural clarification of the onion neck rotting Botrytis species. *Mycotaxon*, **85**: 175-182.

17

Diseases of Amaranthus (*Amaranthus* spp.) and Their Integrated Management

***Munmi Borah*[1] and *Ratul Moni Ram*[2]**

[1]*Department of Plant Pathology, Assam Agricultural University, Jorhat-785013*
[2]*Department of Plant Pathology, A.N.D.U.A & T, Kumarganj, Ayodhya-224229*

Introduction

Amaranthus species were grown as the principle grain crop by the Aztecs 5,000 to 7,000 years ago, prior to the disruption of the South American civilization by the Spanish Conquistadors. Synonyms such as "mystical grains of the Aztecs," "super grain of the Aztecs," and the "golden grain of the Gods" were used to describe the nutritious amaranth grain. Vegetable Amaranthus spp. were and are presently utilized for food from such diverse geographic areas as southwestern United States, China, India, Africa, Nepal, South Pacific Islands, Caribbean, Greece, Italy, and Russia (Stallknecht *et. al.,* 1993). Amaranthus grain is extremely nutritional pseudo-cereal with a higher amount of proteins when compared to true cereals. It is a reasonably well balanced food with functional properties that have been shown to provide medicinal benefits. Amaranth associate underutilized crop and an inexpensive supply of proteins, minerals, vitamin A and C, appears to be a future crop which may substantiate this demand because of its tremendous yield potential and biological process qualities, conjointly recently gained worldwide attention (Topwal, 2019). Amaranth as a unique plant with a long and mysterious history is the object for science and also for business. Its characteristics such as absence of gluten and special composition of oil lead this plant to the position of very important plant for the future. The Advantages of Amaranth for Food Industry: Gluten-free, High-quality protein, Non Food-allergy, Thickening property of amaranth starch. The Advantages of Amaranth oil for Cosmetic Industry: Squalene content has a preserve function for skin and hair against sun and radiation, Available for very sensitive skin, Non- Allergy amaranth oil on skin. Amaranth has a big advantage, because it could be use by two ways-biomass for the bio fuel production and grain for food production (Matusova, 2008). As dyes, the flowers of the 'Hopi Red Dye' amaranth were used by the Hopi (a tribe in the western

United States) as the source of a deep red dye. Also a synthetic dye was named "amaranth" for its similarity in color to the natural amaranth pigments known as betalains (Doring, 2019). The boiled leaves and roots are used as: laxative, diuretic, anti-diabetic, antipyretic, anti-snake ve nom, antileprotic, anti-gonorrheal, expectorant, to relieve breathing in acute bronchitis. It also has anti-inflammatory properties, immunomodulatory activity, anti-androgenic activity and anthelmintic properties (Alegbejo, 2013).

Table 1. Important Diseases of Amaranthus with their causal organism.

S.No	Disease	Causal Organism
1.	Pythium Stem Canker	*Pythium aphanidermatum*
2.	Charcoal Rot	*Macrophomina phaseolina*
3.	Amaranthus Leaf Blight Mosaic	*Rhizoctonia solani*
4.	Bacterial Wilt	*Ralstonia solanacearum*
5.	Soft Rot	*Pectobacterium carotovorum* subsp. *brasilliense*
6.	Leaf curl Disease	*Chilli Leaf Curl*
7.	Papaya Leaf Curl Disease	*Papaya leaf curl virus*
8.	Capsicum Chlorosis disease	*Capsicum Chlorosis Virus*
9.	Amaranthus Mosaic Disease	*Amaranthus Mosaic Virus*

Pythium Stem Canker

Economic Importance

Pythium species are morphologically polymorphic, physiologically unique and ecologically versatile, which make them significant both theoretically and practically. They are ubiquitous in soil and in water, distributed worldwide, and with very diverse host ranges. They include some of the most important and destructive plant pathogens, causing losses of seeds, pre-emergence and post-emergence damping-off, rots of seedlings, roots, basal stalks, decays of fruits and vegetables (Parveen and Sharma , 2015).

Symptoms

Mihail *et. al.,* 1993 reported the symptoms of Pythium stem canker as sunken tan lesions at the soil line which are bordered by narrow bands of black tissue. The cankers did not extend below the soil line and upon excavation the roots appeared white, undecayed and healthy. While on the stem, the interior tissues were wet and soft.

Causal Organism

Pythium aphanidermatum

Disease Cycle

The mode of survival of Pythium in the soil is by oospores, hyphae and sporangia. With the help of oospores, fungi survives the unfavourable soil moisture and temperature for several years. The fungi directly infect the plant by forming germ tube from the oospores or may form a sporangia. The fungi produces motile zoospores. Before encysting and forming a germ tube, the zoospores swim around briefly which cause infection. Either by germinating directly or by formation of zoospores the sporangia develop on plant tissue. With the transportation of infected debris to uninfected areas and by the movement of soil moisture which allows the zoospores to swim , the pathogen is perpetuated (Parker, 2009).

Epidemiological Factors

Parker stated that the fungi *Pythium aphanidermatum* has a very wide host range and occurs world wide in a temperature range between 27-340c . The fungi also requires warm regions and wet conditions (water potential of 0 to -0.01 bars).

Integrated Disease Management

Beckerman, 2010 suggested the following management approaches-

- Surface clean and disinfect all bench surfaces, tools, trays, containers, and equipment that will contact the potting mix.
- Biological control agents such as *Trichoderma harzianum* or *Gliocladiu mvirens* do provide some protection when disease pressures are low.

Charcoal Rot

Economic Importance

Macrophomina phaseolina has a wide host range and is responsible for causing losses on more than 500 cultivated and wild plant species (Khan, 2007).

Symptoms

Mihail *et al.,* 1993 observed stem lesions and small, black microsclerotia typical of *M. phaseolina* adhering to root vascular tissue on excavation of several plants.

Causal Organism

Macrophomina phaseolina

Disease Cycle

Disease cycle of *Macrophomina phaseolina* (Smith *et al.* 2014). (A) *M. phaseolina* overwintering by microsclerotia in crop residue. (B) Infected roots after being in contact or in proximity with microsclerotia. (C) Fungus grows within the stem and root affecting and disrupting the vascular system. More microsclerotia are produced. (D) Abundant microsclerotia presence on lower stem and taproot tissue gives a charcoal-like appearance. Infected residue will become potential source of inoculum for next planted crop.

Epidemiological Factors

Srivastava *et al,* 1979 observed that the mean maximum temperature has a significant and positive correlation with disease incidence.

The severity of the disease is directly related to the population of viable sclerotia in the soil. The pathogen generally affects the fibrovascular system of the roots and basal internodes. The parasitic fitness of a facultative soilborne pathogen before invading the host is linked with its ability to compete for its survival, utilization of organic sources and colonization in the host root rhizosphere by competing other microorganism in the vicinity (Khan, 2007).

Integrated Disease Management

Soil solarization in combination with fungal antagonists and soil amendments has been subjected to evaluation as a potential disease management strategy for the control of charcoal rot (Ibrahim *et al,* 2015)

Amaranthus Leaf Blight Mosaic

Symptoms

The diseases starts off as small straw coloured lesions in the foliage of young plants which later coalesce and turn greenish and often get dislodged from the leaf lamina causing shot holes and disfiguring it entirely.

Causal Organism

Rhizoctonia solani

Disease Cycle

Sclerotia and mycelia in infected plant debris are two primary sources of inoculum. During the cropping season or at harvest, sclerotia fall on the ground and serve as survival structures from one cropping season to other. They survive for long periods in the soil, with up to 2 years in temperate rice production areas, and frequently accumulate in the field over time. Field water movement and irrigation support the dispersal of sclerotia and infected plant debris. Initial infections start with a sclerotium or a piece of infected debris floating on the water surface and coming in contact with the sheath. The fungus gets attracted to the chemical stimuli released by the rice host. Germinating sclerotia or mycelia in debris penetrate the plant tissue either by means of natural openings or by specialized infection structures called appressoria or infection cushions.

The fungus also produces extracellular enzymes that degrade plant cell walls to facilitate colonization. Once the fungus penetrates and colonizes the plant tissue, symptoms are initiated. The fungus grows upwards on the plant, penetrates, and infects upper leaf sheaths, leaf blades, and panicles . The fungus spreads in the field by growing its runner hyphae from tiller to tiller, from leaf to leaf, and from plant to plant, resulting in a circular pattern of damage. The disease frequently starts during the late tillering to joint elongation stages of plant growth and becomes more aggressive as the rice plant shifts to the panicle differentiation (reproductive) stage. Sheath blight also infects weed hosts and causes similar symptoms. Sclerotia are produced on the surface of infected tissues of rice and weed hosts and survive in the soil between crops. (Uppala *et. al.,* 2018).

Epidemiological Factors

The infection spreads most quickly when susceptible varieties are grown under favorable conditions such as warm temperature (28 to 32°C), high humidity (95% or above), and dense stands with a heavily developed canopy.Factors that favor the development of sheath blight include the use of highly susceptible semi-dwarf rice varieties, short rotations with non-host crops, overuse of nitrogen fertilizer, and increased stands that create favorable microenvironments in the canopy.

Integrated Disease Management

Researchers suggested the following management approach for the disease

- Prophylactic spray application of baking soda with turmeric powder at a ratio of 1 : 4 can give rise to systemic induced resistance in the plants.(Dissolve 40 g of Asafoetida (Palkayam) in 10 lit of water. To this add 8 g of sodium bicarbonate and 32 g of turmeric powder with violent agitation . Filter and spray).

- Fungicidal control can be effectively done by the periodic application of mancozeb @ 4g/L using cowdung supernatant as the diluent.
- Spraying 2 % Pseudomonas also effective in managing the disease.

Bacterial Wilt

Symptoms

Sikirou *et al,* 2019 observed wilting in Amaranthus plants after 12 days of inoculation with the pathogen. He further confirmed the disease as the cut stem release whitish bacterial ooze in water.

Causal Organism

Ralstonia solanacearum

Disease Cycle

This soil and water borne bacterium enters the plant roots, multiplies through the xylem, collapses the host and returns to the environment. The bacterium moves to the host roots, attaches to the epidermis, infects the cortex, and colonizes the xylem, resulting in host wilting. After death of the plant, the bacterium is released into the environment, where it seems to survive in reservoir plants, soil and/or water (Alvarez *et. al.,* 2010).

Epidemiological Factors

The Ralstonia solanacearum species complex (RSSC) is a highly diverse cluster of bacterial strains found worldwide, many of which are destructive and cause bacterial wilt (BW) in a wide range of host plants(Ravelomanantsoa *et. al.,* 2018).

Spread of the pathogen from infested to healthy plants occur by infested soil, infested irrigation water, root contact, tools and equipment.(Champoiseau *et. al.,* 2009).

Integrated Disease Management

Champoiseau *et al.,* 2009 suggested the following approaches for the management of the disease

- Crop rotation with non host crops
- Removal of hosts or crop residues when inoculum persists
- Soil treatment by modification of soil pH, solarization

- Application of stable bleaching powder
- Chemical control by fumigation and application of phosphoric acid.

Soft Rot

Symptoms

Typical soft rot symptoms like wilting, defoliation and odd smell were shown by the amaranth plants (Jeeet. *et al.,* 2018).

Causal Organism

Pectobacterium carotovorum subsp. brasilliense

Disease Cycle

Seed tubers are the most important source of inoculum in the blackleg disease cycle. When a contaminated or infected seed potato is planted, one of three things may occur: (1) The blackleg bacteria may move via the vascular bundles directly into the growing plant and result in blackleg disease. If tuber contamination is confined to the lenticels, decay of the seed tuber occurs first and when bacterial populations become great enough, invasion of the growing stem occurs. Both the process of seed tuber decay and the spread of the pathogen into the stem are highly dependent on environmental conditions. Moist, cool conditions favor the disease. (2) When conditions are favorable for growth of the potato plant, no disease may occur even when blackleg-causing bacteria are present. (3) Decay of the seed piece may occur prior to the establishment of a plant, and this, too, is an important manifestation of blackleg.(Boer *et. al.,* 2004).

Epidemiological Factors

The major host of the bacteria include Potato.

Contamination of potato tubers is exacerbated by harvesting and storage operations

The presence of moisture on stored tubers is also conducive to development of decay, since a film of water surrounding tubers causes them to become anaerobic. The lack of oxygen inhibits tuber metabolic activity and prevents them from staging a normal resistance reaction.(Boer *et. al.,* 2004).

Integrated Disease Management

- Use of botanicals to control soft rot of bacteria using extracts of Jute leaf and Cheerotaplant.(Rahman *et. al.,* 2012).

- To minimize decay in storage , injury during harvesting must be avoided. Reducing soil inoculum by roguing out the sof rot infected plants. (Boer *et al.,* 2004).

Leaf Curl Disease

Symptoms

Leaf samples of Amaranthus collected from Banswara district, Rajasthan in India showed symptoms of leaf curling, leaf distortion, leaf crinkling and yellow leaf margins (George *et. al.,* 2014).

Causal Organism

Chilli Leaf Curl Virus

Disease Cycle

The virus is transmitted by White fly, *Bemisia tabaci* in a persistent manner (Thakur *et. al.,* 2018).

Epidemiological Factors

With increase in temperature and relative humidity , white fly population and leaf curl incidence increase. (Mishra *et. al.,* 2018).

Integrated Disease Management

Pandey *et. al.,* 2010 found Use neem seed kernel extract and imidacloprid 17.8 SL (0.003%) most effective against Chilli leaf curl virus.

Papaya Leaf Curl Disease

Symptoms

As compared to healthy ones , the naturally infected ones showed begomovirus like typical symptoms of leaf curl accompanied by stunting .The presence of a whitefly population in the vicinity and begomovirus-like typical leaf curl symptoms indicated the association of begomovirus with the disease (Srivastava *et. al.,* 2015).

Causal Organism

Papaya leaf curl virus

Disease Cycle

The causal pathogen was transmitted through whiteflies *Bemisia tabaci* from infected to healthy *Amaranthus cruentus* seedlings. (Srivastava *et. al.,* 2015).

Epidemiological Factors

It also infects papaya, tomato, tobacco, sunhemp, chilli, *Datura stramonium, Zinia elegans* (Prakas, 2017).

Integrated Disease Management

Prakash, 2017 suggested roguing out of infected plants, keeping the field weed free during monsoon season, use of yellow sticky traps to minimize white fly population in field, regular spraying of neem based insecticides.

Capsicum Chlorosis Disease

Symptoms

Leaf distortion and chlorosis are shown by the Amaranthus leaves infected with Capsicum Chlorosis Virus (CaCV) (Sharma *et al.,* 2014).

Causal Organism

Capsicum chlorosis virus

Epidemiological Factors

The virus systemically infected *Arachis hypogaea, Phaseolus vulgaris, Glycine max, Pisum sativum, Phaseolus mungo, Cassia tora, Sesbania-cannabina, Nicotiana tobacum, N. rustica, N. glutinosa, N.occidentalis, N. benthamiana, Lycopersicon esculentum, Physalis floridama, Petunia hybrida, Datura stramonium, Capsicum annuum* and *Cyamopsiste-tragonoloba* (Haokip *et. al.,* 2016).

Amaranthus Mosaic Disease

Symptoms

Mosaic, Mottling, leaf curling, leaf reduction and chlorotic symptoms are exhibited by the infected leaves (Ehinmore *et. al.,* 2010).

Causal Organism

Amaranthus mosaic virus

Disease Cycle

The virus is transmitted by sap. (Ehinmore *et. al.,* 2010)

Future Approaches

Instead of high nutritional benefits of Amaranthus, the research is limited . The diseases of amaranthus has not been studied widely as compared to its nutritional and economic importance .Due to its high protein and vitamin content , it is can be introduced as a natural food supplement. Moreover, the crop has a potential to be high cash earning crop owing to its requirement in cosmetic and food industry.The crop has a high potential in respect of earnings to the farmers as both the grain and leaf have high nutritional value. It has a high prospective to be a commercially cultivable crop in the future.

Conclusion

The available literature on Amaranthus suggests that more attention must be given to the crop as it has wide scope of development. Recent technologies must be able to exploit the germplasm of Amaranthus for the benefit to the farmers. The crop also requires wide and extensive research on breeding and biotechnological approaches to increase yield , quality and better adaptation.

References

Alegbejo, J. O. (2013). Nutritional value and utilization of Amaranthus (Amaranthus spp.)–a review. *Bayero Journal of Pure and Applied Sciences*, **6**(1): 136-143.

Álvarez, B., Biosca, E. G., & López, M. M. (2010). On the life of Ralstonia solanacearum, a destructive bacterial plant pathogen. *Current research, Technology and Education Topics in Applied Microbiology and Microbial Biotechnology*, **1:** 267-279.

Beckerman, J. L. (2010). Pythium root rot of Herbaceous Plants. *Grower Talks*, **74**(8): 74-76.

Champoiseau, P., Jones, J. B., Allen, C., & Momol, T. (2009). Description and strategies for best management of Ralstonia solanacearum Race 3 biovar 2 as a potential incitant of bacterial wilt of tomato. In *Proceedings of the 24th Annual Tomato Disease Workshop Held on, Nov* (pp. 3-5).

De Boer, S. H., & Rubio, I. (2004). Blackleg of potato. *The Plant Health Instructor*.

Disease: Amaranthus Leaf Blight Mosaic Retrieved from http://www.cpsskerala.in/OPC/pages/AmaranthusDiseaseAmaranthusleafblight.jsp.

Döring M (2019). English Wikipedia - Species Pages. Wikimedia Foundation

Ehinmore, I., & Kareem, K. T. (2010). Effect of Amaranthus mosaic virus on the growth characters of Amaranthus hybridus. *Agriculture and Biology Journal of North America*, **1**(2): 75-79.

George, B., Kumar, R. V., & Chakraborty, S. (2014). Molecular characterization of Chilli leaf curl virus and satellite molecules associated with leaf curl disease of Amaranthus spp. *Virus Genes*, **48**(2): 397-401.

Haokip, B. D., Alice, D., Malathi, V. G, Nagendran, K., & Renukadevi, P. (2016). Detection of Capsicum chlorosis virus (CaCV), An Emerging Virus Infecting Chilli in Tamil Nadu, India. *Vegetos* **29** (4): 25-30.

Ibrahim, M. E., & Abdel-Azeem, A. M. Management of Sesame (Sesamum Indicum L.) Charcoal Rot Caused by Macrophominaphaseolina (Tassi) Goid. Through the Application of Different Control Measure.

Jee, S., Choi, J.G., Hong, S., Lee, Y.G., & Kwon, M. (2018). First Report of Soft Rot by Pectobacteriumcarotovorum subsp. brasiliense on Amaranth in Korea. *Research in Plant Disease*, **24**(4): 339-341.

Khan, S. N. (2007). Macrophominaphaseolina as causal agent for charcoal rot of sunflower. *Mycopath*, **5**(2): 111-118.

.Matušovà, K. (2008). The different utilization of amaranth industry. In *Book of Abstracts of* (Vol. 34).

Mihail, J. D., &Champaco, E. R. (1993). Diseases of Amaranthus spp. caused by Pythium aphanidermatum and Macrophominaphaseolina. *Canadian Journal of Botany*, **71**(9): 1219-1223.

Mishra, R. S., & Chauvey A.N (2018). Chilli Leaf Curl Virus and its Management. *Acta Scientific Agriculture* **23**: 24-28.

Papaya Leaf Curl Virus, (2017, May 9). Retrieved from https://www.biotecharticles. com/ Agriculture-Article/Papaya-Leaf-Curl-Virus-3900.html.

Parker, K. C. (2009). Pythium aphanidermatum. *Soilborne plant pathogens. NC State University. Consulta, 20.*

Parveen, T., & Sharma, K. (2015). Pythium diseases, control and management strategies: a review. *Int. J. of Plant, Animal and Environ. Sci*, **5**(1): 244-257.

Rahman, M. M., Khan, A. A., Ali, M. E., Mian, I. H., Akanda, A. M., & Abd Hamid, S. B. (2012). Botanicals to control soft rot bacteria of potato. *The Scientific World Journal*, 2012.

Ravelomanantsoa, S., Vernière, C., Rieux, A., Costet, L., Chiroleu, F., Arribat, S., & Guérin, F. (2018). Molecular epidemiology of bacterial wilt in the Madagascar highlands caused by Andean (Phylotype IIB-1) and African (Phylotype III) brown rot strains of the Ralstonia solanacearum species complex. *Frontiers in Plant Science*, **8:** 22-58.

Sharma, A., &Kulshrestha, S. (2014). First report of Amaranthus sp. as a natural host of capsicum chlorosis virus in India. *Virus Disease*, **25**(3): 412-413.

Sikirou, R., Dossoumou, M. E., Zocli, B., Afari-Sefa, V., Honfoga, J., Azoma, K., & Bihon, W. (2019). First report of bacterial wilt of amaranth (Amaranthus cruentus) caused by Ralstonia solanacearum in benin. *Plant disease*, **103**(3): 578-578.

Srivastava, A., Jaidi, M., Kumar, S., Raj, S. K., & Shukla, S. (2015). Association of Papaya leaf curl virus with the leaf curl disease of grain amaranth (*Amaranthus cruentus* L.) in India. *Phytoparasitica*, **43**(1): 97-101.

Srivastava, S. K., & Dhawan, S. H. A. S. H. I. (1979). Epidemiology of Macrophomina stem and root rot of Brassica juncea (L.) Czern & Coss in northern India. *Proceedings Indian National Science Academy B*, **45**: 617-622.

Stallknecht, G. F., & Schulz-Schaeffer, J. R. (1993). Amaranth rediscovered. *New Crops. Wiley, New York*, 211-218.

Thakur, H., Jindal, S. K., Sharma, A., & Dhaliwal, M. S. (2018). Chilli leaf curl virus disease: a serious threat for chilli cultivation. *Journal of Plant Diseases and Protection*, **125**(3): 239-249.

Topwal, M. (2019). A review on Amaranth: nutraceutical and virtual plant for providing food security and nutrients. *Acta scientific agriculture*, **3**(1): 9-15.

Uppala, S., & Zhou, X. G. (2018). Rice sheath blight. *Plant Health Instr. http s://doi. org/ 10.1094/PHI-I-2018-0403-01.*

18

Diseases of Spine Gourd (*Momordica dioica* Roxb) and Their Integrated Management

***Joyarani Pegu*[1] and *Pranab Dutta*[2]**

[1]*Department of Crop Protection, College of Horticulture, Assam Agricultural University (AAU), Jorhat, Assam*

[2]*School of Crop Protection, College of Post Graduate Studies in Agricultural Sciences Central Agricultural University (Imphal), Umiam, Meghlaya -793103*

Introduction

Spine gourd/Teasel gourd (English), (*Momordica dioica* Roxb.) is popularly known as Kantola or Kakrol in India. It is a perennial, dioecious, cucurbitaceous climbing creeper that belongs to the cucurbitaceae family (Raj *et al.*, 1993). The spine gourd is also called as janglee karela (Harish, 2008). The crop is cultivated for its fruits, which are used as vegetable (Sastri, 1962). Young twigs and leaves of this crop are also used as vegetable (Bandyopadhyay and Mukherjee, 2006). The fruit is rich in ascorbic acid and contain iodine (Rao, 2001). Karnataka and West Bengal are the two main states that cultivate spine gourd commercially. Indira Kankoda I (RMF 37) is a new commercial hybrid variety of spine gourd developed by Indira Gandhi Agricultural University. This hybrid variety is commercially cultivated in the states of Chhattisgarh, Uttar Pradesh, Orissa, Maharashtra and Jharkhand. Spine gourd can also be grown well up to an altitude of 1500 m in Assam and Garo hills of Meghalaya (Ram *et al.*, 2002). This popular vegetable has high demand in market but still remain underutilized and underexploited mainly due to its vegetative mode of propagation and dioecious nature (Bharathi *et al.*, 2010). Adopting improved genotypes and standardized agro- techniques can increased the yield potential of spine gourd (Tiwari and Tigga, 2015).

Economic Importance

Spine gourd is economically important vegetable plant with high food and medicinal value. The crop is cultivated for its fruits, which is used as vegetable and is known by various local names in India such as Kantola (Hindi), Kankoda

(Rajasthani), Kartoli (Marathi) and Bhat kerela (Assamese). Studies revealed that, per 100g edible fruit is found to contain 84.1% moisture, 7.7 g carbohydrate, 3.1 g protein, 3.1 g fat, 3.0 g fiber and 1.1 g minerals. The fruit also contains small quantities of essential vitamins like ascorbic acid, carotene, thiamin, riboflavin and niacin (Kushwaha *et al.,* 2005). This vegetable has high demand in market because of good nutritional, medicinal value, high keeping quality and ability to withstand long distance transportation, high market price and good export potential (Rasul, 2003).

Important Diseases of Spine Gourd: Powdery mildew, downy mildew, anthracnose, angular leaf spot and mosaic are some of the disease that infects the crop.

Powdery Mildew

Disease Symptoms

The disease appears as pale yellow spots on stems, petioles and leaves. These spots enlarge and white, fluffy mycelium grows over plant surfaces and produces spores, which give the lesions a powdery appearance. Affected leaves become dull and chlorotic. If favorable condition prevails than the diseases may cause significant destruction and loss in the crop production. In severe cases of infection, the pathogen can cause premature death of leaves and reduce yield and fruit quality.

Causal Organism

Podosphaera xanthii (previously known as *Sphaerotheca fuliginea*) and *Erysiphe cichoracearum* are the two fungi that cause powdery mildew. The fungus belongs to the kingdom Fungi, phyllum Ascomycota, subdivision Ascomycotina, class Pyrenomycetes, order Erysiphales and family Erysiphaceae. The fungus produces orange-to-dark brown sexual fruiting body (cleisthothecia) on the diseased plant parts. The conidia of *P. xanthii* are single celled, hyaline and borne singly. Cleisthothecia of the fungus contain a single ascus. Ascospores are single celled, hyaline, spherical and contain 6-10 per ascus. The conidia of *E. cichoracearum* are single celled, hyaline; barrel shaped and occurs in chain. The Cleisthothecia are dark spherical with mycelioid appendages and contains 8-18 per asci.

Disease Cycle

The powdery mildew pathogens are obligate parasites. The pathogens cannot survive in the absence of living hosts except as cleistothecia. The primary initial inoculums come from airborne conidia dispersed potentially long distances from other affected crops. These conidia germinate and the germ tube penetrates the epidermal cells directly. The pathogen has a polycyclic disease cycle and can complete one disease cycle in 3 to 7 days. Several quick cycles in a season results in production of lot of inoculums which causes widespread infection within a short period of time.

Epidemiology

Favorable conditions for occurrence of the disease include dense plant growth and low light intensity. Dryness is favorable for colonization, sporulation and dispersal. The conidia of the fungus contain 52 to 75% water and high lipid content which prevents their desiccations and allows them to germinate in areas of low atmospheric humidity. The fungus requires optimum temperatures of 68-80° F for disease development.Temperature of 100° F or above stop the development of powdery mildew fungus.

Integrated Disease Management

Healthy plants are less susceptible to powdery mildew infection than plants under nutritional stress. Therefore the crop should be planted in sunny areas with proper spacing to provide good air circulation and avoid application of excess fertilizer. Generally application of wettable sulphur @ 0.2% at 20 days interval with the first spray to be given at the time of appearance of the symptoms is recommended. However, powdery mildew can also be controlled effectively by application of Trifloxystrobin- a new strobilurin fungicide @ 7.5-12.5g/100 litre of water (Liguori *et. al.*, 2000). Effective management of powdery mildew can also obtain with milk, sulfur and the mixture sulfur + vegetable oil (Santomauro *et. al.,* 2003).

Downy Mildew

Disease Symptoms

Disease symptoms first appear as small yellow spots or water-soaked lesions on the upper side of the leaves. The centre of this lesion eventually turns tan or brown and dies. Under humid conditions, the lesion often develops a downy growth on the underside of the leaf. The presence of this downy growth on the underside of the lesion is a key to diagnosing this disease. This downy growth is particularly noticeable in the morning after a period of wet weather or when conditions favours dew formation.

Causal Organism

It is caused by the fungus-like water mould *Pseudoperonospora cubensis* (Berk & Curt.) Rostow. It is an obligate parasite. The fungus belongs to kingdom Chromista, phylum Oomycota, subdivision Mastigomycotina, class Oomycetes, order Peronosporales and family Peronosporaceae. The mycelium of the fungus is coenocytic. It remains intercellular with small, ovate to sometimes branched finger-like haustoria. It produces long, branched sporangiophores which come out through stomata on the lower surface of the leaf. The sporangia are borne singly on the pointed tips of sporangiophores that branch at acute angles. Sporangia are lemon shaped thin walled and papillate. Biflagellate zoospores are produced which swim, encyst and germinate. The oospores when produced are thick walled.

Disease Cycle

The fungus can survive both as mycelium and sporangia from one season to another. The fungus produces microscopic sac-like structures called sporangia. These sporangia can be disseminated by irrigation water, splashed by rain or carried by cucumber beetles which on germination produce a germ tube and penetrates the host surface and cause infection.

Epidemiology

The disease is favored by cool, wet and humid conditions. The pathogen can infect the plant between 10-27°C, with optimum day temperature of 25-30°C and night temperature of 15-20°C. However the temperature above 35°C arrests the further invasion of the host. Relative humidity of more than 75% favors the disease development. A film of moisture on the leaf surface is necessary for the infection to occur.

Integrated Disease Management

The crop should be planted with proper spacing between plants. This will prevent development of condition like high humidity and facilitate plants for exposure to sunlight and air currents and ultimately reduce the occurrence of disease. Spraying of mancozeb @ 0.1% and metalaxyl 0.05% mixture starting with the onset of the conducive weather with repetition at 10 days interval is very effective.

Anthracnose

Disease Symptoms

The disease appears as small yellowish or water soaked areas on the leaves, petioles, stem and fruits. These spots enlarge rapidly and turn brown. Infection on the young fruits may result in malformed fruits. When infection occurs in the stem, the whole vine may be killed. Circular black cankers with salmon coloured spores may also be produced on the infected parts.

Causal Organism

The fungus *Colletotrichum orbiculare* (Berk. & Mont.) Arx. cause the anthracnose disease. The fungus produces brown to black stromata on the lesions. The fungus produces acervuli with setae interspersed in the stroma and hyaline conidia are budded one at a time at the tip of the conidiophores.

Disease cycle

The fungus perpetuates in the infected plant debris, in the soil and superficially on seed also. It can also survive on the crop and weeds of other cucurbitaceous family.

Epidemiology

The growth and germination of the fungus occurs at a temperature between 22 and 27°C. Symptoms usually develop within a week when high humidity conditions prevail. The spores formed on the lesions are carried by rain splash and irrigation water leading to secondary infection. The spores may also be carried by beetles and other insects.

Integrated Disease Management

The disease can be managed by removal and destruction of diseased plant debris and cucurbitaceous weeds. The diseased can also be reduced by treating the planting material with with thiram or carbendazim @ 0.2%. Foliar spray with fungicides like zineb, maneb, mancozeb, chlorothalonil at 7days interval @ 0.2% can also gives good control of the disease. Spraying of copper oxychloride and wettable sulfur in combination is also found to be most effective in controlling the disease.

Angular Leaf Spot

Disease Symptoms

The disease symptoms appear as small, water-soaked lesions on leaves and are largely confined by plant veins. Sometimes these lesions may have yellow margins and often have an angular appearance. As the disease progresses, the lesions become dry and turns to tan brown in colour and sometimes dropping out of the leaf may also occur. Lesions may also develop on petioles, stems and fruits. The lesions developed on the fruits are small and water soaked with a light tan center. Fruit infection can also penetrate deeply, causing an internal rot of vascular tissue of the fruit and sometimes deformation of the fruit. It predisposes the fruits to secondary infection by pathogens. As a result it can reduce fruit yield, quality and marketability of the fruits.

Causal Organism

The disease is caused by the bacterium, *Pseudomonas syrinage* pv. *lachrymans.*

Disease Cycle

The bacterium survives in contaminated seed and infested crop debris. The disease cycle begins when seedborne inoculum or bacterial cells in infested crop debris colonize cotyledons upon germination. The bacterium infects hosts through natural openings and wounds, multiplying in and on leaves. The pathogen can be disseminated within and among fields by irrigation water, splashing rain, insects, workers, contaminated equipment, and by wind as aerosols.

Epidemiology

The bacterial pathogen is seed borne. It can also survive in soil on plant debris upto 2 years. The disease is favored by prolonged leaf wetness for infection. The bacterium enters the leaf through stomata, hydathodes or wounds.

Integrated Disease Management

Planting pathogen-free seed is the primary control strategy for angular leaf spot. Contaminated seed can be treated with hot water to reduce, but not eliminate, the number of bacteria in and on the seed. Crop rotation with non host crop. The disease can be checked by application of copper oxychloride.

Mosaic

Disease Symptoms

The infected plants are severely stunted and there is a presence of distinctive yellow mosaic on the foliage. Infected leaves of the plant curl downwards and the size of the leaf is much smaller than normal leaves. Infected plants produces deformed flowers and distorted fruits and are small in size. Fruit produced is often discolored.

The spine gourd is commercially propagated by tuberous roots and stem cuttings. However the emergence of begomovirus infection in some cucurbit crops is a major constraint for the cultivation of the crop. Infected plants exhibit severe mosaic, leaf curl and mottling symptoms. As reported by Venkataravanappa *et.al.* 2019, around 50-60 % of disease incidence is reported from different farmer's field at Varanasi and Uttar Pradesh state of India.

Causal Organism

The disease is caused by Cucumber mosaic virus (CMV).

Disease Cycle

Cucumber mosaic virus (CMV) is a plant pathogenic virus belongs to the family Bromoviridae. It has a worldwide distribution and a very wide host range.

Epidemiology

CMV is mainly transmitted by aphids. This virus has an acquisition period of five to ten seconds and an inoculation period of about a minute. After two minutes, the probability of inoculation largely decreases. Moreover, CMV can overwinter in perennial plants and weeds, as it can survive the winter in the roots of the plant and move to the aerial parts in spring, where it can be transmitted by aphids to other plants.

Integrated Disease Management

The management of the virus is largely dependent on the control of the aphid vectors. Provision for reflective mulches can deter aphid feeding on the plant.

Future Approaches

Spine gourd is one of the important cucurbitaceous crops grown across the world for vegetable and medicinal purposes. Diseases caused by various microorganisms are becoming the major limiting factors for the production of spine gourd, reducing its potential yield. For commercial cultivation of spine gourd, most of the growers used tuberous roots and stem cuttings as a propagation material. This in turn results in an increased occurrence of various diseases. The crop also suffers from various viral diseases. So, there is a need for understanding the causal agent through characterization of which will lead to the designing management strategies for the spine gourd viral disease control.

Conclusion

The commercial cultivation of the spine gourd by most growers depends on the tuberous roots and stem cutting. The occurrence of various diseases in spine gourd gives an alarming signal against utilization of such infected plant materials in the crop breeding and improvement programs. Use of disease free vegetative propagating material is considered as one of the most important methods for controlling several diseases.

References

Bandyopadhyay S, Mukherjee S. K. (2006). Traditional medicine used by the ethnic 'communities of Koch Bihar district (West Bengal-India). *J. Trop. Med. Plants.* **7**(2): 303-312.

Bharathi, L.K, Munshi, A.D, Joseph John, K, Vishal, N and Bisht, I.S. (2010). Genetic resources of spine gourd (*Momordica dioica* Roxb.). ex-Wild.) An underexplored nutritious vegetable from tribal regions of eastern India. *Plant Genetic Resources*. **8**(3): 225-28.

Kushwaha, S.K., Jain, A., Gupta, V.B., and Patel, J.R. (2005). Hepatoprotective activity of the fruits of *Momordica dioica, Nigerian Journals of Natural Product and Medicine*, **(9)**: 29-31.

Raj, N. M., Prasanna, K. P., Peter, K. V. (1993). Momordica spp. In kallo G. Berge Bo (Eds). Genetic improvement of vegetables crops. Pergamon press; Oxford, 239-243.

Ram D., Kalloo G., and Banerjee M.K., (2002). Popularizing kakrol and kartoli: the indigenous nutritious vegetables. *Indian Horticulture*, pp 6-9.

Rao, M. K. (2001). Flora of Maharashtra State, Dicotyledons vol. 2.

Rasul, M. G., Hiramatsu, M., and Okubo, H. (2003). Morphological and physiological variation in kakrol (*Momordica dioica* Roxb.), *J. Fac. Agr Kyushu Univ,* **49**(1): 1-11.

Sastri B.N. (1962). Wealth of India-Raw Materials. Council of Scientific and Industrial Research, Delhi, 406-407.

Tiwari, J.K. and Tigga, K. (2015). Genetic architecture and correlation analysis for fruit yield in different genotypes of spine gourd (*Momordica dioica* Roxb.). Progressive Research – *An International Journal.* **10** *(Special-IV)* : 2425-2428.

Venkataravanappa B., Lakshminarayana Reddy C. N., Shankarappa K. S. , M. and Krishna Reddy. (2019). Association of Tomato Leaf Curl New Delhi Virus, Betasatellite, and Alphasatellite with Mosaic Disease of Spine Gourd (*Momordica dioica* Roxb. Willd) in India. *Iranian J Biotech.* **17**(1): e2134.

19

Diseases of Cucurbitaceous Crops and Their Integrated Management

Partha Sarathi Nath

Department of Plant Pathology, Bidhan Chandra Krishi Viswavidyalaya Mohanpur- 741252, West Bengal

---

Cucurbits are major crops of countries having a wide variety of vegetables. Cucurbits include bottle gourd, bitter gourd, pumpkins, squashes, ridge gourd, sponge gourd, and cucumbers. Cucurbits are a large, diverse crop group that are susceptible to over 200 diseases (Zitter *et al.*, 1996). A wide range of pathogens affect the productivity of cucurbits. The diseases may be caused by fungi, bacteria, Cucurbits crops are warm-season large group of crops that are cultivated and harvested over different seasons like spring, summer, and winter. These crops as a group are an important part of a diverse and nutritious diet. Melons are eaten as fruits, whereas summer and winter squashes, cucumber, and pumpkin are eaten as vegetables. In addition, cucurbits are used for fiber, utensils, decoration, ceremonial and medicinal purposes.

A wide range of pathogens affect the productivity of cucurbits, which constitute over 200 diseases (Zitter *et al.,* 1996). The diseases may be caused by fungi, bacteria, viruses, or mycoplasma- like organisms. The disease may be soilborne, seedborne (carried by seed), spread by wind, or transmitted by insect vectors. The moisture content of the seed, storage period, prevailing temperature, and degree of invasion influence the development of seedborne fungi (Anjorin and Mohammed 2009). The pathogen may cause seed abortion and rot, necrosis, reduction, or elimination of germination capacity as well as seedling damage at later stages of plant growth, resulting in the develop- ment of the disease as systemic or local infection (Khanzada *et al.,* 2001). The diseases cause heavy loss in terms of yield and quality. The diseases caused by pathogen are controlled by fungicides against fungal diseases, antibiotics for bacterial diseases, or insecticide in case of viruses transmitted by insect vectors. The diseases of cucurbetaceous crops can be managed by adopting different cultural, biological, and chemical methods of disease management. The following management practices are required to be taken into consideration for raising disease-free cucurbitaceous crops:

Survey and surveillance should be done for early detection of the disease and disease management in the initial stages which may help to reduce the chemical pesticides load and shall be very economical.Certified seed should used and it should collected from authenticated sources. Growing of tolerance varieties must be taken into consideration. Before sowing Seedsshould be treated with either proper chemicals or botanicals or bio-control agents before sowing.If necessary crop rotation with crops other than cucurbits should be done. A Minimum of 3-year crop rotation should followed.Field sanitation should be maintained, that is, discard plant debris and keep the cropping area free from weeds as they may serve as alternate host to disease-causing organisms.Proper drainage facility should be ensured especially during rainy season.Overcrowding of crop plants shouled be avoided. Maintain desired plant population. Avoid injury during intercultural operations, harvesting, and handling. Discard the diseased portion carefully in a way that inoculum does not spread.Use clean irrigation water and well-sterilized tools.Control insect vector to check viral diseases.Do not spray in the daytime as pollinators are quite active in the day time.Do not use a single chemical to combat disease repeatedly.Adjust harvesting and spray schedule in a way that the waiting period required for a par-ticular pesticide is followed.

Table 1. The Major diseases of cucurbits and their causal organisms.

S.No.	Name of the Disease	Causal Organism
1.	Anthracnose	*Colletotrichum orbiculare*
2.	Downy Mildew	*pseudopronospora cubnsis*
3.	Powdery Mildew	*Erysiphe cichoracearum*
4.	Alternaria leaf blight	*Alternaria cucumerina*
5.	Scab or Gummosis	*cladosporium cucuerinum*
6.	*Cercospora* Leaf Spot	*Cercospora citrullina*
7.	Damping-off of Seedlings and fungal root rots	*Pythium spp, Fhizoctonia spp., and Fusarium spp.*
8.	Angular leaf spot	*Pseudomonas syringae* pv. *lachrymans*
9.	Bacterial leaf spot	*Xanthomonas campestris* Pv. *cucurbitae)*
10.	Cucumber mosaic virus	CMV
11.	Squash mosaic virus	Squash mosaic virus
12.	Zucchiniyellow mosaic virus	Zucchiniyellow mosaic virus
13.	Poty virus	Poty virus
14.	Papaya Ring spot virus	Papaya Ring spot virus
15.	Watermelon mosaic virus	Watermelon mosaic virus

Anthracnoce (*Colletotrichum orbiculare*)

Symptoms

The one of the most important disease in cucurbitaceous family is Anthracnose (*Colletotrichum orbiculare* [(Berk & Mont.) Arx.] and is both air borne and seed borne fungal disease that occur on most cucurbits throughout the year. The characteristics symptoms appear on all the aboveground parts of the plants. The primary symptoms appear as water-soaked spots on the leaves, as the disease progress the water soaked spots which in turn yellowish circular spots and later as brown to black in colour. The symptoms are different in different cucurbitaceous plant. In water- melon The spots are irregular and the colour of the foliage is deep brown or black, while on cucumber and muskmelon the spots are brown and enlarge considerably. The stem, petioles and fruit pedicel, also become infected, resulting in defoliation of leaf, shriveling of the fruit and death of the plant. On the stem dark spots enlarge in size.

The lesions on stem cause wilting of plants in muskmelon. The most characteristics symptoms are found on the fruits. The fruits may turn black, shrivel when it is small in size and when the pedicels are infected fruit dropping is noticed. The symptoms appear as circular, water-soaked and sunken spots with dark borders on mature fruits. The size of the spots may vary from 5 to 11 mm in diameter and 5 mm deep on watermelon. Small, black fruiting body acervuli are formed. Under humid conditions, black stromata with a gelatinous mass of salmon pink spores may be observed on muskmelon, watermelon, and cucumber. Cankers lined with this characteristic color can never be mistaken for any other disease (Zitter 1987). The disease is primarily promoted under humid and moist weather conditions. Seedborne anthracnose infection results in drooping and wilting of cotyledons, and lesions may appear on the stem near the soil line. Huge losses in storage or shipment can occur if infected fruits are packed.

The fungus overwinters on infected residue of the previously grown cucurbit crop and may also be passed on to cucurbit seeds. In the spring, under humid conditions, the fungus releases conidia (airborne spores) that infect foliage of cucurbit vines. High humidity and temperature of 24°C is optimum for disease multiplication though conidia do not germinate below 4.4°C or above 30°C or under nonhumid conditions. The pathogen needs water to liberate the conidia from their sticky covering. Anthracnose usually establishes midseason after the plant canopy has developed. The fungus can penetrate leaves directly and does not require natural openings (e.g., stomata) or wounds.

Management

Growing resistant/tolerant varieties, which should be the utmost important step in managing any diseases of cucurbitaceous crops and can greatly, reduce yield losses particularly due to anthracnose disease. Two distinct races (and perhaps as many as seven), which vary in their ability to infect a range of cucurbit genera, species, and cultivars, have been reported (Goldberg 2004).

Monitor crops closely to identify the disease early in its cycle, especially on the young plants. The management of early infections and protecting young plants from infection prevent inoculums buildup in the field and reduce losses (Thompson and Jenkins 1985).

Follow good sanitation practices such as cleaning up crop debris and wild hosts at the end of the growing season. Also, cucurbit volunteers and alternative hosts in and around the field should be destroyed.

Carry out deep plowing immediately after the final harvest to destroy all infected cucurbit plants and debris in the field (Palenchar *et al.,* 2012).

Seeds should be treated with Thiram or Captan @ 3 g/kg seed.

Spraying of Difolatan (0.2%), Dithane M-45 (0.2%), or Bavistin (0.2%) has been found to be effective (Rai *et al.,* 2008). Fungicide applications generally are not required within 2–3 weeks of the final harvest.

Downy Mildew (*Pseudoperonospora cubensis*)

Downy mildew is a fungal disease and one of the most important disease of the cucurbitaceous family caused by *Pseudoperonospora cubensis* [(Berk. Curt.) Rostow zew]. The disease is airborne in nature. It is particularly serious in warm weather conditions or tropical environment.

Symptoms

The visible symptoms first appear as pale green areas on the upper surface of leaves that changes to bright yellow angular spots. Spots on the leaf are irregular in appearance and are delimited by leaf veins. When the disease appeared in very severe form, the lesions expand and coalesce, tissues turn necrotic and brown, and results in shriveling and death of large areas of leaf surface giving a scorched appearance. Sometimes yellowing may be seen around the edges of the spots. When the symptoms appeared in early stages of plant growth, a reduction in photosynthetic activity that results in stunted plant growth and reduction of yield, especially in cucumber (Colucci and Holmes 2010). When humidity is high, the lower surface of infected leaves shows water-soaked lesions, slightly sunken with profuse growth of light to dark gray or purple sporulation that is also delimited by leaf veins, which is evident as a fuzzy or downy growth. Premature defoliation may also result in sunscalding of fruits due to direct exposure to sunlight. The disease progresses aggressively and kills the plant quickly through rapid defoliation.

High humidity, fog, and heavy dew favor the disease development (Keinath 2014). After infection, downy mildew continues to get worse even under dry weather conditions. In contrast to powdery mildew, spores of downy mildew are dark purplish gray and appear only on the underside of leaves (Mc Grath 2006).

Symptoms on watermelon and cantaloupe are not as distinctive as on cucumber and squash and are mostly mistaken for other diseases such as anthracnose, target spot, *Alternaria* leaf spot, or gummy stem blight (Colucci and Holmes 2010). The symptoms of downy mildew infection are variable in different cucurbit crops, for example, the lesions are angular and are limited by the leaf veins on cucumber and squash, whereas these are typically irrregularly shaped on the foliage of watermelon and cantaloupe and turn a brown rapidly. As the disease progresses, the lesions expand and multiply, causing the field to have a brown and brittle look (Celetti and Roddy 2010).

It is an obligate parasite that requires living host tissue to grow and reproduce. The organism overwinters on the infected cucurbit plants in areas without hard frost. Sporangia are the primary source of inoculum that spread from the infected plants through wind currents or rain splashes to local or distant places. Once the sporangia reach a susceptible host, they germinate and directly infect the leaf within an hour. The optimum temperature for sporulation is 15°C–20°C with at least 6 h of high humidity. The plants with yellow lesions have high sporulation ability. The symptoms appear after 3–12 days of infection, depending upon the inoculum load, temperature, and relative humidity. High temperatures of more than 35°C are not favorable though cooler nights favor disease development.

Management

Resistant/tolerant varieties shoud be planted. With the development of new races, different varieties, especially cucumber, do not exhibit a high level of resistance as they did to previous races. However, resistant varieties are still a valuable component of downy mildew management.

Early in its cycle monitor crops to identify the disease.

Prophylactic measures should be followed on the basis of the weather forecast as the disease typically coincides with the onset of rains.

Grow cucurbits in environments where humidity can be manipulated to manage downy mildew. For example, vining cultivars should be tied to the trellises so that the dew on leaves dries quickly. Early planting can also avoid the disease (Keinath 2014).

Neem oil derived from *Azadirachta indica* is a botanical control for both downy and powdery mildews on cucurbits, but negative impacts have been reported as it has been found to be toxic to ladybeetles (Banken and Stark 1998) and harmful to bees (so, apply when they are not active in the field). Therefore, it should not be used without clear need and caution. Biocontrol agent *B. subtilis* is available in a wettable powder formula- tion that can be used for downy mildew control (Fravel 1999). However, Keinath (2014) reported that no organic-approved fungicides or biofungicides prevent or control cucur- bit downy mildew.

Contact fungicides provide some degree of control. Systemic fungicides are recommended in the initial stages when there is forecast of occurrence or symptoms have just appeared. Applications of systemic fungicide at 5–7-day intervals depending on disease severity. Spraying of Ridomil (0.3%), Blitox (0.2%), or Mancozeb (0.2%). The application of Dithane M-45 (0.3%) at a 15-day interval is good to control the disease (Rai *et al.,* 2008). Rotate fungicide products to reduce the risk of resistance to fungicides.

Alternaria **Leaf Spotor** ***Alternaria*** **blight (*Alternaria cucumerina* and *Alternaria alternate*)**

The disease also known as target leaf spot or *Alternaria* blight, caused by *Alternaria cucumerina* [(Ellis & Everh.) Elliot] and *Alternaria alternate* [(Fr.) Keisslar]. The disease is seed borne and soil borne.It appears on bottle gourd, watermelon, muskmelon, cucumber, snake gourd, and vegetable marrow. High humidity and temperature in the range of 15°C–32°C is quite favorable for its development. The symptoms appear as small circular tan spots 1–2 mm in diameter on the upper surface of the older or crown leaves that may be surrounded by a yellow halo. These spots later spread to the younger leaves toward the tips of the vines. Lesions formed on the lower leaf surface tend to be more diffused. These spots later coalesce to form large lesions, which may be more than 10 mm in diameter. These lesions show a target-like pattern of rings that is typical of most *Alternaria* spp. These lesions bear concentric rings that may cause severe leaf drop. Dark sunken brown spots initiate the disease on fruits, which later develop dark powdery lesions. It causes damage by defoliating the vines and leads to reduction in fruit yield, size, and quality. Partial defoliation may even cause the fruit to sunscald and ripen prematurely. The first symptoms usually were observed at the time of initial fruit development as fruits turn brown and reduce in size, which later turn black and mummified.

The pathogen overwinters as dormant mycelium (saprophyte) in diseased and partly decayed crop debris, in weeds of the cucurbit family, and possibly in the soil. Conidia are produced in the spring and act as the primary inoculum. The inoculum is carried by wind for long distances and splashing water from diseased plants to susceptible tissues (Seebold 2010; Watt 2013). The conidia can survive under warm and dry conditions for several months. The period between infection and the appearance of symp- toms varies from 3 to 12 days (Babadoost 1989). Young plants less than a month old and fruit-bearing plants (70–75 days old) are more prone to the disease infection than midseason plants (45–60 days old). Germinating spores can enter the host directly or through wounds and natural openings.

Management

Effective management depends upon integrating individual control measures that help to reduce leaf spot in the initial stages and thereby slow epidemic development:

- Follow a 3-year crop rotation with noncucurbitaceous crop.
- Use disease-free seeds, resistant varieties/hybrids varieties should be used.
- Seeds should be treated with Captan at 3 g/kg of seed.
- Good sanitation practices such as cleaning up crop debris at the end of the growing season should be followed (Seebold 2010). volunteer cucurbit crop plants and weeds, which may be capable of harboring spores of this disease should be destroyed (Kucharek 2000).
- Apply fungicides that are registered for controlling this disease. Spray Bavistin (0.1%) as soon as the disease appears. Miltox (0.2%) or Dithane M-45 (0.2%) can be used at 15 days interval for effective management (Rai *et al.,* 2008).

Powdery Mildew *(Erysiphe cichoracearum* and *Sphaerotheca fuliginea*)

Powdery mildew disease is a very serious disease in cucurbits affects the crop under field and greenhouse conditions. It is caused by *Sphaerotheca fuliginea* [(Schlect. ex. Fr.) Poll.] and *Erysiphe cichoracearum* (D. ex. Merfat), which is the perfect stage of the fungi. The disease is widely distributed and destructive among cucurbits in most areas of the world and can be a major production problem causing yield losses upto 30%–50% (El-Naggar *et al.,* 2012). The disease first appears on older leaves. Conidia are produced profusely in the white powdery mycelium, and these spores spread quickly through wind to the adjacent leaves or plants as well as travel over long distances.

Powdery mildew appears as powder like colonies on the upper surface of the leaves. As the disease progresses, the entire leaf surface is covered by the pathogen. The disease is most severe after fruit set and in densely planted fields. Symptoms and signs can also develop on stems and fruit. Powdery mass on the leaves decreases the photosynthetic rate (Queiroga *et al.,* 2008), causing reduction in plant growth, premature foliage loss, and consequently reduction in yield. The infected plant parts remain stunted and distorted and may drop prematurely. The infected fruits do not develop fully; as a result, yield and quality of fruits is affected. The yield loss is proportional to the severity of the disease and the length of time that plants have been infected (Mossler and Nesheim 2005). If this disease is not controlled in time, symptoms can be severe enough to cause extensive premature defoliation of older leaves and wipe out the crop (Nunez-Palenius *et al.,* 2012).

Disease development is favored by vigorous plant growth and moderate temperature. The most favorable conditions for disease development are 35°C temperature and high relative humidity of more than 70% (Ali *et al.,* 2013). Temperature and humidity must be examined together since it is the water vapor pressure deficit that has the greatest effect on host–parasite interactions (Jarvis *et al.,* 2002). During periods of intensive dew on leaf surfaces, the severity of this disease is enhanced. However, excessive water on the leaf surface is often detrimental to the development of powdery mildew disease.

Management

Crops should be closely monitored to identify the disease early in its cycle. Even powdery mildew-resistant varieties should be monitored, because there are many races of the pathogen and some of them may break resistance. Summer squash (spring planted) is affected earlier and can be used as an indicator for examination in other crops.

Grow resistant/tolerant varieties. Genetic resistance is used extensively in cucumber and melon and has been incorporated into other cucurbit crops (Mc Grath 2011). However, there are several fungal races of powdery mildew, and hence some resistant cultivars might be susceptible to a specific fungal race (Zitter *et al.,* 1996). According to Konstantinidou- Doltsinis and Schmitt (1998), the need to control powdery mildew disease is one of the reasons for the increased use of fungicides in cucurbits.

Biological control involves the use of fungal spores of *Ampelomyces quisqualis* Ces., which parasitize and destroy the powdery mildew. Similarly, bacteria *Bacillus subtilis* and fungus *Sporothrix flocculosa* (syn. *Pseudozyma flocculosa*) gave promising results (Nunez-Palenius *et al.,* 2012).

Field Sanitation Should be Followed

Neem oil helps in effective management, but the harmful effects on beneficial insects render it impractical. However, cinnamon oil has been found effective (Nunez-Palenius *et al.,* 2012). Most of these natural oils have to be applied repeatedly during the active growing season in order to achieve a reliable level of control. Thus, they cost much higher than the chemicals. Cow's milk sprayed onto the leaves of greenhouse-grown zucchini was found to be effec- tive (Bettiol 1999). White mustard oil at 1% was also found effective (Stoyka *et al.,* 2014).

Fungicides should be applied at an interval of 7–10 days with the appearance of early disease symptoms. Most of the fungicides to control powdery mildew are primarily pre- ventive, that is, they are effective when applied before the disease appears. In addition, powdery mildew fungi can develop resistance to specific fungicides (Brown 2002). Spray sulfur-based fungicides like Karathane (0.05%) or Hexconazole (0.1%) for effective con- trol. Addition of 2.3 mM silicon to the nutrient solution (fertigation) can significantly delay and reduce the incidence of disease (Menzies *et al.,* 1991). Moreover, foliar sprays of chlo- rite mica clay that contains silicon have also shown suppression of powdery mildew in cucumber (Ehret *et al.,* 2001). Follow rotation of protective and systemic fungicides to reduce the chance of fungicide resistance development.

It is caused by seedborne fungi *Cladosporium cucumerinum*. The disease appears on leaves, petioles, stems, and fruits (Ogorek *et al.,* 2012). Numerous water-soaked spots occur on leaves and runners, which eventually turn gray to white and become angular, often with yellow margins. The center of the spots could then drop out to give irregular-shaped holes in the leaves. The lesions on the stem and petiole are comparatively oblong in shape and may be covered with dark-green velvety mold under moist conditions. Lesions on fruit are often confused with *Anthracnose*. The spots might ooze a gummy substance, which could then be invaded by secondary rotting bacteria, and these spots cause foul smell (Yuan 1989; Watson and Napier 2009).

The disease overwinters on crop debris and seed. The spores are blown long distances in moist air and are also spread by insects, implements, and clothes of workers. The spores penetrate the host within 9 h of germination and infection can be seen in 3 days (Zitter 1986). The optimum temperature for disease development is 17°C–27°C along with high humidity. The spread of pathogen is favored by low night temperature, Cercospora leaf spot (*Cercospora citrullina*)

The Cercospora leaf spot disease is spread both by soil and airborne spore, *Cercospora citrullina* (Cooke). The first symptoms can be found found on the foliage but it may also appear on petioles and stems when the disease is severe and conditions are highly favorable for disease development. Dark spots are seen firstonthe older leaves. The spots are circular to irregular in shape with pale or light centers and dark margins. As lesions expand, they coalesce and give a blighted appearance. If the disease is severe, then defoliation occurs and the yield is affected. It can reduce fruit size and quality, but economic losses are rarely severe (Schwartz and Gent 2007).

The pathogen survives on infested crop debris and on weeds between cucurbit crops. The disease infection through its spores (conidia) is readily spread by wind currents, splashing rain, and irrigation water. The infection begins after the deposition of spores onto the leaves and petioles. The disease development is favored by temperatures in the range of 26°C–32°C and high humidity. The new cycles of infection and sporulation occur every 7–10 days during warm and humid weather conditions (Schwartz and Gent 2007).

Management

- Good sanitation practices should be followed by removing old diseased crop residues and cucurbitaceous weeds. Deep plowing may also be practiced immediately after finalharvest.
- Dithane M-45 at 0.2% should be applied as and when disease appeared

Damping-off of Seedlings and Fungal Root Rots (*Pythiumspp, Fhizoctonia* spp., and *Fusarium* spp.)

Although the seeds of cucurbitaceae are directly sown in the field, but the seed and seedlings of cucurbits are affected by a number of soilborne pathogens. Of these, damping-off and root rots caused by a complex of fungi, namely, *Pythium*, *Rhizoctonia*, and *Fusarium*, are most common. Damping-off affects the crop before (pre-emergence damping-off) and after the germination of the seed (postemergence damping-off). Preemergence infection causes the rotting of seed inside the seed coat. The seed may germinate but the radical and cotyledon turn brown and soft and fail to grow further. The initial symptoms of postemergence damping-off appear as yellow to dark brown, water-soaked lesions on the root and hypocotyl tissues. With time, the hypocotyl tissues shrivel, roots decay further, and the seedlings topple down or may wilt and eventually collapse. Plants that survive may show symptoms of root rot. Roots can have a watery gray appearance, particularly the fine feeder roots. Cool temperatures, high soil moisture, and poor aeration favor the disease development. When plastic mulches are used under moist, hot conditions, the roots of even older plants rot. The fungal complex can invade many plants species and can survive even on decaying plant material. Usually, sporadic outbreaks are difficult to control.

Management

- Seed treatment with seed treating fungicides is one of the most important practices followed by the farmers. Seed should be treated with Ceresan or Thiram at 3 g/kg seed. Phosphonate seed treatment is a cost-effective way of protecting cucumber plants from *Pythium* damping-off (Abbasi and Lazarovits 2006).
- Ensure proper drainage and avoid overwatering.
- Drench the soil with Bavistin (0.1%).
- Highly effective biological control of soilborne pathogens can be attained only with the combined application of organic amendments and microbial biocontrol agents (Chang *et al.,* 2007; Singh *et al.,* 2007).

Bacterial Diseases

Angular leaf spot (*Pseudomonas syringae* Pv. *lachrymans*)

The most important and widespread bacterial disease of cucurbits that causes reduction in fruit number, fruit yield, and quality of the fruit is the Angular leaf spot disease (Pohronezny *et al.,* 1977). This disease is caused by the bacteria *Pseudomonas syringae* pv. *lachrymans*. The disease primarily affects cucumber, but it may occur on melons, squashes, and pumpkins. The bacterium infection occurs on all aboveground parts of cucurbit plants. Initially, the symptoms appear on leaves as small, water-soaked lesions that later enlarge. The shape of older lesions tends to be angular as they enlarge and encounter veins. As the infection pro- gresses, the affected tissues often dry and fall, and thus, the leaves are left with torn irregular-shaped holes. The shredded leaves have lower photosynthetic efficiency, causing indirect yield losses. Under severe conditions, the leaves turn yellow, and occasionally, the growing tips of vines become water- soaked and yellow and growth ceases. On fruits, the infection appears as small, circular spots with a yellow halo. Later, the spots turn dull white. Dry cracks may occur, which are a little deeper, rendering the fruits unmarketable. The cracks expose fruits to other secondary infections like soft rot, which usually follows the bacterial infection. Sometimes, the infection progresses up to the seeds.

Angular leaf spot is most active between 24°C and 28°C and is favored by high humidity. Under very humid conditions and warm temperatures, white bacterial ooze may be found on the underside of lesions, which dries to form a thin, white crust. The disease is seed- and soilborne. The bacterium does not have spores adapted to carry them over long unfavorable periods and survive as vegetative cells in seed, soil, diseased plant debris, or weed plants (Leben 1981). The pathogen is disseminated by splashing rain, irrigation water, soil, insects, farm tools, and field workers. Infection occurs through natural openings and wounds. The bacterium is usually trans- mitted by water.

Resistance to *P. lachrymans* is controlled by a large number of recessive genetic factors (Klossowska 1976). However, it has also been reported that the disease is controlled by a single recessive gene "pl" (Dessert *et al.,* 1982). The use of resistant cucumber cultivars is effective in reducing damage and losses due to this disease.

Management

- Used disease-free seeds is the first and most important method of disease control of angular leaf spot of bacterial disease.
- Another important method of angular leaf spot of bacterial disease is seed should be treated with mercuric chloride solution @ (1:1000) for 5–10 min. Also, hot water treatment of seeds at 50°C ± 2°C can reduce the incidence but cannot eliminate.
- Crop rotation with noncucurbitaceous crop for three years can eliminate the disease.
- Maintain field sanitation during the crop-growing period and remove crop debris immediately after final harvest.
- Do not work in the crop when it is wet. Cultivation in dry soil is most effective in reducing bacterial survival (Kritzman and Zutra 1983).
- Also, follow deep plowing and soil solarization in summer months.
- Umekawa and Watanabe (1982) reported suppression to an appreciable extent by managing nighttime humidity (to 80%–90%) with dehumidifiers under protected structures.
- Chemical controls are most effective when integrated with sound cultural control practices. Streptocycline is an effective antibiotic against bacterial plant pathogens at a rate of 400 ppm. Copper-based bactericides are often necessary at an interval of 4–7 days to reduce the severity of the disease (Schwartz and Gent 2007).

The biological control agent pentaphage (lysate of the virulent strain of *P. syringae*) was most effective when applied at high relative humidity (90%) in the morning and evening at intervals of 12–14 days (Korol and Bylinskii 1994). Avagimov and Panteleev (1984)reported reduced incidence and increased yield by using an isolate of the L33 strain of *Pseudomonas geniculata*. Systemic induction of resistance by infection with tobacco necrosis virus on first leaves was reported by Jenns and Kuc (1979).

Bacterial leaf Spot (*Xanthomonas campestris* Pv. *cucurbitae*)

The disease is not common but can occur during persistent warm and humid conditions on different cucurbits. This disease is caused by the seed-and soilborne bacterium *Xanthomonas campestris* pv. *cucurbitae* (Bryan). The initial symptoms appear on seedlings as small brown spots on cotyledons. The disease appears as necrotic spots on leaves and characteristic lesions on fruits. The symptoms appear as small water-soaked or greasy patches on the underside of leaves and as indefinite yellow patches on the upper surface of leaves. As the disease cycle progresses, the patches turn brown, round to angular with translucent centers and a typical yellow halo. The necrotic areas on the leaves do not fall, unlike the angular leaf spot. The disease occasionally occurs on young stems, but young fruits may be infested. The appearance of disease on fruits is erratic and can be correlated with moisture content at rind matu- rity (Goldberg 2012). The affected fruits bear water-soaked, sunken and cracked areas with light-brown exudates. Later, they become distinctly sunken and the rind may crack, leading to fruit rotting and other secondary infections. Viscous liquid drops of a yellowish-brown color are formed on the surface of these spots in damp weather. The infection can progress into the seed cavity of fruits and result in the rotting of flesh and contamination of seed. The disease can be transmitted by seed and infected crop residue, rain splash, and infected tools. The disease development is favored by high temperature in the range of 25°C–30°C and high relative humidity in the range of 90%. The disease incidence increases during the vegetative growth period. Severe disease incidence takes place due to precipitation/overhead irrigation. Yield losses may exceed 50% under humid/moist conditions (Babadoost 2012).

Management

The management practices are similar to the angular leaf spot except for the biological control as there are no reports of biological control for bacterial spot of cucurbits.

Viral Diseases

Cucurbits are very sensitive to viral infection. A complex of viruses is able to infect cucurbits, and over thirty viruses have been reported (Zitter *et al.,* 1996). The majority of these viruses cause huge losses in cucurbit production. The important viruses are cucumber mosaic virus (CMV), squash mosaic virus (SqMV), watermelon mosaic virus (WMV), zucchini yellow mosaic virus (ZYMV), and papaya ring spot virus (PRSV) (Tobias and Tulipan 2002). These viruses are transmitted by aphids except SqMV, which is spread by beetles and is seedborne. The symptoms caused by different cucurbit viruses are very similar, and it is very difficult or even impossible to identify the causal virus. Enzyme-linked immunosorbent assay is widely used for the detection of plant viruses (Clark and Adams 1977). The other viruses that are not very common include cucumber green mottle mosaic virus and melon necrotic spot virus.

Cucumber Mosaic Virus

CMV is probably the most widely distributed and important virus disease of cucurbits (Ferreira and Boley 1992). It infects all cucurbit crop plants and has a very wide range of natural hosts, including many other noncucurbit crop plants and weeds belonging to different crop families.

On cucurbits, symptoms may occur on about 6-week-old plants at vigorous growth stage. The first symptoms appear on young leaves that exhibit mottled mosaic leaf pattern of alternate light-green and dark-green patches with edges that curl downward. The characteristic symptoms are yellow- colored mottling, leaf distortion, and stunted plant growth due to shortening of stem internodes. Older leaves develop chlorotic areas, which turn necrotic along the margins and later cover the entire leaf. Dead leaves either fall off or droop; wilting of the petioles occurs leaving the older vine mostly bare. The new leaves in case of muskmelon and cucumber may wilt and die, while older crown leaves may turn yellow and later dry up, resulting in a slow decline of plant health (MacNab *et al.,* 1983).

The stem end portion of the infected fruits becomes mottled with yellowish-green and dark- green spots, and this mottling pattern gradually covers the whole fruit. The wartlike raised dark spots are usually formed on the fruit, and thus the fruit appears distorted. Fruits produced by the plants in the later stages of the virus infection are somewhat misshapen but have a smooth gray-white color with some irregular green areas, often called white pickle. The infected fruits of cucumber may have a bitter taste and make soggy pickles (Agrios 1978). In squash, there is blotchy appearance of fruits in nonyellow varieties. The infected plants usually produce few runners, flowers, and fruits.

CMV is transmitted through aphid vectors, namely, *Aphis gossypii* and *Myzus persicae*, and also through seeds and several weed hosts. The virus overwinters in many perennial weed sources especially attractive to aphids in spring. It can be transmitted through sap adhered on the hands and clothes of workers harvesting fruit. Agrios (1978) has reported that the entire field of cucurbits sometimes turns yellow due to CMV immediately after the first harvest.

Squash Mosaic Virus

SqMV infects cultivated as well as wild or native cucurbits. However, it does not infect noncucur- bitaceous crops or weeds. The virus causes stunting, chlorotic leaf mottling, green vein banding, and leaf distortion of infected seedlings. There are two strains of this virus, namely, strain 1 (greater effect on melons than squash) and strain 2 (reverse action of strain 1) (Haudenshield and Palukaitis 1998). Older infected plants develop distorted margins, blistering, hardening, and mild to severe dark-green mosaic pattern on leaves. This pattern may be confused with hormonal herbicide effect. Infected fruits may have mottling on the skin and become malformed or distorted (Dukic *et al.,* 2002). The virus is seedborne and spread by beetles (*Acalymma* and *Diabrotica*). Upon feeding, the beetles inject the virus through saliva and infect plants. The virus can also spread readily through a mechanical injury. The virus is carried within the seed and cannot be eliminated by hot water or chemical treatment (Zitter and Banik 1984).

Poty Virus

These include WMV, PRSV, and ZYMV. All three viruses are transmitted by different aphid spe- cies, namely, Aphis spp., *Brachycaudus* spp., *Cavariella* spp., *Myzus* spp., and *Liriomyza* spp. In general, high activity of aphids causes potyviruses in the field, which often spread very fast, and symptoms are localized.

Zucchini Yellow Mosaic Virus

All cucurbitaceous crops are susceptible to ZYMV, and it also infects certain noncucurbitaceous weeds and wild cucurbits. The infection causes blistered, deformed leaves with severe mosaic symp- toms besides reduction in size and stunted plant growth. Melon fruits get discolored with hardening of flesh, seed deformation, and external cracks (Desbiez and Lecoq 1997). Pumpkin and zucchini fruits become discolored and deformed due to knobbiness, while rockmelon fruits have poor sur- face netting (Blua and Perring 1989). This virus is transmitted by a wide range of aphid species ofwhich the green peach (*M. persicae*) and melon (*A. gossypii*) aphids are the most important. The cowpea aphid has been reported to transmit the virus more efficiently than the cotton or melon aphid (Yuan and Ullman 1996).

Papaya Ring Spot Virus

PRSV biotype "W" infects many cucurbit crops but does not infect noncucurbitaceous crops (Bateson *et al.,* 2002). Some wild and native cucurbit species act as infection reservoirs. Characteristic mosaic symptoms appear as profuse mottling, puckering, and deformation of crown leaves, while the symptoms do not occur on lower mature leaves. Leaf distortion and blistering have also been reported. Zucchini squash was the most susceptible to PRSV, followed by watermelon and cucumber (Mansilla *et al.,* 2013). Infected zucchini and pumpkin develop lumpy, distorted fruits, while rockmelon fruits may have poor-quality surface netting. Watermelon fruits may develop uneven surface or typical ring spot patterns on the skin. This virus is spread by a number of aphid species including the green peach and melon aphids (Jensen 1949).

Watermelon Mosaic Virus

The virus infects almost all cucurbit crops, mainly *Cucurbita pepo*, *C. maxima*, and *C. moschata* (Greber *et al.,* 1987), and also wild cucurbits. The diseased vines bear a typical petunia-like appearance, that is, the tips of vines and a proliferation of shoots around the crown extend beyond the general level of the vines. The internodal length of the shoots gets shortened, resulting in crowding of leaves that become rolled, blistered, and small in size with mild mosaic symptoms and have little effect on fruits. It produces less severe symptoms on cucurbit leaves and fruits than PRSV, ZYMV, and SqMV (Coutts 2006). Flowers of the severely infected plants are abnormal in size, shape, and color, resulting in mottled, misshapen small fruits.

Management of Virus Diseases

Identification is very important for the management of viral diseases. Viral diseases can be con- fused with many other diseases or sometimes nutrient deficiency symptoms or physiological disorders. Early identification provides an effective management; otherwise, huge losses may be encountered. The management of the virus transmitting vector is more important than the pathogen itself. Prevention of viral disease is the first strategy that may include using resistant cultivars and pathogen free planting material, preventing spread and overwintering of virus, and reducing patho- gen/vector population. The general control measures are summarized as follows:

Grow resistant varieties: Planting resistant cultivars can serve as an effective control for viral diseases. The most effective and simplest method of controlling viral diseases is to utilize genetic resistance. Even though resistance has been found in some cucurbit species, genetic barrier due to incompatibility among species poses a problem for transferring this resistance to other cucurbits (Zitter and Murphy 2009).

- Use virus-indexed seeds and seedlings.
- Seed treatment of *C. pepo* with hot air (70°C for 2 days) or hot water (50°C for 60 min) prevents the inoculum, but the seed coat prevents systemic inoculum.
- In the early stages of crop growth, monitor the crop carefully and remove the plants with symptoms of virus.

- Grow taller, nonsusceptible barrier crops such as corn that may delay initial infections.
- Remove all weeds and volunteer cucurbit crop plants within and around cucurbit crops as these can harbor aphids and viruses. Eradication of weed hosts is often a difficult task because of the extensive host range of different viruses, but clean cultivation can reduce the incidence and intensity.
- Avoid planting cucurbit crops consecutively in close proximity to each other.
- Destroy the plant debris promptly once the crop is harvested.

Follow long Crop Rotations with Non-cucurbitaceous Crops

Use of super reflective plastic mulch deters vector aphids from landing and thus limits the spread of virus. Floating row covers or reflective mulches may help exclude or repel aphids (Stapleton and Summers 2002; Barbercheck 2014).

The severely infected crop should be burnt.

Parasitoids can be used effectively to manage the aphid vectors such as *Aphidius cole- mani*, *Aphidius matricariae*, *Lysiphlebus fabarum*, and *Binodoxys angelicae* (Kos *et al.,* 2008).

Vector management is not very simple since aphids need only probe the plant to transmit the virus that infects instantly. To control the insect effectively, insecticides would need to be applied at regular intervals, leading to huge monetary inputs and may not be economically feasible. For CMV, spray dimethoate (0.05%), monocrotophos (0.05%), Confidor (0.03%), or Metasystox (0.02%) at weekly intervals. Though managing viruses through insecticides is not a good approach because insecticides do not act fast enough to prevent the rapid spread of the viruses by aphids, once they have well-established colonies, they may increase the disease spread. However, the aphid control can reduce the infection intensity and the losses.

Conclusions

The production of cucurbitaceous crops is being challenged by many new diseases as well as earlier described diseases that are appearing in new areas. The main objective of plant disease management is to reduce the economic damage and to the quality of produce. The indiscriminate use of chemicals has led to the development of resistance causes problem in cucurbits. The IDM approach is essential to cope with the problem. IDM consists of scouting and management strategies, these include, site selection, field preparation, use of resistant cultivars, altering planting time, optimum plant density, optimum irrigation, proper drainage, mulching, optimum use of nutrients and judicious use of pesticides. In addition, monitoring survey environmental factors (temperature, moisture, nutrients, etc.), forecasting diseases, and establishing economic thresholds are important to the management scheme (Maloy 2005).

References

Abbasi, P.A. and G. Lazarovits (2006). Seed treatment with phosphonate (AG3) suppresses *Pythium* damping- off of cucumber seedlings. *Plant Disease.* **90**: 459–464.

Agrios, G.N (1978). *Plant Pathology*, 2nd ed. Academic Press Inc., San Diego, CA, pp. 466–470.

Ali, M., A. Hannan, J. Shafi, W. Ahmad, C.M. Ayyub, S. Asad, H.T. Abbas, and M.A. Sarwar (2013). Nutrient supplement efficacy against powdery mildew of pumpkin (*Sphaerotheca fuliginea*) and its correlation with environmental factors. *International Journal of Advanced Research.* **10**(1): 17–20.

Anjorin, S.T. and M. Mohammed (2009). Effect of seed borne fungi on germination and seedling growth of water melon (*Citrullus lanatus*). *Journal of Agriculture and Social Sciences.* **5**: 77–80.

Babadoost, M. (1989). Alternaria leaf spot or blight of cucurbits. http://web.aces.uiuc.edu/vista/pdf_pubs/918.pdf.

Babadoost, M. (2005). Phytophthora blight of cucurbits. *The Plant Health Instructor.* doi:10.1094/PHI-I-2005- 0429-01. University of Illinois Extension. Department of Crop Sciences, University of Illinois.

Babadoost, M. (2012). Bacterial spot of cucurbits. Report on plant disease, RPD no. 949. extension.cropsci. illinois.edu/fruitveg/pdfs/949_bacterial_spot.pdf.

Bateson, M.F., R.E. Lines, P. Revill, W. Chaleeprom, C.V. Ha, A.J. Gibbs, and J.L. Dale (2002). On the evolu- tion and molecular epidemiology of the poty papaya ringspot virus. *Journal of General Virology.* **83**(10): 2575–2585.

Celetti, M. and E. Roddy (2010). Downy mildew in cucurbits. http://www.omafra. gov.on.ca/english/crops/facts/10-065.htm.

Davis, R.M., T.A. Turini, B.J. Aegerter, and J.J. Stapleton (2012) a. Cucurbits: Charcoal rot-Macrophomina phaseoli. http://www.ipm.ucdavis.edu/PMG/r116101311.html.

Egel, D.S. and R.D. Martyn (2010). *Vegetable Diseases. Mature Watermelon Vine Decline and Similar Vine Decline Diseases of Cucurbits.* Purdue Extension BP-65-W, Purdue Extension Education Store, Purdue University, USA.

Egel, D.S. and R.D. Martyn (2013). *Fusarium* wilt of watermelon and other cucurbits. *The Plant Health Instructor.* © The American Phytopathological Society. pp. 1–11.

Haudenshield, J.S. and P. Palukaitis (1998). Diversity among isolates of Squash Mosaic Virus. *Journal of General Virology.* **79**: 2331–2341.

Hopkins, D.L., J.D. Cucuzza, and J.C. Watterson (1996). Wet seed treatments for the control of bacterial fruit blotch of watermelon. *Plant Disease.* **80**: 529-532.

Jarvis, W., W.G. Gubler, and G.G. Grove (2002). Epidemiology of powdery mildews in agricultural ecosys- tems. In: Belanger, R., W.R. Bushnell, A.J. Dik, and T.L.W. Carver (eds.), *The Powdery Mildews: A Comprehensive Treatise.* The American Phytopathological Society, St. Paul, MN, pp. 169–199.

Jenns, A. and J. Kuc. (1979). Graft transmission of systemic resistance of cucumber to anthracnose induced by *Colletotrichum lagenarium* and tobacco necrosis virus. *Phytopathology.* **69**: 753–7563.

Jensen, D.D. (1949). Papaya virus diseases with special reference to papaya ring spot. *Phytopathology.* **39**: 191–211.

Keinath, A.P. (2014). Cucurbit downy mildew management for 2014. *Clemson University Extension Information Leaflet*, p. 90.

Khanzada, K.A., M.A. Rajput, G.S. Shah, A.M. Lodhi, and F. Mehboob (2001). Effect of seed dressing fun- gicides for the control of seed borne mycoflora of wheat. *Asian Journal of Plant Sciences*. **1**: 441–444.

Kritzman, G. and D. Zutra (1983). Systemic movement of *Pseudomonas syringae* pv. *lachrymans* in the stem, leaves, fruits and seed of cucumber. *Canadian Journal of Plant Pathology*. 273–278.

Kucharek, T. (2000). *Alternaria* leaf spot of cucurbits. http://plantpath.ifas.ufl.edu/extension/fact-sheets/pdfs/pp0032.pdf.

Latin, R.X. and D.L. Hopkins (1995). Bacterial fruit blotch of watermelon: The hypothetical exam question becomes reality. *Plant Disease*. **79**: 761–765.

Leben, C. (1981). Survival of *Pseudomonas syringae* pv. *lachrymans* with cucumber seed. *Canadian Journal of Plant Pathology*. **3**: 247–249.

Li, H.N. (2014). Anthracnose of cucumber. www.ct.gov/caes.

Louws, F.J., G.J. Holmes, and K.L. Ivors (2008). *Cucurbits-Phytophthora Blight*. North Carolina State University, North Carolina. http://www.cals.ncsu.edu/plantpath/extension/fact_sheets/Cucurbits_-_Phytophthora_blight.html.

MacNab, A.A., A.F. Sherf, and J.K. Springer (1983). *Identifying Diseases of Vegetables*. The Pennsylvania State University, Pennsylvania.

McGrath, M.T. (2004). Diseases of cucurbits and their management. In: Naqvi, S.A.M.H. (ed.). *Disease of Fruits and Vegetables*, Vol. 1. Kluwer Academic Publishers, Dordrecht, Netherlands. pp. 455–510.

McGrath, M.T. (2006). *Update on Managing Downy Mildew in Cucurbits*. http://vegetablemdonline.ppath.cornell.edu/NewsArticles/Cuc_Downy.htm.

McGrath, M.T. (2011). Powdery mildew of cucurbits. *Vegetable Crops*. Fact Sheet Page: 732.30. http://vegeta-blemdonline.ppath.cornell.edu/factsheets/Cucurbits_PM.html.

Menzies, J.G., D.L. Ehret, A.D.M. Glass, T. Helmer, C. Koch, and F. Seywerd (1991). Effects of soluble silicon on the parasitic fitness of *Sphaerotheca fuliginea* on *Cucumis sativus*. *Phytopathology*. **81**: 84–88.

Miller, S.A., R.C. Rowe, and R.M. Riedel (1996). *Phytophthora* blight of pepper and cucurbits. http://ohioline.osu.edu/hyg-fact/3000/pdf/3116.pdf.

Ogorek, R., A. Lejman, W. Pusz, A. MiB uch, and P. Miodynska (2012). Characteristics and taxonomy of *Cladosporium* fungi. *Mikologia Lekarska*. **19**(2): 80–85.

Owen, J.H. 1955. *Fusarium* wilt of cucumber. *Phytopathology*. **45**: 435–439.

Palenchar, J., D.D. Treadwell, L.E. Datnoff, A.J. Gevens, and G.E. Vallad (2012). Cucumber anthracnose in Florida. Publication #PP266. http://edis.ifas.ufl.edu/pp266.

Pohronezny, K., P.O. Larsen, D.A. Ematty, and J.D. Farley (1977). Field studies of yield losses in pickling cucumber due to angular leafspot. *Plant Disease Reporter*. **61**: 386–390.

Queiroga, R., M. Puiatti, P.C.R. Fontes, and P.R. Cecon (2008). Produtividade e qualidade de frtos de meloeir- ovariandonumero de frutos e de folhasporplanta. *Horticultura Brasileira*. **26**: 209–215.

Rai, M., S. Pandey, and S. Kumar (2008). Cucurbit research in India: A retrospect. In: Pitrat, M. (ed.), *Proceedings of the IXth EUCARPIA Meeting on Genetics and Breeding of Cucurbitaceae*. INRA, Avignon, France.

Roustaee, A., M.K. Reyhanb, and M. Jafaric (2012). Study of interaction between salinity and charcoal rot dis- eases of melon (*Macrophomina phaseolina*) in Semnan and Garmsar. *Desert*. **16**(2): 111–180, Article 11.

Schaad, N., W.G. Sowell, R.W. Goth, R.R. Colwell and R.E. Webb (1978). *Pseudomonas pseudoalcaligenes*sp. *citrulli* subsp. nov. *International Journal of Systematic Bacteriology.* **28**: 117–125.

Schaad, N.W., E. Postnikova and P.S. Randhawa (2003). Emergence of Acidovorax avenae subsp. citrulli as a crop threatening disease of watermelon and melon. In: Iacobellis, N.S., A. Collmer, S.W. Hutcheson.

Schwartz, H.F. and D.H. Gent (2007). Cercospora leaf spot (Cucumber, Melon, Pumpkin, Squash, and Zucchini). http://wiki.bugwood.org/uploads/CercosporaLeafSpot-Cucurbits.pdf.

Sharma, Akhilesh, Viveka Katoch, and Chanchal Rana.2016. Important Diseases of Cucurbitaceous Crops and Their Managementhttps://www.researchgate.net/publication.

Shimizu, M., S. Yazawa, and Y. Ushijima (2009). A promising strain of endophytic *Streptomyces* sp. for bio- logical control of cucumber anthracnose. *Journal of General Plant Pathology.* **75**(1): 27–36.

Singh, A., S. Srivastava, and H.B. Singh (2007). Effect of substrates on growth and shelf life of *Trichoderma harzianum* and its use in biocontrol of diseases. *Bioresource Technology.* **98**: 470–473.

Srinon, W., K. Chuncheen, K. Jirattiwarutkul, K. Soytong, and S. Kanokmedhakul (2006). Efficacies of antag- onistic fungi against *Fusarium* wilt disease of cucumber and tomato and the assay of its enzyme activity. *Journal of Agricultural Technology*. **2**(2): 191–201.

Stapleton, J.J. and C.G. Summers (2002). Reflective mulches for management of aphids and aphid-borne virus diseases in late-season cantaloupe (*Cucumis melo* L. var. *cantalupensis*). *Crop Protection*. **21**: 891–898.

Stoyka, M., L. Tsvetana, V. Nikolay, and V. Georgi (2014). Botanical products against powdery mildew on cucumber in greenhouses. *Turkish Journal of Agricultural and Natural Sciences*. **2**: 1707–1712.

Thompson, D.C. and S.F. Jenkins (1985). Influence of cultivar resistance, initial disease, environment, and fungicide concentration and timing on *anthracnose* development and yield loss in pickling cucumbers. *Phytopathology.* **75**: 1422–1427.

Tobias, I. and Tulipan M. (2002). Results of virological assay on cucurbits in 2001. *Novenyvedelem*. **38**: 23–27.

Umekawa, M. and Y. Watanabe. 1982. Relation of temperature and humidity to the occurrence of angular leaf spot of cucumber grown under plastic house. *Annals of the Phytopathological Society of Japan*. **48**: 301-307.

Watson, A. and Y.T. Napier (2009). Diseases of cucurbit vegetables. Primefact 832. http://www.dpi.nsw.gov.au/_data/assets/pdf_file/0003/290244/diseases-of-cucurbit-vegetables.pdf.

Watt, B.A. (2013). Alternaria leaf blight of cucurbits. Pest Management Fact Sheet #5086. http://extension.umaine.edu/ipm/ipddl/publications/5086e.

Yang, Q.Y., K. Jia, W.N. Gena, R.J. Guo, and S.D. Li. (2014). Management of cucumber wilt disease by *Bacillus subtilis* f. sp. *cucumerinum* in rhizosphere. *Plant Pathology Journal*. **13**: 160–166.

Yuan, C. and D.E. Ullman (1996). Comparison of efficiency and propensity as measures of vector importance in Zucchini yellow mosaic potyvirus transmission by *Aphis gossypii* and *A. craccivora*. *Phytopathology*. **86**: 698–703.

Zitter, T.A. (1986). Scab of cucurbits. *Vegetable Crops*. Fact Sheet Page 732.50. http://vegetablemdonline.ppath.cornell.edu/factsheets/Cucurbit_Scab.html.

Zitter, T.A. (1987). Anthracnose of cucurbits. *Vegetable Crops*. Fact Sheet Page 732.60. http://vegetablemdon-line.ppath.cornell.edu/factsheets/Cucurbit_ Anthracnose.html.

Zitter, T.A. (1998). Fusarium diseases of cucurbits. http://vegetablemdonline. ppath.cornell.edu/factsheets/Cucurbits_Fusarium.html.

Zitter, T.A. and M.T. Banik (1984). Vegetable crops: Virus diseases of cucurbits. Fact Sheet Page: 732.40. Vegetable MD online. Cornell University. Cornell, New York. http://vegetablemdonline.ppath.cornell. edu/factsheets/Viruses_Cucurbits.htm.